AF572585

CERAMIC FILMS AND COATINGS

CERAMIC FILMS AND COATINGS

Edited by

John B. Wachtman and Richard A. Haber

The Center for Ceramics Research
Rutgers, the State University of New Jersey
Piscataway, New Jersey

Library of Congress Catalog Card Number: 92-540
ISBN: 0-8155-1318-6
Printed in the United States

Published in the United States of America by
Noyes Publications
Mill Road, Park Ridge, New Jersey 07656

10 9 8 7 6 5 4 3 2 1

Library of Congress Cataloging-in-Publication Data

Ceramic films and coatings/edited by John B. Wachtman, Richard A. Haber.
p. cm.
Includes bibliographical references and index.
ISBN 0-8155-1318-6
1. Ceramic coating. 2. Thin films. I. Wachtman, J.B., 1928- . II. Haber, Richard A., 1960- .
TS695.9.C46 1992
667'.9--dc20 92-540
CIP

MATERIALS SCIENCE AND PROCESS TECHNOLOGY SERIES

Editors

Rointan F. Bunshah, University of California, Los Angeles *(Series Editor)*
Gary E. McGuire, Microelectronics Center of North Carolina *(Series Editor)*
Stephen M. Rossnagel, IBM Thomas J. Watson Research Center *(Consulting Editor)*

Electronic Materials and Process Technology

DEPOSITION TECHNOLOGIES FOR FILMS AND COATINGS: by Rointan F. Bunshah et al

CHEMICAL VAPOR DEPOSITION FOR MICROELECTRONICS: by Arthur Sherman

SEMICONDUCTOR MATERIALS AND PROCESS TECHNOLOGY HANDBOOK: edited by Gary E. McGuire

HYBRID MICROCIRCUIT TECHNOLOGY HANDBOOK: by James J. Licari and Leonard R. Enlow

HANDBOOK OF THIN FILM DEPOSITION PROCESSES AND TECHNIQUES: edited by Klaus K. Schuegraf

IONIZED-CLUSTER BEAM DEPOSITION AND EPITAXY: by Toshinori Takagi

DIFFUSION PHENOMENA IN THIN FILMS AND MICROELECTRONIC MATERIALS: edited by Devendra Gupta and Paul S. Ho

HANDBOOK OF CONTAMINATION CONTROL IN MICROELECTRONICS: edited by Donald L. Tolliver

HANDBOOK OF ION BEAM PROCESSING TECHNOLOGY: edited by Jerome J. Cuomo, Stephen M. Rossnagel, and Harold R. Kaufman

CHARACTERIZATION OF SEMICONDUCTOR MATERIALS—Volume 1: edited by Gary E. McGuire

HANDBOOK OF PLASMA PROCESSING TECHNOLOGY: edited by Stephen M. Rossnagel, Jerome J. Cuomo, and William D. Westwood

HANDBOOK OF SEMICONDUCTOR SILICON TECHNOLOGY: edited by William C. O'Mara, Robert B. Herring, and Lee P. Hunt

HANDBOOK OF POLYMER COATINGS FOR ELECTRONICS—Second Edition: by James J. Licari and Laura A. Hughes

HANDBOOK OF SPUTTER DEPOSITION TECHNOLOGY: by Kiyotaka Wasa and Shigeru Hayakawa

HANDBOOK OF VLSI MICROLITHOGRAPHY: edited by William B. Glendinning and John N. Helbert

CHEMISTRY OF SUPERCONDUCTOR MATERIALS: edited by Terrell A. Vanderah

CHEMICAL VAPOR DEPOSITION OF TUNGSTEN AND TUNGSTEN SILICIDES: by John E.J. Schmitz

ELECTROCHEMISTRY OF SEMICONDUCTORS AND ELECTRONICS: edited by John McHardy and Frank Ludwig

(continued)

HANDBOOK OF CHEMICAL VAPOR DEPOSITION: by Hugh O. Pierson

DIAMOND FILMS AND COATINGS: edited by Robert F. Davis

ELECTRODEPOSITION: by Jack W. Dini

Ceramic and Other Materials—Processing and Technology

SOL-GEL TECHNOLOGY FOR THIN FILMS, FIBERS, PREFORMS, ELECTRONICS AND SPECIALTY SHAPES: edited by Lisa C. Klein

FIBER REINFORCED CERAMIC COMPOSITES: by K.S. Mazdiyasni

ADVANCED CERAMIC PROCESSING AND TECHNOLOGY—Volume 1: edited by Jon G.P. Binner

FRICTION AND WEAR TRANSITIONS OF MATERIALS: by Peter J. Blau

SHOCK WAVES FOR INDUSTRIAL APPLICATIONS: edited by Lawrence E. Murr

SPECIAL MELTING AND PROCESSING TECHNOLOGIES: edited by G.K. Bhat

CORROSION OF GLASS, CERAMICS AND CERAMIC SUPERCONDUCTORS: edited by David E. Clark and Bruce K. Zoitos

HANDBOOK OF INDUSTRIAL REFRACTORIES TECHNOLOGY: by Stephen C. Carniglia and Gordon L. Barna

CERAMIC FILMS AND COATINGS: edited by John B. Wachtman and Richard A. Haber

Related Titles

ADHESIVES TECHNOLOGY HANDBOOK: by Arthur H. Landrock

HANDBOOK OF THERMOSET PLASTICS: edited by Sidney H. Goodman

SURFACE PREPARATION TECHNIQUES FOR ADHESIVE BONDING: by Raymond F. Wegman

FORMULATING PLASTICS AND ELASTOMERS BY COMPUTER: by Ralph D. Hermansen

HANDBOOK OF ADHESIVE BONDED STRUCTURAL REPAIR: by Raymond F. Wegman and Thomas R. Tullos

CARBON-CARBON MATERIALS AND COMPOSITES: edited by John D. Buckley and Dan D. Edie

Preface

Ceramic films and coatings are both active fields of research and widely used areas of technology. The relatively high hardness and inertness of ceramic materials make ceramic coatings of interest for protection of substrate materials against corrosion, oxidation and wear resistance. The electronic and optical properties of ceramics make ceramic films and coatings important to many electronic and optical devices.

This book presents a series of reviews of many of the most active and technically important areas of ceramic films and coatings. The chapters are intended to be useful to a B.S. or higher level person who is not necessarily an expert in the area. The book is introduced by a survey of the uses and methods of preparation of ceramic films and coatings. The following chapters each focus on an area of application of a type of film with outstanding properties. Each chapter typically considers the processing, properties, and applications of the subject area of film technology. Each of these chapters can be read on a stand-alone basis, but groups of chapters are related, as described below, and reinforce each other by providing perspective and alternate viewpoints.

Four chapters focus on coatings used for protection. An example of a field in which ceramic coatings have become indispensable is treated first: coated cutting tools. A related research area of great promise is taken up in the next chapter: creating an in situ wear resistant film by ion implantation. Chemical protection is then considered. A widely used and important area of technology is the use of ceramic enamel films for protection. Incidentally, decorative aspects of such use are of high commercial importance also. Finally in the area of protective films thermal protection is considered. Such films have become essential in many high temperature devices including combustors and parts of jet engines.

A chapter on synthetic diamond films also relates to the use of films for protection of a substrate. However, these films also have many other potential applications for their electronic, thermal, and optical properties. Diamond thin films are still in a stage of rapid advance. The state of the art in controlling their structure and microstructure is summarized.

Several chapters relate to films used for their optical and electronic properties. Inorganic, nonmetallic thin films for microoptic devices are discussed. Electronic thin films are treated from three standpoints. First, electronic films made by an organic precursor route are treated. Second, ceramic thick film technology for insulators, conductors, and special electrical functions is treated. Third, superconducting thin films are treated in a separate chapter. These latter materials present extreme challenges to thin film technology because of the critical importance of achieving a very narrow range of crystal structure and microstructure required to give the best superconducting properties.

Finally, two areas are taken up which cut across thin film technology. Sol–gel preparation techniques offer a wet chemical route to many types of thin films. This approach has many advantages, but also has limitations. The procedures used for sol–gel film making and the types of microstructures that can be achieved are surveyed.

Characterization of thin films is a requirement for research as well as for quality control in production. The major characterization techniques are brought together and their capabilities and limitations are treated in a single chapter.

The editors thank the authors for their perserverance in preparing chapters and updating them in rapidly moving areas (e.g., superconductivity) even as this book was being produced.

Piscataway, New Jersey
September, 1992

John B. Wachtman
Richard A. Haber

Contributors

Christopher C. Berndt
Department of Materials
Science & Engineering
State University of New York
Stony Brook, NY

Robert Caracciolo
ITT Avionics
Clifton, NJ

David G. Coult
AT&T Bell Laboratories
Solid State Technology Center
Breinigsville, PA

Brian D. Fabes
Department of Materials
Science & Engineering
University of Arizona
Tucson, AZ

Edward N. Farabaugh
Materials Science &
Engineering Laboratory
National Institute of
Standards and Technology
Gaithersburg, MD

Albert Feldman
Materials Science &
Engineering Laboratory
National Institute of
Standards and Technology
Gaithersburg, MD

Richard A. Haber
The Center for Ceramics
Research
Rutgers, the State University
of New Jersey
Piscataway, NJ

Thomas E. Hale
Carboloy, Inc.
Warren, MI

Herbert Herman
Department of Materials
Science & Engineering
State University of New York
Stony Brook, NY

Arun Inam
Bellcore
Red Bank, NJ

Frank A. Kuchinski
Technology Partners, Inc.
Landisville, PA

Carl J. McHargue
Metals and Ceramics Division
Oak Ridge National Laboratory
Oak Ridge, TN

Lawrence H. Robins
Materials Science & Engineering Laboratory
National Institute of Standards and Technology
Gaithersburg, MD

Daniel J. Shanefield
The Center for Ceramics Research
Rutgers, the State University of New Jersey
Piscataway, NJ

Donald R. Uhlmann
Department of Materials Science & Engineering
University of Arizona
Tucson, AZ

T. Venkatesan
Physics Department
University of Maryland
College Park, MD

Robert W. Vest
Potter Engineering Center
Purdue University
West Lafayette, IN

John B. Wachtman
The Center for Ceramics Research
Rutgers, the State University of New Jersey
Piscataway, NJ

Hougong Wang
Sherritt Gordon Ltd.
Fort Saskatchewan, Alberta
Canada

Xin Di Wu
Physics Department
University of Maryland
College Park, MD

Brian J.J. Zelinski
Department of Materials Science & Engineering
University of Arizona
Tucson, AZ

Contents

1

Ceramic Films and Coatings - An Overview

John B. Wachtman and Richard A. Haber

1.0 INTRODUCTION

The processing, study and use of ceramic films and coatings is done by people with various technical backgrounds. The wide range of materials, techniques for preparation and types of application make this an inherently interdisciplinary field. The present general overview attempts to introduce the subject to the worker who is interested in this field but whose expertise is in one of the related disciplines and who is not an overall expert in films and coatings.

Films and coatings are used for an enormous and diverse set of applications including electronic and optical devices (1)-(3)(3a), protection at high temperatures (4), cutting tool enhancement, and large-scale architectural and automotive use (5). Many of these applications require the properties associated with inorganic, nonmetallic materials; i.e., with ceramics.

The special physical properties of ceramics derive from their fundamental bond type (6). Characteristically ceramics are compounds with bonds that are primarily of a mixed ionic/covalent type rather than a metallic type. As a result, most ceramics have completely filled electronic valence bands separated by a wide forbidden band from completely empty electronic conduction bands causing them to be electrical insulators and to be transparent. Because of the availability of the wide range of ceramic compounds and the ability to introduce additives into their structures, their electronic and optical properties can be tailored to make them semiconductors and electro-optic materials useful as wave guides, modulators, and detectors.

Because of the basic bonding and structural features of ceramics, some possess large and useful amounts of ferroelectricity, ferromagnetism, piezoelecticity, and pyroelectricity. Many ceramics have the high bond strength between atoms that leads to great hardness, stiffness and strength. Many also have good resistance to corrosion and oxidation at high temperatures.

The special properties of ceramics lead to a wide range of applications. Continuing reduction in the scale of microelectronics and increases in the complexity of microelectronic devices have greatly extended thin film technology. Lines and other geometric features of one micron lateral scale or less are used. Films are needed for insulators, conductors, and hermetic seals. Integrated optics technology requires sources, transmission lines, modulators, and detectors made by thin film technology.

Development of new materials has affected ceramic film and coating technology. The discovery that multiple layer coatings can have extraordinary mechanical properties including very high values of elastic moduli is one example. The development of procedures for growing true diamond thin films has set off a whole field of research. The discovery of the high Tc superconducting ceramics which have their best properties as highly-oriented thin films has caused intense efforts to process these materials into good thin films and to combine them with normal conductors to make devices.

Modern ceramic film technology has been strongly affected by many lines of progress in related technologies. This progress includes developments in vacuum technology, film processing, film characterization, materials science of ceramics, semiconductor device technology, optical technology, and cutting tool technology. The following sections summarize the major areas of application, discuss the major processing techniques, discuss the major characterization techniques, and give some perspective on exciting trends toward new ceramic films and new applications.

2.0 AREAS OF APPLICATION OF CERAMIC FILMS AND COATINGS

Table 1 gives a brief summary of areas of application with some typical examples of ceramics for each (6). Practical ceramic films vary greatly in function and thickness. Some have been in use for a long time and others are just coming into use. At one extreme are the naturally-forming oxide films which act as the oxidation barrier on stainless steels. These are typically as thin as 10 nm or less. At the other extreme are the porcelain enamel films as thick as 1 mm or more which are used to protect steel from corrosion. There is thus a long-standing ceramic film and coating technology with its roots in practical needs.

Table 1. Uses of Ceramic Films and Coatings (6)

Use	Typical Ceramic Material
Wear Reduction	Al_2O_3, B_4C, Cr_2O_3, $Cr B_2$ $Cr Si_2$, Cr_3Si_2, DLC*, Mo_2C $MoSi_2$, SiC, TiB_2, TiC, WC
Friction Reduction	MoS_2, BN, BaF_2
Corrosion Reduction	Cr_2O_3, Al_2O_3, Si_3N_4, SiO_2
Thermal Protection	Ca_2Si_4, $MgAl_2O_4$, MgO, ZrO_2 (Mg or Ca stabilized)
Electrical conductivity	In_2O_3/SnO_2, $YBa_2Cu_3O_{7-x}$
Semiconductors	GaAs, Si
Electrical Insulation	SiO_2
Ferroelectricity	$Bi_4Ti_3O_{12}$
Electromechanical	AlN
Selective optical transmission and reflectivity	BaF_2/ZnS, CeO_2, CdS, CuO/Cu_2O, Ge/ZnS, SnO_2
Optical wave guides	SiO_2
Optical processing (electrooptic, etc.)	GaAs, InSb
Sensors	SiO_2, SnO_2, ZrO_2

* DLC = Diamond-like carbon

3.0 PROCESSING OF CERAMIC FILMS AND COATINGS

It can be argued that most if not all of the current film production techniques are merely extensions of processes already studied more than 50 years ago. Such techniques can be grouped into four categories: *(i)* atomic deposition processes, *(ii)* particulate deposition processes, *(iii)* bulk coating, and *(iv)* surface modification (6). The principal techniques in these categories are listed in Table 2. The degree of extension of these basic techniques in recent decades is so great in many cases that it constitutes a revolution in process control and in the type and quality of films and coatings which can be made. For example, the degree of process control now possible allows the growth of epitaxial layers with desirable properties in many systems (1)(2)(7)-(15). A relatively new process is laser ablation which uses very short-pulse lasers to transfer complex compounds to a substrate with little or no change in composition (8).

Progress in vacuum technology in the 1950s and 1960s made it possible to operate at pressures of 10^{-11} torr instead of the previous limit of about 10^{-6} torr (16). This makes it possible to study the chemistry and structure of surfaces without significant contamination. Also, these vacuum techniques can be used to lower the contamination during film production. Even when the film production process operates at a higher pressure, the background contamination can be kept down.

The interaction of depositing ions with the surface is complex. Takagi (17)(18) has emphasized the importance of the kinetic energy of the depositing ions in affecting the processes which occur. Figure 1 from Takagi shows the range of energy per ion and incident flux density of ions corresponding to characteristic operating conditions for deposition, etching, and implantation. Also shown on the figure are the energy ranges that correspond to significant interactions. He notes that an ion energy of a few hundred eV would be very useful but is difficult to achieve because of the space charge repulsion effect. The technique of ionized cluster beam (ICB) deposition is used to avoid space charge repulsion. In this technique, films are deposited by clusters of 500 - 2000 atoms with a small charge per ion compared to individual ions.

The energy of the deposited ion after it has come to equilibrium with the temperature of the surface is also important. Metastable films can be formed if the temperature is too low. The temperature of the substrate is critical to the crystal structure (or lack of it) and the texture of the film. For example, formation of films of superconducting $Ya_1Ba_2Cu_3O_{7-x}$ with good crystallinity and highly preferred orientation requires both a suitable substrate and a deposition temperature above 600°C as discussed in the chapter on superconducting thin films in this volume (Ch. 11).

Some film processing techniques involve a rapid drop in energy of the atoms just before and just after attachment to the substrate. Non-equilibrium phases, both metastable crystalline phases and amorphous phases, can be formed. These typically remain frozen up to temperatures

Table 2. Materials Coating Techniques (6)

- ATOMIC DEPOSITION
  - Chemical Vapor Environment
    - Chemical Vapor Deposition
    - Reduction
    - Decomposition
    - Plasma Enhanced
    - Spray Pyrolysis
  - Electrolytic Environment
    - Electroplating
    - Electroless Plating
    - Fused Salt Electrolysis
    - Chemical Displacement
  - Plasma Environment
    - Sputter Deposition
      - Diode
      - Triode
      - Reactive
    - Evaporation
      - Direct
      - Activated Reactive
    - Ion Plating
      - Hot Cathode Discharge
      - Reactive
    - Diffusion
  - Vacuum Environment
    - Vacuum Evaporation
    - Ion Beam Deposition
    - Ion Implantation
    - Molecular Beam Epitaxy

- PARTICULATE DEPOSITION
  - Fusion Coatings
    - Electrostatic
    - Electrophoretic
    - Sol-Gel
  - Impact Plating
  - Thermal Sprating
    - Plasma Spraying
      - Low Pressure Plasma Spraying
      - Laser Assisted Plasma Spraying
    - Flame Spraying
    - Detonation Gun
    - Electric Arc Spraying

- BULK COATINGS
  - Mechanical
    - Coextrusion
    - Explosive Cladding
    - Roll Bonding
    - Electromagnetic Impact Bonding
  - Electrostatic Spraying
    - Printing
  - Spin Coating
  - Overlaying
    - Laser Glazing
    - Brazing
    - Weld Coating
      - Oxyacetylene
      - Powder Welding
      - Manual Metal Arc
      - Metal Inert Gas
      - Tungsten Inert Gas
      - Submerged Arc
  - Diffusion
    - Diffusion Bonding
    - Hot Isostatic Pressing
  - Wetting Processes
    - Dipping
    - Enameling
    - Painting
    - Spraying
    - Thick Film

- SURFACE MODIFICATION
  - Chemical Conversion
  - Chemical (liquid)
    - Oxidation
  - Chemical (vapor)
    - Thermal
    - Plasma
  - Electrolytic
    - Anodization
    - Fused Salts
  - Leaching
  - Ion Implantation
  - Mechanical
    - Shot Peening
  - Sputtering
  - Surface Enrichment
    - Diffusion From Bulk
  - Thermal
    - Laser Alloying
    - Quenching
    - Diffusion

of about 30% of the melting temperature (10). For very hard materials with melting points above 2500°C this would give a service temperature of 500°C or so. Evidently metastable phases might be useful for some hard material applications.

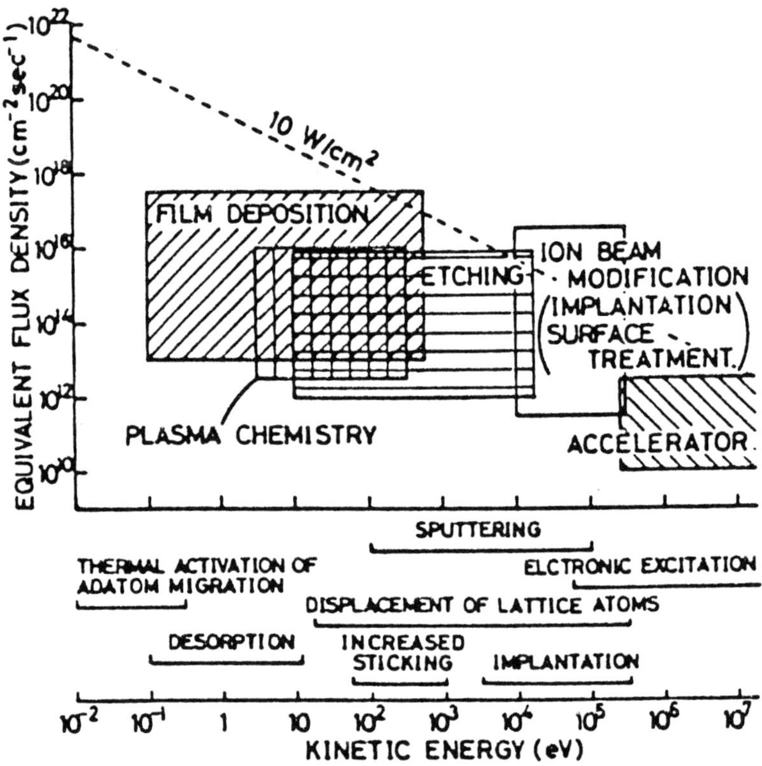

Figure 1. Ranges of kinetic energy and equivalent flux density of incident atoms, corresponding to various engineering applications which include ion-surface or vapor-surface interactions. Kinetic energy ranges of atoms where significant ion-surface or vapor-surface interactions occur are also shown.

4.0 CHARACTERIZATION OF CERAMIC FILMS AND COATINGS

Progress in instrumentation for process control and characterization of films after processing has greatly facilitated modern film research and development. Instruments based on electron-beam technology give compositional and structural information to a very fine scale, approaching atomic dimensions in some cases (19)(20). Transmission electron microscopy allows microanalysis on the scale of 5 nm using energy dispersive x-ray spectroscopy (EDS) and to 20 nm using electron energy loss spectroscopy (EELS) while selective area electron diffraction allows phase analysis on a fine scale. X-ray photoelectron spectroscopy (XPS or ESCA) allows surface analysis of 0.5 mm diameter to a depth of only a few atomic layers. Auger electron spectroscopy (AES) has similar capabilities and allows element-specific images to be formed. Rutherford backscattering spectroscopy (RBS) has become a routine tool for non-destructive depth profiling near a surface.

The principal instrumental microanalytical techniques commonly used to characterize thin films are listed in Table 3 (19). The sensitivity and accuracy of these techniques vary with the material and circumstances so the table entries should be taken as typical generic values rather than precise values. Such characterization is generally concerned with composition and structure (including defect structure). Electron beam instruments operating in spectroscopic or imaging modes (or in combination) have become essential tools for determining elemental composition, phase composition, and microstructure.

An interesting round robin study of chemical analysis techniques was recently reported (21). Titanium nitride films were analyzed by electron probe microanalysis (EPMA), Auger electron spectroscopy (AES), and X-ray photoelectron spectroscopy (XPS). A wet-chemical gravimetric analysis was also performed. Standards were used, the spectra were obtained by operators familiar with titanium nitride, and the data was carefully analyzed to correct for background. The EPMA and AES results for N:Ti ratio agreed well with the wet-chemical analysis results. The XPS data showed some unexpected deviations. The results indicate that the surface analysis techniques can be used to give accurate results but that experience, standards and care are needed.

Semiconductor wafer topography presents special characterization needs. As summarized by Biddle (22), wafer topography includes the aspects of wafer flatness, film thickness, linewidths and spaces, surface profile, and surface roughness. Common film/substrate combinations used in semiconductor wafer technology are summarized in Table 4. A summary of the capabilities of many common techniques for film thickness measurement has been prepared by Biddle and is summarized in Table 5. Again, these values are typical; the references cited by Biddle should be consulted for details.

Table 3. Microanalytical Techniques Useful for Film Characterization (19) (see Appendix A for abbreviations)

Name	Input	Output	Lateral Resolution	Depth Resoulution	Detection Limit	Lightest Element	Imaging	Other Information	Advantages
SEM, EDS	e^- (2 to 30 keV	SE (50 eV)	5 nm	5 nm			yes		routine, easy sample prep, and analysis
		BSE (E_o)	2 μm	2 μm			yes		
		X-rays (>1 kV)	2 μm	2 μm	1000 ppm	Na	yes		
EMPA, WDS	e^- (5 to 50 keV)	X-rays(>100 eV)	2 μm	2 μm	50 ppm	Ba	yes		quantitative analysis
TEM	e^- (100 keV to 1 MeV)	TE (E_o)	0.2 nm	<100 nm			yes	SAED	high resolution, electron diffraction
AEM	e^- (100 to 400 keV)	TE (E_o)	0.2 nm	<100 nm			yes	CBED	analytical resolution
AEM/STEM		SE (50 eV)	3 nm	3 nm			yes		
AEM/EDS		X-rays (>1 kV)	5 nm	<100 nm	1000 ppm	Na	yes		
AEM/EELS		electrons	20 nm	<10 nm	1000 ppm	Li	no		small amts of light elements
AES	e^- (1 to 3 keV)	e^- (>200 eV)	0.2	<3 nm	1000 ppm	Li	yes		quant. surf. analysis
XPS/ESCA	X-rays (1 to 1.5 keV)	e^-(>10 eV)	0.5 nm	<3 nm	1000 ppm	He	no	binding energy	chemical state info.
SIMS	ions (4 to 15 keV)	secondary ions	5	0.1 nm	1 ppb	H	some	depth profile, mass	analytical sensitivity for surface & light elements
PIXE	H^+, HE^{++} (3 MeV)	X-rays	2 μm		1 ppm	Na	some		analytical sensitivity
RBS	H^+, H^{++} (2 to 15 MeV)	H^+, H^{++} (<E_o)	2 μm	10 nm	1 to 1000 ppm	Li	no	depth profile, mass	non-destructive depth profile
LIMA/LMMA	UV (250 nm)	ions	2 μm	2 μm	1 ppm	H	yes	mass	
Micro Raman	vis. light	vis. light	2 μm				some	molecular spectrum	phase identification
Micro FTIR	IR light	reflected or transmitted IR	10 μm				no	molecular spectrum	phase and functional group identification

Table 4. Common Films in Wafer Fabrication (22)

Film Material	Typical Film Thickness (micrometers)	Substrates(s)/ Underlayer(s)
SiO_2	0.01 - 3.0	Silicon
Si_3N_4	0.04 - 1.0	Si, SiO_2
Polysilicon	0.04 - 1.0	SiO_2
Photoresist	0.4 - 3.0	SiO_2
Photoresist	0.05 - 3.0	Silicon
Aluminum	1.0 - 2.0	SiO_2
W, Pt, & Pd	0.1 - 0.5	SiO_2
Au	0.1 - 1.0	Various
Epi Silicon	0.1 - 0.5	Silicon
PSG	0.05 - 1.0	SiO_2

Table 5. Comparison of Film-Thickness Measurement Techniques (22)

Measurement Technique	Thickness Range (micrometer)	Vertical Accuracy (micrometer)	Horizontal Resolution (micrometer)
Beta-backscatter	0.1 - 60	+-2nm to +- 2%	100 - 300
Eddy current	0.1 - 1000	1-2%	2500 - 40000
Ellipsometer	0.02 - 5	0.1nm to 10 nm	25 - 3000
FTIR	0.5 - 1000	Sample dependent	20 - 250
Micro-spectrometry	0.01 - 4	Sample dependent	3.5
Stylus Profiling	0.005 - 160	±6% - ±1.2nm	0.04 - 50
X-Ray Fluorescence	0.05 - 300	Sample dependent	125 - 250

5.0 TRENDS IN CERAMIC FILMS AND COATINGS

Several trends in the development and use of ceramic films and coatings are evident. The already vast use of polymers is increasing and many applications need hard protective coatings and diffusion barriers. Processing techniques compatible with the relatively low temperature limit for most polymers are needed. Electronic and optical technology needs very thin films patterned with small lateral dimensions. The need for sensors for industrial processes and for medical applications has opened a broad range of special requirements for films. The desire to tailor the properties of films has lead to efforts to control processing at the atomic level and to allow for the building up of controlled microstructures on a nanometer scale.

A comparison of market estimates for ceramic thin films was prepared by Richardson (23) and is given in Table 6.

Table 6. The Market for Ceramic Coatings as Estimated by Various Sources. Estimates are in 1985 $ million (23).

	U.S.		World	
Source	1985	1995	1985	1995
Gorham	360	1640	—	—
SRI	700 - 800	—	—	—
Kline	—	—	1100	3000
BCC	240	585	—	—

Some especially interesting trends in ceramic thin films are *(i)* the development of synthetic diamond films, *(ii)* the development of high Tc superconducting films, *(iii)* the development of the sol-gel method for making thin films, *(iv)* the important improvements in cutting tool performance resulting from the use of ceramic thin films, *(v)* the extensive use of thin films in the manufacture of semiconductor integrated circuits, *(vi)* the use of very large area thin films on architectural and automotive glass.

Subsequent chapters treat many of these areas in detail. A brief introduction is given here to indicate major directions and the potential for growth in the use of ceramic coatings.

5.1 Diamond Coatings

Synthesis of diamonds has presented a major challenge to scientists (24)(25). A process using high pressure and high temperature in combination was announced by General Electric in 1955. The resulting grains are useful in cutting, grinding and polishing but the process does not lend itself to producing coatings or to working with materials which would be destroyed by the temperatures and pressures involved. As early as 1958, some success in producing diamond films at relatively low pressures with very slow growth rates was reported. Within the last decade great progress has been made in Russia, then Japan, and subsequently in the U. S. and Europe, in growing true diamond films from the vapor phase on substrates at 800 to 1000°C. Several vapor phase processes have successfully produced diamond films, but it appears that all require activation of the gas to give appreciable growth rates and the presence of atomic hydrogen for efficient growth. The potential applications of diamond films are summarized in Table 7 which is taken from Spear (24) who expanded it from one given by Nishimura, Kobashi, Kawate, and Horiuchi.

Table 7. Properties and Applications of Diamond Coatings (24)

Properties	
Hardest known material	Electrical insulator
Low coefficient of friction	High band gap semiconductor (either p- or n-doped)
High thermal conductivity	Low dielectric constant
Low thermal expansion	High hole mobility
Heat resistive	Visible and IR transparent
Acid resistive	Large refractive index
Radiation resistive (to x-ray, ultraviolet, gamma)	
Applications	
Coatings for cutting tools	Radio-frequency electronic devices
Abrasive coatings	High-frequency electronic devices
Coatings for bearings	Sensors for severe environments
Heat sinks for electronic devices	Window and lens materials
High-power microwave devices	Electro-optic devices

5.2 High Tc Superconducting Ceramic Thin Films

The processing of high Tc films presents a special challenge. The most widely used high Tc ceramic is $Y_1Ba_2Cu_3O_{7-x}$ where x<<1. Processing of thin films of this material is treated in more detail in Chapter 11 of this book. Here we briefly note that the promise of applications of thin films of this material is high, the film quality required is quite high, the processing difficulties are great, but that films of useful quality have been produced (26)(27). The major techniques used include pulsed laser deposition, electron beam evaporation, sputtering (single target and multi-target), and metal-organic chemical vapor deposition.

5.3 The Sol-Gel Method for Making Ceramic Thin Films

Sol-gel (or solution-gelation) technology is the process of reacting liquid state precursors to form a porous unfired ceramic shape. The most common case involves the reaction of a metal alkoxide with water to form an oxide. The term sol-gel is somewhat imprecisely defined but is increasingly taken broadly to include liquid state chemical preparation and to include chemical routes to produce carbides and nitrides as well as oxides (12)-(14).

Preparation of a ceramic by a sol-gel technique usually involves three stages: *(i)* hydrolysis and condensation to form individual particles or polymers, *(ii)* linking of these to form a high viscosity gel, and *(iii)* aging and drying to remove all or part of the solvent and form a shrunken and porous ceramic precursor. This precursor is then heated to a temperature which depends upon the degree of consolidation wanted.

Ceramic thin films are made by applying the sol or gel to a substrate at a point in the sol-gel processing before the viscosity has become too high. The sol-gel process is complicated in detail with many competing reactions occurring in parallel. Considerable variation in microstructure and properties is possible with a given chemical system through process variables (14)(15)(28). Applications of sol-gel coatings are subject to competition from coatings made by other processes. Most applications are emerging from the research stage with a few, such as coating of large sheets of window glass, having become a commercial reality. This technology is still in an early stage of growth. Pierson (29) lists recent applcations as follows:

- Contrast enhancement coatings for computer monitors and terminals
- Passivation coatings for solar cells and optical discs
- Indium tin oxide and other coatings for electroluminescent panels
- Oriented birefringent films
- Achromatic coatings
- Color effect filters
- Coatings of plastic for improved surface properties such as abrasion resistance

- Multilayer antireflection films or painting on art glass, stained glass, TV screens, monitors, glass cases and the like
- Magneto-optic materials of yttrium garnet for optical waveguides
- Transparent cathodoluminescent coatings for cathode ray tubes

Paquette (30) discusses the potential for sol-gel processing to produce:

- Diffusion of oxidation barriers for metals and composites
- Fiber reinforcement barrier coatings for composites
- Advanced fiber development
- Matrices for low dielectric constant composites

5.4 Ceramic Thin Films on Cutting Tools

The use of ceramic films on tools for metal cutting is one of the greatest success stories for ceramic thin films (30)-(32). Titanium carbide coatings on tungsten carbide tools were introduced in 1969 and gave dramatic improvements in tool performance (31). An order of magnitude improvement in machining productivity can result from increased cutting speeds, feeds and deeper cuts (32). High hardness and good chemical stability at high temperature are important film properties. Good adhesion to the substrate and good microfracture toughness are needed. Complex multilayer coatings (up to four layers) are now in use. Clavel lists TiN/TiC/TiCN/TiN, TiN/TiC/Al_2O_3, TiN/Al_2O_3/TiN, and TiN/TiCO/Al_2O_3 as the most widely used.

Hard coatings on cemented carbide cutting tools are so successful that the technology might be considered mature. About 67% of all metal cutting inserts in the U.S. and Europe are now coated (32). Several factors may cause further change. These include: *(i)* use of even more layers, *(ii)* use of PVD rather than CVD to give harder coatings, and *(iii)* use of diamond and boron carbide coatings.

5.5 Ceramic Thin Films in Semiconductor Integrated Circuits

Capasso (34) concludes that materials science techniques, particularly epitaxial growth of semiconductor thin films, is one of the three areas which together are driving developments toward nanoscale and ultrafast electronic devices. The other two areas are the physical understanding of electronic transport and device fabrication technologies such as nanolithography.

Wilson (35) indicates that state-of-the-art integrated circuits require 100 or more processing steps of which 20% or more involve the deposition of conducting and dielectric films. As many as 20 or more films may be required. Chemical vapor deposition and plasma-enhanced chemical vapor deposition are favored techniques. According to Wilson the commonly used dielectrics are:

Oxide:	CVD silicon dioxide
	plasma silicon dioxide
Silicates:	Phosphosilicate glass
	Borosilicate glass
	Borophosphosilicate glass
Nitrides:	Plasma silicon nitride
Oxynitrides:	Plasma silicon oxynitride
	Spin-on-glass

This film technology is clearly at the heart of today's semiconductor devices and will probably be vital to the next generation (36)(37). Most of the thin film processing techniques covered in other chapters of this book are pertinent even if discussed in other contexts.

The properties of thin films can differ appreciably from those of bulk material of the same composition and these are becoming important as the thickness (and sometimes other dimensions) of films continues to be reduced as electronic devices are further miniaturized. The electrical conductivity decreases as the film becomes thinner (38). Various quantum effects become important at small dimensions. For short times (less than one picosecond) the electron velocity can exceed its drift value (34). Layer structures made up of successive thin films (composition-modulated foils) can have elastic modulus values three to five times the bulk value (39). In such composition-modulated foils, the short diffusion distances can cause changes in properties with time (40).

5.6 Ceramic Thin Films on Architectural and Automotive Glass

In contrast to the ultraminiaturization of electronics, there is a growing class of applications requiring very large areas of uniform, high quality thin films. Use of coated glass on a large scale to control solar heat gain began in the mid 1960's (41). Early glass panels were made by pyrolytic coating and then by chemical vapor deposition or electron-beam deposition. Sputtering deposition began in 1968 and horizonatal sputtering deposition in 1974. One cannot escape seeing the results on the sides of modern buildings in the cities of the developed world. The technology is a triumph of applying an optically uniform coating over a large area (up to 321 cm by 600 cm) in enormous volume (over 2 million square meters per year from a single coating line). A development coming into use is the coating of curved glass. One application is for electrically heated windshields to melt ice and act as defoggers. A potential future application of very large area coatings is for solar cells.

6.0 CONCEPT OF THE PRESENT BOOK

Activity in thin films in general and ceramic thin films in particular is accelerating (42). Most of the papers being published report on progress in specific processes and compositions. Such papers typically discuss details of how process variables affect the composition and structure of the films in particular systems. Such papers are essential to progress in specific films, but assume considerable background knowledge. They are accordingly of limited use to the reader wishing to get an overall picutre of a broad area of ceramic film technology. The present book is intended to present a series of major areas of ceramic thin film technology and to facilitate the reading of papers on specific compositions and processes. A list of abbreviations and acronyms used in this book is included in Appendix A.

APPENDIX A. List of Abbreviations and Acronyms

AE	1) Auger Electron, 2) Acoustic Emission
AEM	Analytical Electron Microscopy
AES	Auger Electron Spectroscopy
APS	Atmospheric Plasma Spraying
ARE	Activated Reactive Evaporation
BE	Binding Energy
BSE	Backscattered Electrons
CBED	Convergent Beam Electron Diffraction
CMA	Cylindrical Mirror Analyzer
CTE	Coeffieient of Thermal Expansion
CVD	Chemical Vapor Deposition
DRM	Dynamic Recoil Mixing
DTA	Differential Thermal Analysis
EDAX	Emergy Dispersive Analysis of X-rays
EDS	Energy Dispersive Spectroscopy
EELS	Electron Energy Loss Spectroscopy
ELS	Electron Loss Spectrometer
EPMA	Electron Probe Micro Analyses
ESCA	Electron Spectroscopy for Chemical Analysis, generally used as another name for XPS but sometimes used in a more general sense to include Auger as well
FTIR	Fourier Transform Infrared Spectroscopy
HIC	Hybrid Integrated Circuit
HREELS	High Resolution Electron Energy Loss Spectroscopy
HSA	Hemispherical Analyzer
HTSC	High Temperature Superconductor
IAC	Ion Assisted Coating
IAD	Ion Assisted Deposition
IBAD	Ion-Beam Assisted Deposition
IBED	Ion-Beam Enhanced Deposition
ICB	Ionized Cluster Beam
ISS	Ion Scattering Spectroscopy
IVD	Ion Vapor Deposition
KE	Kinetic Energy
LAMMA	Laser Microprobe Mass Analyzer = LIMA
LEIS	Low Energy Ion Scattering
LIMA	Laser Ionization Mass Spectrometry = LAMMA

LMMA	Laser Microprobe Mass Spectrometry = LAMMA
LPCVD	Low Pressure Chemical Vapor Deposition
LPE	Liquid Phase Epitaxy
LWP	Long Wave Pass
MBE	Molecular Beam Epitaxy
MOD	Metallo-Organic Decomposition
OMCVD	Organometallic Chemical Vapor Deposition
PIXE	Particle (usually proton) Induced X-ray Emission
PLD	Pulse Laser Deposition
PVD	Plasma Vapor Deposition
QMS	Quadrupole Mass Spectroscopy
RBS	Rutherford Backscattering Spectroscopy
RED	Radiation Enhanced Diffusion (Deposition)
RIBED	Reactive Ion-Beam Enhanced Deposition
RIS	Radiation Induced Segregation
SAED	Selected Area Electron Diffraction
SEM	Scanning Electron Microscopy
SIMS	Secondary Ion Mass Spectroscopy
SIS	Superconductor-Insulator-Superconductor
SQUID	Superconducting Quantum Interference Device
STEM	Scanning Transmission Electron Microscope
SWP	Short Wave Pass
TBC	Thermal Barrier Coating
TCR	Thermal Coefficient of Resistance
TEM	Transmission Electron Microscopy
TGA	Thermogravimetric Analysis = TG
TTBC	Thick Thermal Barrier Coating
UPS	Ultraviolet Photoelectric Spectroscopy
WDS	Wavelength Dispersive Spectroscopy
XPS	X-ray Photoelectron Spectroscopy = ESCA

REFERENCES

1. *Metallurgical Coatings 1987,* 4 vols. (R. C. Krutenat, ed.), Elsevier Applied Science (1987)

2. Pulker, H. K., "Coatings on Glass," *Thin Films Science and Technology,* Vol. 6, Elsevier (1984)

3. Chapra, D. L. and Kaur, I., *Thin Film Device Applications,* Plenum Press (1983)

3a. Sayer, M., and Sreenivas, K., *Science,* 247:1056-1060 (2 March 1990)

4. *Coatings for High Temperature Applications,* (E. Lang, ed.), Applied Science Publishers (1983)

5. Nyce, Andrew C., organizer, *The Global Business and Technical Outlook for High Performance Inorganic Thin Films and Coatings, Monterey, California,* Gorham Advanced Materials Institute, Gorham, Maine (Oct. 30 - Nov. 1, 1988)

6. Wachtman, J. B., Jr. and Haber, R. A., *Chemical Engineering Progress,* pp. 39-46 (January 1986)

7. Vossen, J. L., and Kern, W., *Thin Film Processes,* Academic Press (1978)

8. See Chapter 11 on superconducting thin films in this volume

9. Zaat, J. H.. *Annual Reviews of Materials Science,* 13:9-42 (1983)

10. Hersee, S. D., and Duchemin, J. P., *Annual Reviews of Material Science,* 12:62-80 (1982)

11. Reinberg, A. R., *Annual Reviews of Materials Science,* 9:341-372 (1979)

12. Hess, D. W., *Annual Reviews of Materials Science,* 16:163-183 (1986)

13. Tu, K. N. and Rosenberg, R., *Treatise on Materials Science and Technology,* Vol. 24, Academic Press (1982)

14. Klein, L. C., *Annual Reviews of Materials Science,* 15:227-248 (1985)

15. Klein, L. C., *Sol-Gel Technology for Thin Films, Fibers, Preforms, Electronics and Specialty Shapes,* Noyes Publications, Park Ridge, NJ (1988)

16. Stuart, R. V., *Vacuum Technology, Thin Films, and Sputtering - An Introduction,* Academic Press (1983)

17. Takagi, T. *Thin Solid Films,* 92:1-17 (1982)

18. Takagi, T., *J. Vac. Sci. Technol.* pp. 382-388 (April-June 1984)

19. Friel, J. J., Princeton Gamma Technology, private communication with the authors

20. Young, W. S., McVay, G. L., and Pike, G. E., *Ceramic Transactions,* Vol. 5, The American Ceramic Society (1989)

21. Perry, A. J., Strandberg, C., Sproull, W. D., Hofman, S., Ernsberger, C., Nickerson, J., and Chollet, L., *Thin Solid Films,* 153:169-183 (1987)

22. Biddle, D A., *Microelectronic Manufacturing and Testing,* pp. 15-17 (March 1985)

23. Richardson, R. J., *The Global Business and Technical Outlook for High Performance Inorganic Thin Films and Coatings, Monterey, California,* Gorham Advanced Materials Institute, Gorham, Maine (Oct. 30 - Nov. 1, 1988)

24. Spear, K. E., *J. Am. Ceram. Soc.,* 72(2):171-191 (1989)

25. Messier, R., Badzian, A. R., Badzian, T., Spear, K. E., Bachmann, P, and Roy, R., *Thin Solid Films,* 153:1-9 (1987)

26. Simon, R., *Superconductor Industry,* pp. 22-27 (Spring 1989)

27. Ford, R. G., *Materials and Processing Report,* Vol. 3, No. 11/12 (1989)

28. Brinker, C. J., *The Global Business and Technical Outlook for High Performance Inorganic Thin Films and Coatings, Monterey, California,* Gorham Advanced Materials Institute, Gorham, Maine (Oct. 30 - Nov. 1, 1988)

29. Pierson, H. O., *The Global Business and Technical Outlook for High Performance Inorganic Thin Films and Coatings, Monterey, California,* Gorham Advanced Materials Institute, Gorham, Maine (Oct. 30 - Nov. 1, 1988)

30. Paguette, E. L., *The Global Business and Technical Outlook for High Performance Inorganic Thin Films and Coatings, Monterey, California,* Gorham Advanced Materials Institute, Gorham, Maine (Oct. 30 - Nov. 1, 1988)

31. Clavel, A., *The Global Business and Technical Outlook for High Performance Inorganic Thin Films and Coatings, Monterey, California,* Gorham Advanced Materials Institute, Gorham, Maine (Oct. 30 - Nov. 1, 1988)

32. Quinto, D. T., *The Global Business and Technical Outlook for High Performance Inorganic Thin Films and Coatings, Monterey, California,* Gorham Advanced Materials Institute, Gorham, Maine (Oct. 30 - Nov. 1, 1988)

33. Sarin, V. K., *The Global Business and Technical Outlook for High Performance Inorganic Thin Films and Coatings, Monterey, California,* Gorham Advanced Materials Institute, Gorham, Maine (Oct. 30 - Nov. 1, 1988)

34. Capasso, F., *Physics Today,* pp. 22-23 (February 1990)

35. Wilson, S. R., *The Global Business and Technical Outlook for High Performance Inorganic Thin Films and Coatings, Monterey, California,* Gorham Advanced Materials Institute, Gorham, Maine (Oct. 30 - Nov. 1, 1988)

36. Ghandhi, S. K., *VLSI Fabrication Principles,* Wiley-Interscience (1983)

37. Sze, S. M., *VLSI Technology,* McGraw Hill (1983)

38. Eckertova, L., *Physics of Thin Films,* Plenum Press (1986)

39. Tsakalakos, T., and Jankowski, A. F., *Annual Reviews of Materials Science,* 16:293-313 (1986)

40. Tu, K. N., *Annual Reviews of Materials Science,* 15:147-176 (1985)

41. Hall, D. S., *The Global Business and Technical Outlook for High Performance Inorganic Thin Films and Coatings, Monterey, California,* Gorham Advanced Materials Institute, Gorham, Maine (Oct. 30 - Nov. 1, 1988)

42. "High Performance Ceramic Films and Coatings," *Materials Science Monographs,* (P. Vincenzini, ed.), 67, Elsevier Science Publishers (1991)

2

CVD Coated Cutting Tools

Thomas E. Hale

1.0 INTRODUCTION

CVD coated cutting tools were first introduced in 1969 and were an instant hit due to their significant impact on metal cutting productivity (1). Typical performance gains of 100 - 300% in tool wear or 50% in cutting speed were realized with the initial commercial offerings of TiC coated products, representing the most significant single improvement in cutting performance obtained since the introduction of cemented carbides. Within 10 years, about 40% of the cutting tools sold were coated and today the coated usage stands at about 70%. Today every cutting tool manufacturer in the world offers CVD coated tools and most of them make their own coatings, making the carbide cutting tool industry one of the largest users of the CVD coating process.

In the first decade of coatings usage, significant improvements were made in the quality and performance of coated tools through improved control of the CVD process to reduce coating porosity and improve coating thickness uniformity. Improvements were also made in the control of the metallurgy of the coating-substrate interface, resulting in improved coating adhesion and suppression of strength-degrading interface reactions. Additional improvements resulted from the modification of the cemented carbide substrate compositions to optimize the two most important substrate properties, fracture toughness and thermal deformation resistance. TiN and Al_2O_3 coatings were added to the coatings offering and achieved major commercial success.

The 1980's, the second decade of CVD coated cutting tool usage, featured the rise to prominence of multiple layered coatings using layer combinations such as TiN-TiC, TiN-TiCN-TiC, TiN-Al_2O_3-TiC, and alternating layer sequences having as many as 15 or more layers. The multiple layer coatings provided additional improvements in metal cutting performance by combining the best features of the different coating types and providing improved control of the coating grain structure. Significant modifications of the cemented carbide substrates, featuring surface region enrichment of the cobalt matrix phase, also gained acceptance during this period, resulting in improved combinations of toughness and cutting speed capability. The improvements made during this period permitted coatings to be very successfully applied to the more severe interrupted type machining operations, such as milling, whereas they were previously confined to lower impact straight turning operations due to toughness limitations.

In this chapter, the types of coatings used for cutting tools are described along with some details of the CVD processes used. The relationships between coating composition, thickness, and metal-cutting wear enhancement are also described.

2.0 TiC COATINGS

TiC, the first commercially successful CVD coating used on a cutting tool, was introduced in late 1969 and was found to yield substantial improvements in wear resistance when used to machine steel and cast iron. Figure 1 shows steel machining wear resistance as a function of cutting speed for a commercial uncoated cemented carbide of the type used to machine steel vs. the same material with an optimal quality TiC coating. The nearly order-of-magnitude improvement obtained under the laboratory conditions used is not usually observed in "real" shop operations due to the higher impact loading obtained machining "real" parts, but wear resistance improvements of 200% - 400% are common in shop operations for this coating when used to machine steel and cast iron at the speed normally used for uncoated cemented carbide grades. An even greater production benefit is obtained by increasing the cutting speed to 50% - 100% higher than used with the uncoated material, yielding the same tool life as the uncoated material at the lower speed, but significantly improving the production rate and overall cost of the operation.

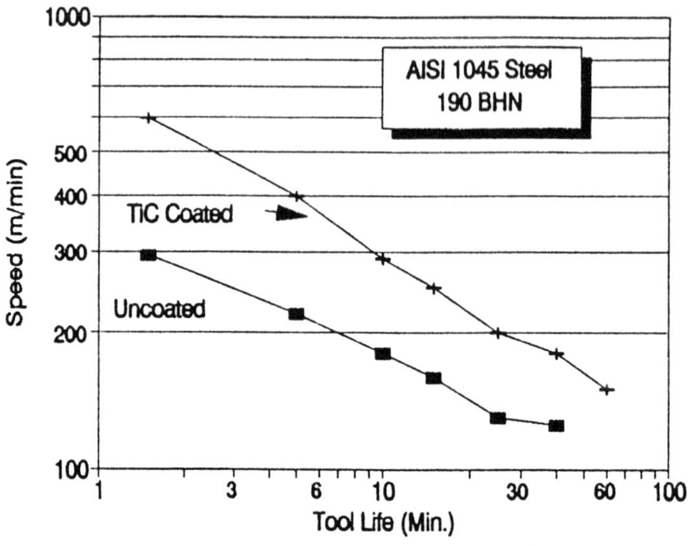

Figure 1. Cutting speed vs. tool life showing the effect of TiC coating.

2.1 CVD Process Conditions for TiC Coatings

Figure 2 illustrates a typical CVD reactor used for coating cutting tool inserts. The reactor is a hot wall type usually constructed from Inconel or high chromium stainless steel, capable of operating at temperatures up to about 1200°C. The commercial CVD reactors used for the production coating of cutting tools are sized to handle about 5,000 to 20,000 cutting inserts per run. The inserts are set on screens or ventilated shelves and stacked vertically in the reactor. Since all corners and edges of most inserts are used in machining operations, each insert must be uniformly coated top, bottom, and all sides. The coating conditions used should thus be surface catalyzed and surface reaction controlled in order to obtain such coating uniformity. For TiC coatings, these requirements are easily met using coating temperatures between about 950°C and 1100°C and a vertical flow of reactant gases consisting of hydrogen as the carrier, 1 - 5 vol% $TiCl_4$ for the titanium source, and 2 - 5 vol% CH_4 for the carbon source. The $TiCl_4$ gas is generated either by bubbling part of the carrier gas through liquid $TiCl_4$ or by direct vaporization of liquid $TiCl_4$. System pressures ranging from about 100 torr to atmospheric are used with average gas flow velocities ranging from about 1 cm/sec to about 10 cm/sec translated to atmospheric pressure (STP). Under this range of conditions, the TiC coating is deposited with excellent "throwing power,"

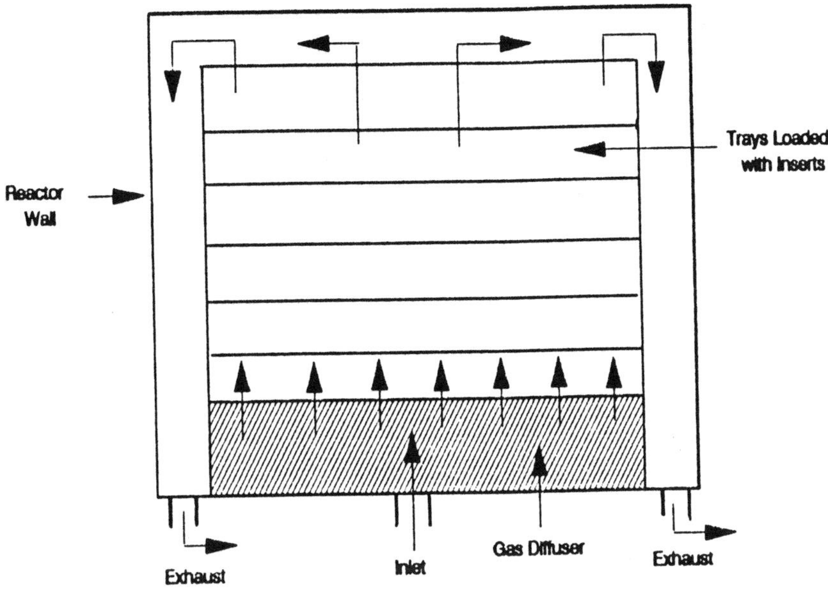

Figure 2. Schematic showing CVD reactor for tools.

readily coating the bottoms of the inserts sitting on the shelves and even penetrating any fine cracks that might be present on the inserts (not present, of course, on good quality inserts). Care must be taken in the design of the reactor gas flow paths to avoid recirculation of exhaust products, especially HCl, which can pick up reactor wall materials (Fe, Ni, and Cr) and deposit them on the cutting inserts. Such deposits can form soft metallic underlayers which cannot resist the shear forces of metalcutting, causing the coating to spall off in service.

TiC Deposition Rate. Within the general range of deposition conditions described above, the deposition rate increases with increasing temperature and system pressure (2). At low CH_4 concentrations (CH_4:$TiCl_4$<1), a significant portion of the carbon supplied to the coating is derived by diffusion from the substrate, causing the formation of the ternary compound, eta phase (W_6Co_6C), at the coating-substrate interface. In this deposition regime, the deposition rate also depends upon the carbon activity of the cobalt matrix phase, showing significantly higher rates at higher carbon activities (3). The deposition rate increases with increasing CH_4 concentration up to at least 12 vol% CH_4 (4). The deposition rate for commercial coatings ranges from about 1 to about 5 microns/hr.

TiC Coating - Substrate Interface. Control of the coating-substrate interface is critical to the success of coated cutting tools. There is probably no other application of CVD coatings that places such high shear stresses on the coating as occurs in metal cutting. In the first few years of coated tool usage, poor interface metallurgy was frequently the reason why one coated insert performed worse than another in the same application. The key ingredients for good interface control are optimization of coating-substrate interactions, elimination of interface porosity, and minimization of deposition of unwanted materials from the reactor environment.

The principal coating-substrate reaction that can occur during the deposition of TiC coatings is formation of the brittle ternary carbide, eta phase (W_6Co_6C), which occurs when the gas phase carbon activity is relatively low, causing carbon from the substrate to be used as a major part of the carbon source for the coating. The tendency for eta phase formation is also enhanced by low initial carbon contents of the cobalt matrix phase of the substrate. By proper adjustments of the gas phase CH_4:$TiCl_4$ ratio and substrate carbon content, eta phase formation can be completely eliminated, if so desired. There is evidence, however, that coating adhesion is promoted and toughness is not significantly degraded by the presence of a slight amount of eta phase (5), so most commercial processes are carefully balanced to avoid excessive eta phase formation and still obtain excellent coating adhesion.

In addition to eta phase formation, there is some diffusion of substrate elements, mainly tungsten (also tantalum and niobium, if present), into the coating, such that most TiC coatings show the presence of these elements when carefully analyzed. While the presence of low levels of these elements is not known to be deleterious to coating performance, excessive diffusion is avoided since it often causes pore generation in the substrate and interfacial regions with resultant degradation of performance. Reasonable control of substrate diffusion results from use of deposition temperatures no greater than about 1050°C and limiting deposition times to no more than a few hours.

Minimization of porosity in the interface and adjacent substrate regions is the most difficult coating process control issue. Dirty or contaminated substrate surfaces prior to being placed in the reactor and subsequent downfall of old deposits and chemicals from improperly cleaned reactors are frequent causes of coating and interface porosity issues. Diffusional porosity can result from use of conditions promoting excessive substrate-coating interactions, including the formation of excessive amounts of eta phase (greater than about 2 microns thickness) and excessive movement of substrate elements into the coating. With proper control, there should be at most only very slight porosity evident in the interface or immediately adjacent substrate regions.

3.0 TiN COATINGS

CVD TiN coated cutting tools made their appearance in the early 1970's and were distinguished by their pleasing gold color and claims of reduced metal cutting friction coefficient and improved crater wear resistance relative to TiC coatings. Metal cutting data showed TiN coatings have less flank wear resistance and equal or somewhat greater crater wear resistance than TiC coatings of the same thickness (6)(7). TiN coatings have superior resistance to galling and metal buildup in soft steel cutting operations and there is some indication that TiN coated tools have greater edge toughness than their TiC coated counterparts, perhaps leading to the popularity of TiN coatings for low speed steel turning, threading, and milling operations. The most popular usage of TiN coatings is as a component of multilayer coatings where TiN is used as the top layer and sometimes in the subsurface layers. In multilayer coatings, the top TiN layer improves the appearance of the tool with its pleasing gold color and it minimizes metal buildup and sticking problems associated with machining soft, gummy steels.

3.1 CVD Process Conditions for TiN Coatings

CVD TiN coatings are deposited using the same basic reactor design and process conditions as used for TiC coatings, substituting N_2 gas for the CH_4 used in TiC coatings. The deposition rate is not as sensitive to N_2 concentration as the TiC coat process is to CH_4 content and N_2 contents ranging from about 10% to about 75% or more are used. The deposition rate, crystal morphology, and uniformity are influenced by $TiCl_4$ concentration, gas flow velocity, total pressure, and temperature, leading to more variation in the commercial process conditions used for TiN coatings than for TiC coatings. The deposition rate can be dominated by either mass transport or surface reaction rates, depending upon the processing conditions employed (8). Typical commercial deposition rates are from about 1 to about 5 microns/hr.

4.0 Al_2O_3 COATINGS

The Al_2O_3 coating was the third commercially successful coating that emerged for use on cutting tools in the 1970's. The potential of Al_2O_3 for superior metalcutting performance was well known and exploited with some success in solid ceramic tools. The outstanding wear resistance of these tools when used for high speed machining of cast irons and steels was attributed to the superior thermodynamic stability of Al_2O_3 relative to the carbides and nitrides normally used in bulk compositions and coatings. The Al_2O_3 resisted the chemical dissociation and dissolution reactions that dominate the metalcutting wear process at higher cutting speeds (9). Solid Al_2O_3 based ceramic tools were, however, much more brittle than cemented carbide tools, limiting their usage to only light duty, low stress cutting operations. On the other hand, when applied as a coating of optimal thickness, the superior wear resistance of Al_2O_3 was obtained and the superior toughness of the cemented carbide tool was maintained.

Unlike TiC and TiN coatings, which are metallurgically compatible with cemented carbide substrates, the Al_2O_3 coating has no such compatibility, imposing much greater technical challenges to obtain good adhesion and avoid deleterious reactions with the substrates. The coating is basically deposited by the reaction of H_2O with $AlCl_3$ (or Al_2Cl_6) in the presence of H_2 carrier gas and the timing of reactant gas introduction into the reactor is critical to avoid unwanted oxidation reactions with the substrate surface. When properly controlled, a good coating can be directly deposited upon a cemented carbide substrate, but adhesion to the substrate is usually poor, insufficient for commercial metalcutting operations.

Two basic approaches to solve the challenge of providing commercially useful Al_2O_3 coated carbide tools were developed in the early 1970's. The first method (10)(11) formed a thin non-metallic underlayer between the alpha Al_2O_3 coating and the substrate which was composed of a complex phase containing elements of both the coating and the substrate. This underlayer bonded well to the carbide substrate and to the coating, thus providing the needed adhesion.

The second method developed for Al_2O_3 coatings (12) utilized a first layer of TiC and then deposited at reduced pressure the Al_2O_3 layer which under the conditions employed formed significant amounts of the kappa phase rather than the alpha form deposited by the first method and which had good adherence without any apparent underlayer. The adherence in this case was attributed to development of an epitaxial relationship between the non-surface-oxidized TiC layer and the Al_2O_3 layer (13).

Other methods have now been developed which form bonding underlayers of complex oxides, oxycarbides, oxynitrides, oxycarbonitrides, etc. The use of bonding underlayers is generally necessary for alpha phase coatings thicker than about 2 microns.

4.1 CVD Process For Al_2O_3 Coatings

The CVD reactors used to make Al_2O_3 coatings are of the same design as described previously for TiC and TiN coatings. The input gases are usually a mixture of H_2 (carrier), 1 - 5 vol% $AlCl_3$, and 1 - 10 vol% CO_2, but HCl and CO gases may be optionally added to help moderate the deposition rate and uniformity. The range of conditions to make good coatings on cemented carbide substrates is:

Temperature:	950 to 1100°C
Pressure:	30 torr to 1 atm.
$AlCl_3$:	1 - 5 vol%
CO_2:	1 - 10 vol%
HCl:	0 - 10 vol%
CO:	0 - 10 vol%
Flow velocity:	0.5 - 5 m/min @ STP

As the input gases heat up on their way to the deposition zone of the reactor, the CO_2 reacts with H_2 (the well known "water-gas" reaction) to form H_2O and CO. The H_2O then hydrolizes the $AlCl_3$ to form the Al_2O_3 phase. Deposition is mainly mass transport rate controlled and is significantly influenced by reactant concentrations (H_2O, $AlCl_3$, HCl), system pressure, and gas flow velocity. If deposition conditions are not carefully chosen, significant amounts of Al_2O_3 dust can form in the reactor and on the substrates due to gas phase or "homogeneous nucleation." "Throwing power" is relatively poor, necessitating careful attention to placement of the substrate inserts in order to obtain good gas flow around all critical wear surfaces and thus assure uniform coating coverage of these surfaces. The initial reactant gas introduction sequence and timing have significant effects on the final oxidation state of the substrate and the initial nucleation conditions for the coating and every cutting tool manufacturer has its own proprietary "recipe" to obtain the proper interface composition and structure. Despite all of these difficulties, almost all of the modern coated grades used for high speed steel and cast iron machining contain a layer of Al_2O_3 in order to have the requisite wear resistance to be commercially competitive.

5.0 MULTI-LAYER COATINGS

The TiC, TiN, and Al_2O_3 coatings described above are all used commercially as single layer coatings and they each have metalcutting applications in which they excel. The TiC coating has unequaled low speed abrasion resistance and thus has superior flank wear resistance when used for relatively low speed (100 - 200 m/min) steel or cast iron machining operations. The Al_2O_3 coating, on the other hand, has outstanding resistance to chemical wear and thus displays superior crater wear resistance and even superior flank wear resistance at high cutting speeds (200 - 600 m/min) where chemical dissociation triggered wear mechanisms dominate. Figure 3 illustrates the differences in wear resistance between TiC and Al_2O_3 coatings when used to machine steel. Both coated tools had superior wear resistance to the uncoated tool, but have different slopes of tool life vs. cutting speed, resulting in superior life for the TiC coating at speeds less than about 200 m/min while the Al_2O_3 coating was superior at speeds above about 200 m/min. Similar crossover speeds are observed in all steel and cast iron turning tests, but the value of the crossover speed changes with the machinability rating of the workpiece. When TiN coatings are run in such tests, they show similar life vs. speed

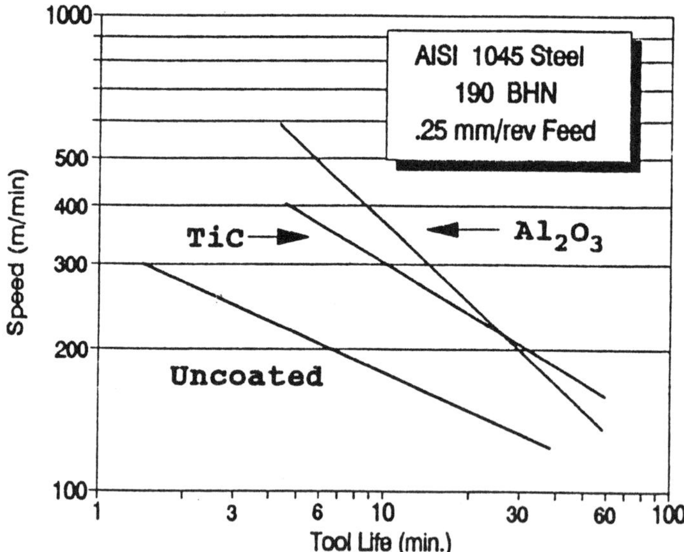

Figure 3. Plot of cutting speed vs. tool life showing the superiority of oxide coating at high speeds.

slopes as TiC coatings, but at lower overall life values. The virtue of TiN shows up in "real world" machining operations where it displays superior metal buildup resistance and toughness.

Since each coating has advantages, it is easy to conceive the concept of multi-layer coated tools having a layer of each type and thus obtain a multi-purpose tool with broad range capabilities. While there are many processing problems involved in making multiple layers and there are thickness constraints imposed by toughness considerations, multi-layer tools have been developed which do display some of the desired broad range attributes. In Fig. 4, the tool life curve for a double layer coating of Al_2O_3 over TiC is compared with monolayer TiC and Al_2O_3 data, showing intermediate life values between the Al_2O_3 and TiC coatings, thus a broader range of speed capability.

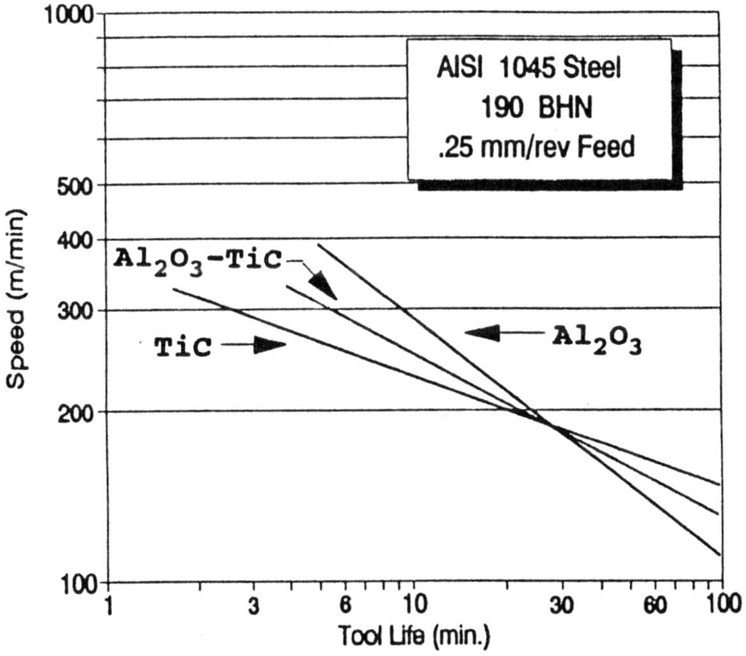

Figure 4. Multi-coat wear resistance.

5.1 TiN-TiC Type Multi-Layers

Tools with coatings containing TiN, TiC, and, sometimes, titanium carbonitride (TiCN) constitute one of the two major classes of multi-layer coated tools in current commercial service. These tools always use a top layer of TiN which has the most attractive appearance and color and which imparts superior resistance to metal sticking and buildup during machining operations. Additionally, the light gold color contrasts sharply with the dark appearance of a used edge, improving the ability of the machine operator to distinguish used from unused edges. TiC layer(s) are used in this coating class to obtain the superior wear resistance imparted by TiC. A layer of TiCN, which has intermediate characteristics between TiN and TiC, is sometimes used instead of TiC or is placed between TiN and TiC layers with the claim of providing a good transition between the two. The total number of layers used in commercial TiN-TiC type multi-layer coatings varies from two to about eight or more, sometimes using alternating combinations of TiN, TiC, and, in some cases, TiCN. The use of alternating layers tends to produce finer grain structure and smoother coatings, with potential toughness advantage. Total thickness is maintained in the range of about 3 - 15 microns due to the constraint imposed by the effect of total thickness upon toughness.

5.2 Al_2O_3 Layer Multi-Coatings

The second major class of multi-layer coatings is distinguished by containing one or more layers of Al_2O_3. The simplest coating sequence used in this class consists of a top layer of Al_2O_3 and then a base layer of TiC. The many variations of this sequence involve the use of multiple TiC-TiN layers in the base layer, complex transition layers between the Al_2O_3 and the other layers, and multiple Al_2O_3 layers, some containing 10 or more Al_2O_3 layers separated by thin layers of TiN or TiC. A TiN top layer is often employed in this class of coatings. The Al_2O_3 layer tends to take the kappa phase form when thinner than about two microns and when deposited directly on TiC or TiN layers. It consists mainly of the alpha form when thicker and when bond enhancing underlayers are employed. One commercial coating on the market uses two different Al_2O_3 layers consisting of an upper layer of kappa and a lower layer of alpha, separated by a proprietary transition layer. The kappa form is softer, has a finer grain structure, and has fewer micro-pores than the alpha phase (13), imparting potentially higher toughness to the kappa layer coatings, while the alpha phase would be expected to have superior wear resistance. Both types impart superior high speed wear resistance relative to other coating types when the wear mode is dominated by chemical dissociation reactions (high speed steel and cast iron machining). The total thickness of Al_2O_3 multi-coatings generally ranges from about 7 - 15 microns.

5.3 CVD Processes For Multi-Layer Coatings

Multi-layer coatings are made in the same type of CVD reactors used to make single layers and the complete sequence of layers is usually deposited in one run. Since the deposition parameters must be varied widely for each layer and transition periods are often needed to switch from one layer type to another, a very complex sequence of cycle steps is required, which can only be accomplished reliably and with good reproducibility using computer controls to sequence the changes in gases, temperature and system pressure. The practicality of multi-layer coatings has thus only recently been established with the advent of improved reactor controls using fully automated cycle sequencing.

6.0 COATING THICKNESS OPTIMIZATION

The performance of CVD coated cutting tools is strongly affected by the thickness of the coating layer(s). If the coating is too thick, it excessively weakens the cutting edge, causing the tool to be unnecessarily brittle with attendant high risk of fracture in service. Excessively thick coatings also tend to have a very coarse crystal structure and many cracks due to thermal expansion mismatch with the substrate, both of which also contribute to poor metal-cutting performance. On the other hand, coatings that are too thin do not provide sufficient wear resistance enhancement when compared with coatings of optimal thickness. Determination of optimal coating thickness is thus a very important element in the design of a commercially successful coated cutting tool grade and, once determined, development of reliable CVD manufacturing techniques to maintain the thickness tolerances is one of the principle factors determining the quality of the coated grade.

6.1 Cutting Tool Wear Modes

Figure 5 shows a schematic drawing of a cutting tool illustrating the two principle wear regions that occur during machining operations. The crater wear region is the wear spot that forms when the hot chip from steel, cast iron, or nickel alloy workpieces rubs the top or rake face of the tool. Crater wear is predominately caused by chemical breakdown and dissolution reactions and is strongly retarded by the coating. Flank wear occurs on the edge or side of the tool and is caused by the rubbing contact between

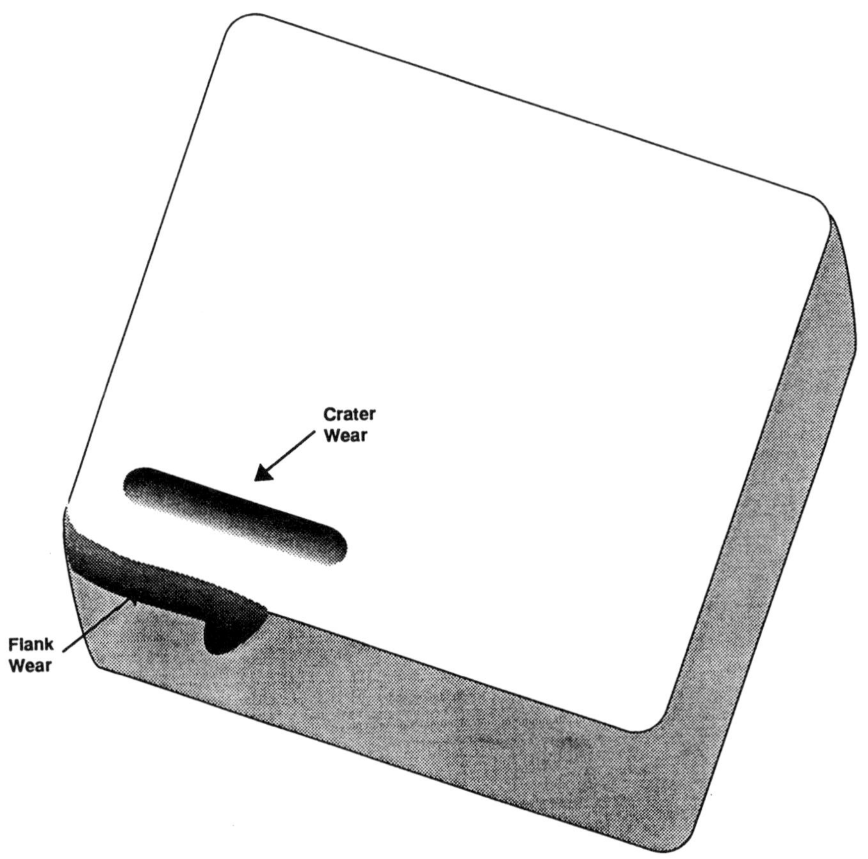

Figure 5. Metal cutting wear patterns.

the tool and the workpiece being machined. Flank wear is dominated by mechanical abrasion at lower cutting speeds and then transitions to chemical wear domination at high speeds, hence the crossover wear responses observed in Fig. 3 where the TiC coating with its outstanding mechanical abrasion resistance was superior at the lower speeds and the Al_2O_3 coating was best at the higher speeds.

6.2 Influence of Thickness Upon Flank Wear Resistance

Figure 6 shows the influence of coating thickness upon flank wear resistance for an Al_2O_3 coating used to machine alloy steel and for a TiC coating machining a medium carbon steel. In both cases, the flank wear resistance increases rapidly with increasing coating thickness up to about 4 - 6 microns and then levels off with further thickness increase. Similar thickness response has been observed for all of the other commonly used coatings when used to machine steels or cast irons. It is thus apparent that a coating thickness of about 4 - 6 microns is optimum for flank wear resistance.

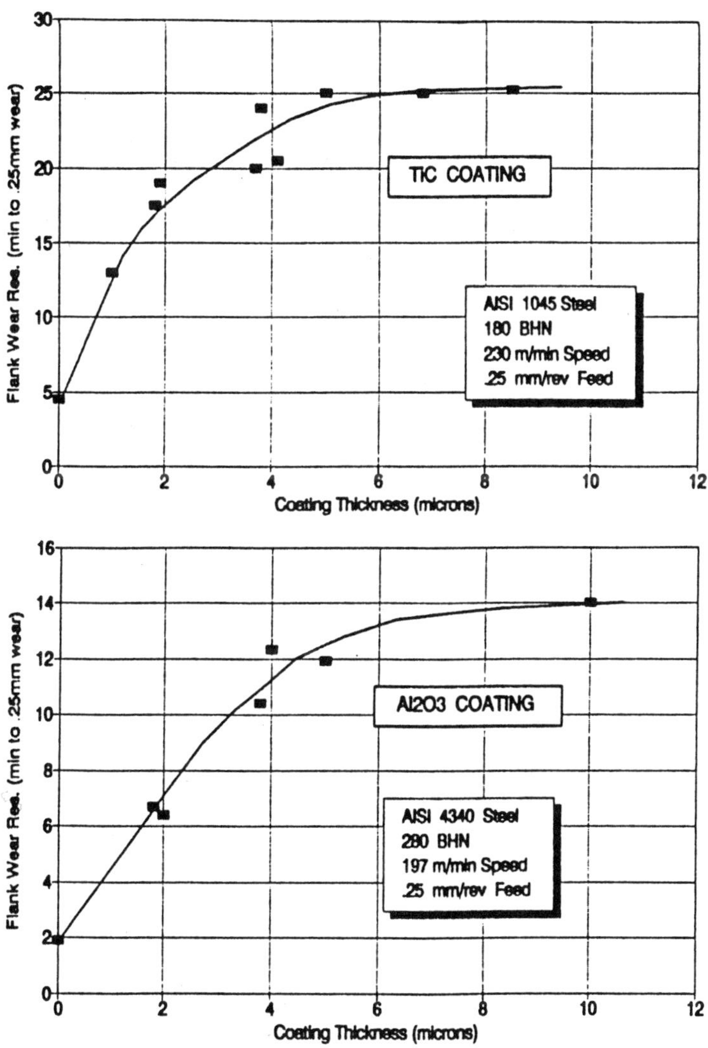

Figure 6. Flank wear resistance vs. coating thickness.

6.3 Thickness Influence On Crater Wear Resistance And Strength

In Fig. 7, the crater wear resistance of the three most common coatings is seen to increase linearly with increasing coating thickness. In order to minimize crater wear, the coating should thus be very thick. In Fig. 8, however, it is seen that the strength of the coated tool is decreased with increasing coating thickness and thus the coating should be very thin to maintain good cutting edge toughness. The optimum thickness with regard to the two types of wear and edge toughness is thus generally greater than four microns in order to maximize flank wear resistance and less than about 15 microns to obtain sufficient crater wear resistance to let flank wear be the dominant failure mode and to avoid excessive weakening of the cutting edge. In general, tools designed to operate at low to medium cutting speeds (100 - 300 m/min), where flank wear dominates and a moderate degree of toughness is required, are manufactured with coatings about 4 - 7 microns thick while tools designed for higher speed operations, where crater wear is dominant, will use thicker coatings, up to about 15 microns total. For high impact operations, such as milling, thinner coatings (2 - 4 microns) are often used to maintain higher edge strength.

6.4 Thickness of Multi-Layer Coatings

The thickness of the individual layers in multi-coated inserts is generally at least about one to three microns for each major wear resisting layer, usually TiC and/or Al_2O_3, and a few tenths to about 2 microns for transition layers and a top TiN layer. Since the edge strength is determined by the total thickness, the total thickness of multi-coated inserts is usually limited to about 7 - 15 microns.

7.0 OTHER COATINGS

The coatings TiC, TiN, and Al_2O_3 are used in the overwhelming majority of CVD coated cemented carbide cutting tool applications. While these coatings were among the first to be tried for cutting tool applications, they were by no means the only coatings evaluated, but have instead reached their position of dominance by an evolutionary development process. Some of the other coating compositions which have either demonstrated their value or appear to have potential are briefly discussed in this section.

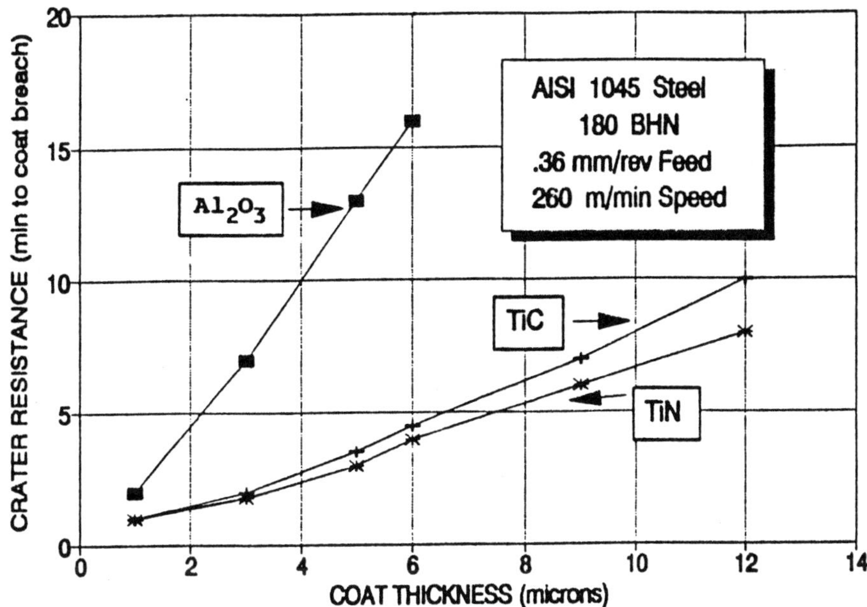

Figure 7. Crater resistance vs. coating thickness.

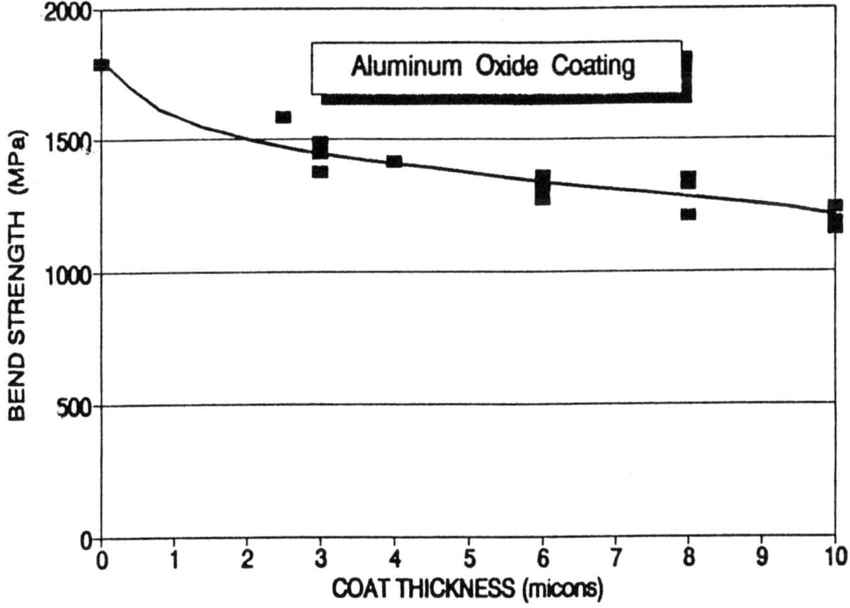

Figure 8. Coating thickness vs. bend strength.

7.1 Hafnium and Zirconium Based Coatings

As might be expected by their position in the periodic table of the elements, hafnium and zirconium form carbides, nitrides, and carbonitrides that have the same structure and general properties as their titanium analogues and coatings can be deposited using analogous gaseous reactants (hafnium and zirconium chlorides, hydrogen, methane, and nitrogen). Formation of hafnium and zirconium chloride vapors is more complex than for titanium tetra-chloride since the Zr and Hf chlorides are solid at room temperature and vaporize at higher temperatures by sublimation. Zirconium and hafnium chloride vapors are most readily formed by direct chlorination of the metal, a procedure requiring some safeguarding to prevent any accidental mixing of Cl_2 and hydrogen with potentially explosive results.

The general conditions used to make good zirconium or hafnium based coatings are:

Temperature:	1000 - 1200°C
Pressure:	20 torr - 1 atm.
$Zr/HfCl_4$:	1 - 10 vol%
N_2:	10 - 80 vol%
CH_4:	1 - 10 vol%
Flow velocity:	2 - 10 m/min @ STP
Thickness:	1 - 10 microns

Influence of Coating Composition on Machining Performance of Zr/HfCN Coatings. Unlike TiCN coatings, which display the best metal-cutting flank wear resistance with 0% nitrogen (pure TiC), HfCN coatings have optimal flank wear resistance at a C:N ratio of about 2.5 and ZrCN coatings are best at a C:N ratio of about 0.45 (14). The crater wear resistance of ZrCN and HfCN coatings is superior to that of TiCN coatings at any C:N ratio, although not as good as Al_2O_3 coatings of the same thickness. Additional enhancement of metal-cutting performance was observed (15) by use of thin underlayers of TiN/TiC to enhance the edge strength of the coated insert.

Commercial Uses of Zr/Hf Based Coatings. HfN coatings are offered commercially by one carbide supplier and ZrN is used as the top layer of a multi-coated insert by another supplier. The HfN coating provides good performance relative to TiN-TiC type coatings (16), but is generally outperformed by multi-coated products, especially those containing a layer of Al_2O_3. There are no known commercial products using carbonitride coatings of Hf or Zr, probably due to the cost and complexity of manufacturing such coatings and due to the equivalent or superior performance of the currently available multi-coated products.

7.2 TiB_2 Coatings

The excellent hardness and abrasion resistance of TiB_2, better than any of the current coatings used on cemented carbide substrates, makes this compound appear very attractive for applications where purely mechanical abrasive wear modes predominate (wear parts, dies, molds, coal cutters, etc.). On the other hand, TiB_2 does not have good potential for machining iron group metal alloys due to the strong reaction between TiB_2 and such metals, forming iron group metal borides. TiB_2 coatings have been developed and evaluated in several laboratories and have demonstrated good mechanical wear resistance enhancement (17)(18), but there are no known commercial products so far.

CVD Process For TiB_2 Coatings. TiB_2 coatings can be made in the same type CVD reactor previously described for TiC/TiN coatings, using H_2, $TiCl_4$, and BCl_3 as the principle reactants:

Temperature:	700 - 1000°C
Pressure:	50 torr - 1 atm.
$TiCl_4$:	1 - 10 vol%
BCl_3:	1 - 10 vol%
Flow Velocity:	1 - 5 m/min @ STP
Thickness:	1 - 10 microns

Due to the strong reaction tendency between boron or borides and the cobalt cementing phase of the cemented carbide substrate, the direct deposition of TiB_2 onto the substrate will usually result in the formation of a considerable quantity of brittle boride phases in the interface region of the substrate. This problem can be controlled by first depositing an underlayer of TiN or TiC to suppress diffusion of boron into the substrate. If properly controlled, a low level of boron diffused to a depth of about 10 to 30 microns into the substrate, followed by the TiB_2 layer (with or without a TiC/TiN underlayer), has been found to enhance abrasive wear resistance (18).

7.3 Tungsten Carbide Coatings

Due to its high hardness, relatively high toughness, and low thermal expansion coefficient, WC coatings have the potential for superior performance in applications requiring a good combination of mechanical abrasion resistance and impact resistance. However, there are no known commercial products utilizing CVD tungsten carbide coatings.

Coatings of W_2C and W_3C were CVD deposited on steel substrates and demonstrated good wear characteristics (19) and a tungsten-carbon coating with 0.6 - 2.0% C was made which demonstrated outstanding strength characteristics (20). Tungsten carbide coatings would be expected to be inferior to TiC and Al_2O_3 coatings in all machining operations involving iron group workpieces due to the poor chemical reaction resistance of WC. WC coatings would also probably be inferior to TiB_2 coatings in operations requiring only mechanical abrasion resistance but might excel when a combination of good mechanical abrasion resistance and edge toughness is required.

CVD Process Conditions for Tungsten Carbide Coatings. Tungsten carbide coatings are formed at temperatures ranging from 500 to 1100°C using mixtures of H_2, WCl_6 or WF_6, and CH_4 or a more complex hydrocarbon such as C_6H_6, at pressures ranging from 5 - 500 torr. At lower temperatures mixed deposits of tungsten metal and lower carbides (W_2C, W_3C) form with little or no WC. WC deposits can be formed on cemented carbide substrates at temperatures above 900°C and coatings of the lower carbides can be converted to WC by use of carburizing heat treatments.

REFERENCES

1. Wilson, R., *Iron Age* (November 1970)

2. Lee, C. et al, *Thin Solid Films* 86:64-71 (1981)

3. Hara, A. et al, *9th Plansee Seminar* (1977)

4. Stjernberg, K. et al, *Thin Solid Films* 40:81-88 (1977)

5. Sarin, V. et al, *6th Int. Conf. on CVD* (1977)

6. Hale, T., *Int. Machine Tool Show Technical Conf.* (Sept. 1982)

7. Peterson, J., *J. Vac. Sci. Tech.,* 11:4 (July/Aug 1974)

8. Kim, M. et al, *Proceedings of the Fourth European Conference on Chemical Vapour Deposition* (May/June 1983)

9. Suh, N., *The Carbide Journal*, pp. 3-9 (Jan/Feb 1977)

10. Hale, T. U.S. Patent #3,736,107 (May 1973); reissues #32,093 (March, 1986) and #32,110 (April 1986)

11. Hale, T., U.S. Patent #4,018,631, (April 1977)

12. Lindstrom, J. et al, U.S. Patent 3,837,896 (Sept 1974)

13. Chatfield, C. et al, *Journal De Physique, Colloque C5,* supplement au n°5, Tome 50 (mai 1989)

14. Hale, T., U.S. Patent #3,854,991 (Dec 1974)

15. Hale, T., U.S. Patent #4,268,569 (May 1981)

16. Leverenz, R., *Mfg. Eng.* (July 1977)

17. Pierson, H. et al, *Thin Solid Films* 54:119-128 (1978)

18. Hale, T., U.S. Patent 4,268,582, (May 1981)

19. Archer, N., *Wear* 48:237-250 (1978)

20. Glaski, F., *The Carbide Journal* (Jan-Feb 1975)

3

Wear Resistant Thin Films by Ion Implantation

*Carl J. McHargue**

1.0 INTRODUCTION

The removal of material from a solid surface by mechanical forces is influenced by material parameters (hardness, fracture toughness, yield strength, surface free energy) as well as system parameters (force, loading velocity, environment). Since the processes involved in material removal occur at the surface, changes in the surface mechanical properties, in the surface composition, and in the phases present should strongly affect the tribological response.

In this discussion, wear processes will be divided into four major types: *(i)* adhesive wear, *(ii)* abrasive wear, *(iii)* corrosive or oxidative wear, and *(iv)* surface fatigue and fracture. Each of these processes may involve several steps which depend upon material properties that may be modified by surface processing treatments. Mathematical models proposed to describe these wear mechanisms frequently contain hardness and fracture toughness as the primary material parameters. In many real systems, two or more wear processes may occur simultaneously.

Friction is also an important parameter since it determines the manner in which an applied tangential force is transferred between contacting moving components. Generally, the lower the friction, the lower is the resultant wear. The coefficient of friction can be strongly influenced by surface chemical and phase composition.

---

*Research sponsored by the Division of Materials Sciences, U.S. Department of Energy, under contract DE-AC05-84OR21400 with Martin Marietta Energy Systems, Inc.

Ion beam processing can be used to either alter the microstructure and composition of a near-surface layer or to synthesize a new compound on or near the surface of a suitable substrate. In the present discussion, attention will be given to both approaches—the formation of a modified surface on a ceramic substrate and the synthesis of ceramic compounds on metallic and ceramic substrates.

The ion beam processes to be covered herein are ion implantation, ion beam mixing (IBM), and ion beam assisted deposition (IBAD). These processes are characterized by the use of high energy ions (tens to thousands of keV). In comparison, techniques such as ion plating, ion nitriding, and ion cluster beam deposition use ions of much lower energy (in the range of 1 keV or less).

Bombarding the surface of a material with ion beams can serve several purposes. First, it allows the introduction of impurity or alloying elements in a highly controlled and reproducible manner that is not limited by the usual thermodynamic constraints. The injection of the ions is a physical process that depends upon the kinetic energy of the accelerated ion and may produce compositions and microstructures that cannot be obtained by other techniques. Thus, solid solutions can be made from elements which would ordinarily be considered to be completely immiscible in the solid state.

Second, the ions deposit their kinetic energy into the near-surface region of the target. This deposited energy may be in the form of large numbers of defects (vacancies, interstitials, defect clusters, dislocations), as intense ionization of local regions, or as local "hot-spots" or thermal "spikes." The resultant microstructure is largely governed by the number and kind of defects that remain after the bombardment. Since the spatial extent of each ion's influence is limited to a few hundred or a few thousand atoms and the lifetime of the cascade events is short, less than a nanosecond, the results often are not characteristic of equilibrium thermodynamics. Such effects cannot be achieved by simply raising the temperature of the surface.

Additionally, the ion beam causes sputtering at the surface. Sputtering limits the composition attainable by ion implantation. It also continuously cleans the surface of deposited impurities during ion beam assisted deposition.

In the following chapter, we will first discuss the characteristics of the wear of ceramics to identify the properties that might be altered by ion beam processing. Then, the ion beam techniques will be described and examples of their application to the production of wear-resistant ceramic surfaces will be given.

2.0 WEAR PROCESSES IN CERAMICS

2.1 Friction

By 1800 Coulomb had described the role of surface roughness on the friction between two moving and contacting bodies. The "laws of friction" recognized that the frictional force is proportional to the normal load applied to the sliding contact and that the frictional force is independent of the area of the sliding surface. However, frictional processes involve the dissipation of energy, a fact not accommodated by the simple roughness model. Nevertheless, the concept of surface irregularities is central to today's understanding of friction.

The modern view of friction, as summarized by Tabor (1), considers the important factors to be: *(i)* the true contact areas, *(ii)* the type and strength of bonds formed at the contact areas, and *(iii)* the deformation and rupture at and around the contact areas.

As two clean pieces of a metal are brought into contact, bonds first form due to van der Waals forces. These increase as the separation distance decreases, leading to screened van der Waals bonds. Much stronger metallic bonds form at a separation distance corresponding to about one interatomic distance. The bonding between dissimilar metals has often been discussed in terms of mutual solubility. The suggestion that mutually soluble pairs show strong adhesive bonding while insoluble pairs show weak adhesive bonding is useful as a general guideline, but does not hold rigorously.

Since strong bonds form only at interatomic distances, the presence of a monolayer of absorbed gas may reduce the interaction to that of the residual van der Waals forces. Thus the presence of impurities or thin oxide films strongly affect the friction.

Considerable attention has been given to the deformation and rupture of the bonds formed at the contact areas. The simplistic view considers failure to occur at the weakest point, randomly in both surfaces for pairs of similar material or in the weaker component of dissimilar couples. If the interface junction is brittle, it fractures with little or no plastic deformation, and the work to break the bond should be close to the theoretical bond strength. If the junction is ductile, the work involved in plastically deforming the surrounding region must be considered, and the work-hardening characteristics of the materials become important.

Following the arguments of Tabor (1), friction arising from interfacial bonds, in the absence of deformation or ploughing, is given as

Eq. (1) $$F = A_o\tau = \frac{W}{\sigma_o}\tau$$

where F = friction force, A_o = true contact area, W = normal force, σ_o = yield strength, and τ = interfacial shear strength. The coefficient of friction is then

Eq. (2) $$\mu = F/W = \tau/\sigma_o$$

If there is a ploughing deformation, an additional term must be added to account for the work of deformation. The important factor for the present discussion is the dependence of friction on the material parameters of shear strength and yield strength.

2.2 Adhesive Wear

Adhesive wear results from the formation of junctions at contacting asperities on the two surfaces and subsequent rupture of the junctions. It is characterized by the transfer of one material to the surface of the other. Figure 1 contains a photograph of a wear track made by a diamond stylus sliding on a TiB_2 disk. The diamond adhered to the TiB_2 as a carbonized carbon layer and exhibited a high wear rate, whereas only a small amount of the TiB_2 was removed (2). A profilometer trace across the wear track shows the buildup of the adhered layer.

Junction formation is often attributed to welding due to local temperature increases caused by the sliding of rough surfaces or to local adhesive bonding. Because of the small tip size of asperities and the high local loads, the instantaneous temperature can be quite high. Junction formation due to solid state bonding (adhesion) is the same as discussed above for friction. Thus, adhesive wear should decrease as one progresses from couples of the same material and those that form solid solutions to insoluble couples and to metal-ceramic pairs.

Again more attention has been given to the breaking of junctions than in their formation. As a basis for discussing the effect of ion beam processing on the wear properties, it is instructive to examine the wear models in order to determine which properties may be altered by the processing. Among the proposed models are *(i)* shearing of the weaker material (3)(4), *(ii)* minimization of surface energy (5), *(iii)* critical strain for crack growth, i.e., fracture toughness (6), and *(iv)* the delamination process (7).

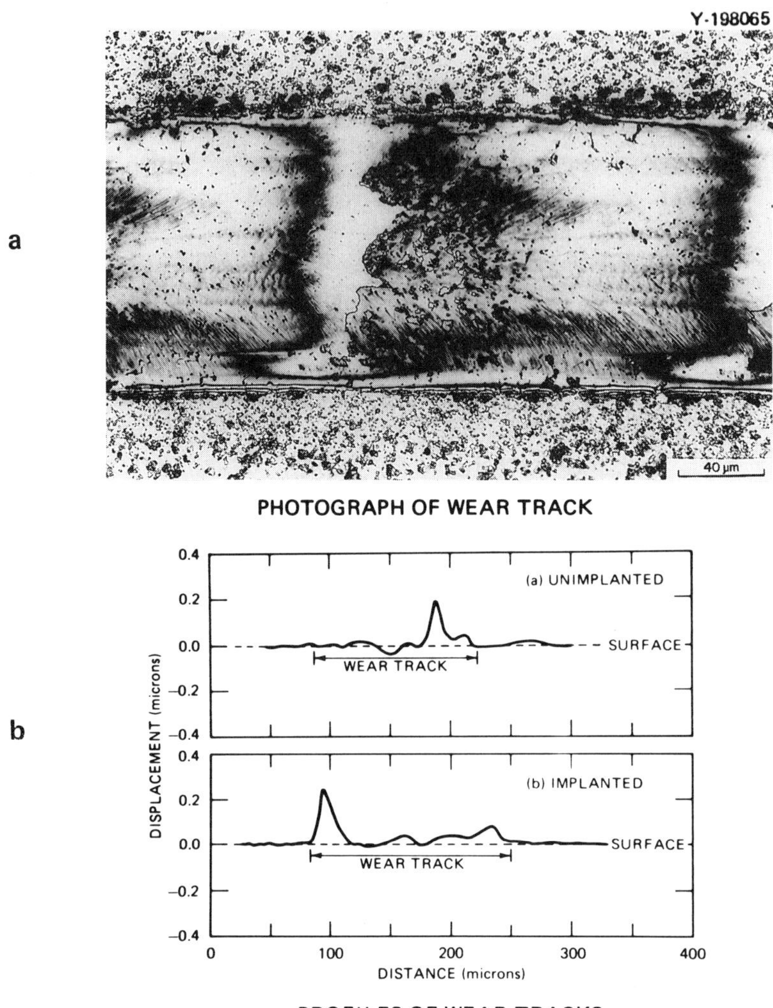

Figure 1. (a) Optical photograph of an adhesive wear track made by diamond sliding on TiB_2. (b) Profilometer trace of (a). (Ref. 2.)

Starting with the premise that sliding causes the rupture of junctions formed due to the contact of asperities, Archard (3) obtained the wear rate, W, as

Eq. (3) $$W = KL/\sigma_y \quad \text{or} \quad W = K'L/H$$

where K or K′ is the probability that a junction between two surfaces leads to the formation of a wear particle (i.e., is removed), L is the applied load,

σ_y is the yield stress, and H is the hardness. The hardness is assumed to be three times the yield stress in the latter expression. This is a simple relationship between the wear rate and the hardness of the deforming (weaker) material.

Rabinowicz (5) proposed that the size of the particles removed or transferred from one surface to the other is determined by balancing the elastic volume energy against the surface free energy. This approach leads to the expression for wear rate

$$\text{Eq. (4)} \qquad W = Nn\pi \left[\frac{3\Gamma_{ab}}{\nu^2 Y \varepsilon^2} \right]^2$$

where N = number of atoms removed per interaction, n = number of contacts, Γ_{ab} = surface free energy of the a-b interface, ν = Poisson's ratio, Y = Young's modulus, and ε = maximum elastic strain. This expression relates the wear volume of interfacial free energy and the elastic properties of the solid.

Hornbogen (6) modified Archard's expression to express the wear rate in terms of fracture toughness for conditions under which crack growth determines the removal of material. In this case,

$$\text{Eq. (5)} \qquad W = K' \frac{\Gamma^{3/2}\, m^2 Y \sigma_y}{K_I^2\, H^{3/2}}$$

where K' and Γ are empirical constants, m is the work-hardening coefficient, σ_y is the yield strength, K_I is fracture toughness, and H is hardness.

Perhaps the most detailed description of the material removal mechanism is given by the delamination model of Suh (7). According to this model, the asperities on the softer material are quickly removed by fracture and further wear occurs by asperity (harder material) contact on a plane (softer material). This type of wear introduces subsurface deformation which accumulates with repeated loading. Eventually cracks form and propagate approximately parallel to the surface. When such cracks emerge at the surface, long thin sheets of material "delaminate." This model leads to an expression of the form

$$\text{Eq. (6)} \qquad W = \frac{\mu}{H\, K_I}$$

where μ is the coefficient of friction.

Certain of the material properties that appear in the above relationships can be altered by ion bean processing. As we will see, ion implantation may change the hardness, yield stress, and fracture toughness of ceramics. The surface energy may be changed by changing the surface composition.

2.3 Abrasive Wear

Abrasive wear is characterized by the removal of material from one of two surfaces in relative motion caused by the presence of hard protuberances or hard particles either between the surfaces or embedded in one of them. The action is primarily cutting or ploughing and the worn surface contains grooves or scratches. Following the arguments of Archard (3) and Mulhearn and Samuels (8), the following expression for volume wear rate is obtained:

Eq. (7) $$W = \frac{BF\sigma_a}{H}$$

where σ_a = applied stress, B = the fraction of contacts that actually remove material, and F = fraction of groove volume removed.

Again, there is a simple inverse relationship between volume of material removed and the hardness. However, although the data of Kruschov (9) show that such an expression may hold for a given class of material, there are marked deviations among different classes (e.g., cold-worked metals vs. annealed metals). Evans and Wilshaw (10) found that the wear resistance of brittle solids varied as fracture toughness and hardness.

2.4 Surface Fracture

Material removal due to direct surface fracture is important for brittle solids and may occur during abrasive or adhesive wear conditions. Figure 2 shows the groove made in SiC during a scratch test (11). There are many radial cracks which extend into the surrounding material for distances comparable to the groove width. Much of the material was removed in large pieces due to the lateral cracks caused by the tensile component of the stress which is present during unloading as the stylus passed the sample. Since the material is removed by the linking of cracks, the wear rate should be inversely proportional to fracture toughness and hardness. Since the fracture mode of brittle solids is sensitive to the details of the stress state, being particularly susceptible to tensile stresses, the residual surface stresses introduced by ion implantation can affect the wear response.

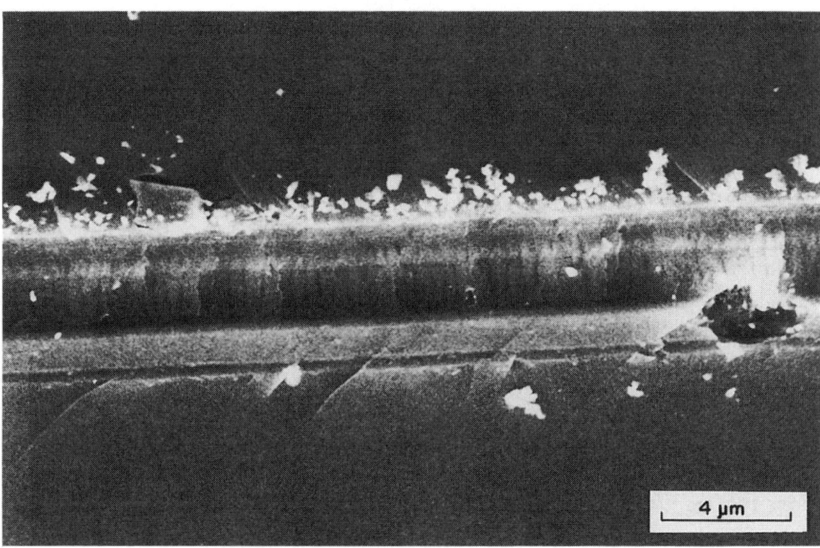

Figure 2. Scanning electron micrograph of surface cracking on and around the wear track made in α-SiC. (Ref. 11.)

3.0 FUNDAMENTAL PROCESSES IN ION-SOLID INTERACTIONS

The ion beam processes described herein consist of bombarding the surface of a material in a vacuum chamber by an electrostatically accelerated beam of ions. The process begins by forming a plasma that contains the element to be implanted. The desired ions are extracted from the plasma and accelerated to an energy typically in the range of 10 to 400 keV. For ion implantation, there seems to be a trend to higher energies and recent studies have used energies of 1 MeV or more. The ion beam may be passed through bending magnets in order to extract impurities by allowing only one isotope to pass.

Because of their kinetic energy, the ions become embedded to a depth controlled by the incident ion energy for a given ion/target material combination. This range is generally less than a micrometer. The ion comes to rest by dissipating its kinetic energy in displacing target atoms (ions) from their normal lattice sites and by ionizing them, thus producing large numbers of defects before coming to rest as an impurity, dopant, or alloying element.

The nature of the process allows us to introduce any element into the near-surface region of any solid in a controlled and reproducible manner that is independent of most equilibrium constraints. Since the process is non-equilibrium in nature, compositions and structures unattainable by conventional methods may be produced.

In metals and ceramics, the concentrations required for property alteration are in the 0.1 to 20 at.% range; hence, damage levels are high and the resultant defect structure may be complex. The structure and property changes may be a result of the damage as well as the composition changes.

Many features of the damage microstructure in metals and alloys are well understood because of the large body of literature generated by studies of radiation damage induced by neutron bombardment in nuclear reactors. In many cases there are reasonably accurate theoretical models to describe the evolution of the microstructure due to the generation of point defects.

Implantation- or radiation-induced damage in ceramics is much more complex and less studied. In the displacement cascade, one must deal with at least two sublattices that have different atomic masses and may have different displacement energies. The types of defects that are produced are strongly influenced by requirements of local charge neutrality, the local stoichiometry, and the nature of the chemical bonding of the particular lattice. In addition, ionizing effects may be significant in producing lattice defects, whereas, in metals such effects are unimportant.

3.1 Range of Incident Ions

An ion incident on a target continuously loses energy by collisions with the nuclei and electrons of the target atoms until it reaches thermal energies. The time required for an ion to come to rest is of the order of 10^{-14} sec. The major mechanisms of energy loss are *(i)* direct collisions with the screened nucleus of a target atom, and *(ii)* excitation of electrons bound to such atoms. Each process is energy-dependent and makes different contributions to the energy loss along the ion's path. To a reasonable approximation, these may be considered to be independent processes so that the linear rate of energy loss is given by the sum of the two contributions. The nuclear stopping usually dominates at lower energies and electronic stopping dominates at higher energies.

Both processes make significant contributions in the region typically of interest for ion implantation and ion beam mixing. The nuclear stopping is important in determining ion ranges, displacement damage, and sputtering. The electronic stopping is important for excitation phenomena which result in secondary electron emission and can produce defects in insulators. The energy transfer processes are well understood theoretically (12)-(14) and there are a number of computer programs (15)-(17) or tabulations (18)(19) that can be used to calculate the profiles of the final ion distribution and the deposited energy.

To a first approximation, the distribution of ions or deposited energy is given by a Gaussian function. However, in some cases, a Type IV Pearson distribution with four moments gives a better fit to experimental observations than does the Gaussian distribution (20)(21). The Gaussian usually underestimates the peak concentration on highly skewed distributions, i.e., for higher energies and lighter ion masses.

The solutions of the LSS theory (12) for stopping power and ranges presented in published tables are normally limited to ions entering single element system. To a reasonable approximation, the stopping cross-section of multi-element targets (compounds, alloys) are linearly additive. According to Bragg's rule, the energy loss is the sum of the losses in the constituent elements weighted proportionately to their concentration in the target. Such estimates are generally accurate to within 10%. Reference 22 contains a tabulation of ion ranges in a number of ceramics.

3.2 Defect Production and Retention

The nuclear component of the energy loss is dissipated in elastic collisions which may cause atom displacements in a crystal for energy transfers greater than some threshold value. This threshold for displacement, E_d, lies in the range of 20 to 40 eV for most metallic materials. In the case of compounds (ceramics), the displacement energy may be different for each ion species. For example, in Al_2O_3, the displacement energy for Al is 18 eV and that for oxygen is 76 eV (23). Table 1 contains a compilation of the measured or estimated displacement energies for a number of ceramics.

Since the target atom initially struck by the incident ion may receive an energy that is considerably greater than E_d, it may cause further displacements along with the incident ion. Thus a cascade of collisions and defect production ensues. Figure 3 is a schematic view of the resulting energy transfer-defect production. Since the recoil atom loses its energy by both nuclear (collisions) and electronic (ionizing, excitation) processes, the damage energy deposited is less than the total nuclear stopping of the primary ion.

The basic unit of radiation-induced defects in crystals is the Frenkel pair-one vacancy and one self-interstitial. In the low-density cascade regime, this defect production can be related to the deposited damage energy, $S_D(x)$, by the modified Kinchin-Pease relationship (24) as:

Eq. (8) $$dpa(x) = 0.8\ (\tau)\ S_D(x)/(N\ 2E_d)$$

Table 1. Displacement Energies for Some Ceramics

Material	E_d	(eV)	Reference
Al_2O_3	Al	18	23
	O	76	
MgO	Mg	64	25
	O	60	
$MgAl_2O_3$	Mg	86	26
	Al	77	
	O	130	
UO_2	U	40	27
	O	20	
ZnO	Zn	57	28
	O	57	
Si_3N_4	Si	60	29
	N	60	

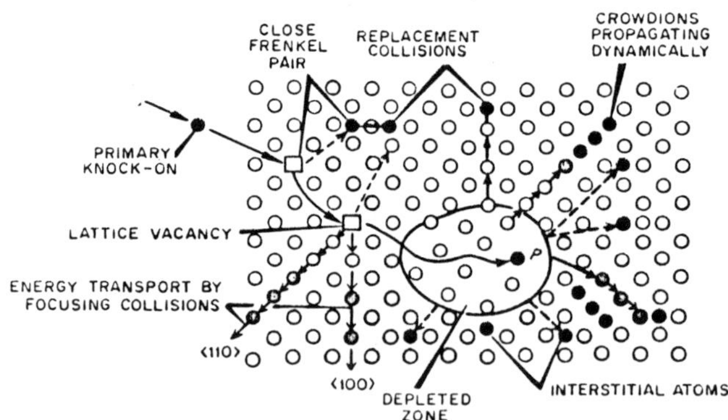

Figure 3. Schematic representation of energy loss and defect production for energetic ion incident on a solid target.

where dpa indicates displacements per atom, (τ) is the fluence of bombarding ions, N is the number density of target atoms, E_d is the target displacement energy, and $S_D(x)$ is the deposited damage energy at depth x. In the first part of the ion's path, the vacancy and its interstitial are separated by several interatomic distances. However, in the dense cascades and at end-of-range, there is a large concentration of both, consisting of a vacancy-rich "core" surrounded by a "shell" of interstitials.

The amount of "radiation damage" or defect production occurring during implantation is substantial. As an illustration of the number of defects generated, consider the number of vacancies or displacements per target atom calculated from Eq. (8) for 10^{17} N/cm^2 incident on an iron target at 200 keV. The peak damage corresponds to 27 dpa (displacements per atom). That is, each iron atom has been displaced from a lattice site an average of 27 times. This is similar to the radiation damage caused in one year to structural materials located in the core of a typical nuclear reactor.

The energy partitioned to electronic excitation can also produce defects in insulators. See Ref. 30 for a recent review of the types and number of defects generated during ion implantation. The depth profile for electronic energy losses (sometimes called inelastic losses) can be calculated in the same manner as for nuclear processes. Since electronic stopping dominates at higher energies and nuclear stopping at lower energies, the deposited electronic energy is maximum at the surface and falls off rapidly. The relative amount of energy deposited in the two processes varies strongly with incident energy and with both ion and target atom atomic numbers. Thus, for very light ions, the majority of the energy is lost in electronic processes.

It is the defects that survive the "cooling down" period after the cascade, of the order of 10^{-10} sec, that cause the microstructural changes. Many of the vacancy-interstitial pairs recombine and annihilate each other during this period. However, observations by electron microscopy indicate that defect clusters may be directly formed in the volume of the cascade. Interstitials can form clusters in the form of dislocation loops while vacancies may form dislocation loops, stacking fault tetrahedra, or even three-dimensional cavities.

4.0 ION IMPLANTATION OF CERAMICS

Implantation is the ion beam process that uses the bombarding ion beam to alter the surface composition and/or microstructure. It produces non-equilibrium microstructures which may range from high concentrations of point defects to supersaturated solid solutions to new phases. The microstructure of implanted ceramics depends upon the implantation parameters of fluence, ion species, and substrate temperature, and the material parameter of chemical bonding type (31). The effects on mechanical properties and wear resistance will be discussed first in terms of microstructural changes and then in terms of phase synthesis.

4.1 Microstructural and Property Changes in Ion Implantation

The microstructure of as-implanted materials depends upon the damage left after dynamic recovery processes annihilate most of the defects produced in the collision cascade and rearrange the remaining defects into metastable configurations such as dislocation loops, stacking faults, etc. At low fluences or low temperatures where recovery is inhibited, defects accumulate as the fluence is increased. If recovery is sufficiently suppressed, a defect concentration may be reached where the long-range order of the crystal lattice is destroyed and an amorphous state is produced. There is evidence that some implanted species are more effective than others in stabilizing the disordering defects that cause amorphization, suggesting a "chemical" effect in addition to the damage energy effect (32). The temperature at which significant recovery occurs is governed primarily by the type of chemical bonding present; directional covalent bonds are more difficult to reform than ionic bonds.

The microstructure of oxide ceramics after implantation with relatively low fluences or in the temperature range where dynamic recovery prevents amorphization is characterized by point defect clusters (bounded by dislocation loops). The microstructure of Al_2O_3 determined by transmission electron microscopy (TEM) shows a large concentration of "black spots" similar to that seen after high-energy electron or neutron bombardment. It is likely that these are dislocation loops bounding stoichiometric interstitial clusters of aluminum and oxygen faulted with respect to the cation (Al) sublattice (23).

Some oxides become amorphous due to damage accumulation after high fluence implantation at room temperature or at lower fluences at low temperatures. The fluence of chromium necessary to amorphize Al_2O_3 at 77 K is about 200 times lower than at room temperature (31). The critical damage level at 77 K is a deposited energy of about 2.5×10^{22} keV/cm^3.

The covalently bonded ceramics SiC and Si_3N_4 are easily amorphized at room temperature. The critical damage level for SiC corresponds to a damage energy density of 2 to 10×10^{21} keV/cm^3, more than two orders of magnitude lower than for Al_2O_3 at the same temperature. The critical damage energy density for silicon nitride is about a factor of two higher than that for silicon carbide (33). Implantation at elevated temperatures (≈1050 K) does not produce an amorphous structure in SiC for damage levels as high as 16 dpa (energy densities greater than 5 x 10^{22} keV/cm^3).

Most of the published data on the hardness of implanted ceramics comes from low load (10 g [0.098 N] to 50 g [0.49 N]) Knoop or Vickers micro-indentation tests. In these cases, the depth of the indentation is equal to or exceeds the thickness of the implanted zone. Thus, the reported changes in hardness do not give the true value but do define the direction of change.

A tabulation of published relative hardness values, taken from Ref. 34, is given in Table 2. The values are given as relative ones, i.e., hardness of implanted sample divided by hardness of the unimplanted material. The test method used in each case is listed. The data should be the most accurate for the ultra-low load (ULL) technique and least accurate for the Vickers tests.

The hardness of the implanted ceramic increased in each instance that the sample remained crystalline. Figure 4 shows the relative hardness as a function of fluence for Al_2O_3 (c-axis orientation) implanted with 280 keV Cr^+ at room temperature (31).

From measurements of the lattice disorder and the location of the implanted Cr ions, it was estimated that defect hardening (radiation damage) accounted for essentially all the increase in hardness at fluences below about 2 x 10^{16} Cr/cm^2 and more than half of it at 1 x 10^{17} Cr/cm^2. Bull (47) has also concluded that radiation-induced defects are responsible for most of the hardening in Al_2O_3.

Implantation of Al_2O_3 at higher temperatures (≈640 K) produced less disorder near the specimen's surface than did room temperature implantation (36). The hardness was less for the former implants, reflecting the lesser amount of residual damage (due to enhanced recovery).

All studies of implantation-induced amorphization show that the crystalline-to-amorphous transformation results in a significant softening of the material. Data for Al_2O_3 (32)(33)(37)(39), MgO (41), ZrO_2 (42)(43), and SiC (35)(45)(46)(48) indicate the hardness of the amorphous state is 50 to 60% that of the respective crystalline state.

The transverse rupture strength as measured in three- or four-point bending increases with implantation in a number of ceramics. Increases of 10 to 100% have been reported for Al_2O_3 (39)(46)(49) and SiC, Si_3N_4, and ZrO_2 (46).

Table 2. Hardness of Ion Implanted Ceramics (34)

Material	Implantation Conditions Species	Fluence (ions/cm^2)	Energy (keV)	Temp.	Hardness method*	Relative Hardness** [implanted / unimplanted]	Comments	Ref.
Al_2O_3								
c-axis	Cr	10^{16} - 10^{17}	280	RT	K-15	1.27-1.55		(35)
	Cr	$4x10^{16}$	280	640°K	K-15	1.1		(36)
	Cr	$3x10^{15}$	280	77°K	K-15	0.6	amorphous	(32)
	Al+O	[$4x10^{16}$ Al	90	77°K	ULL	0.45	amorphous	(37)
		$6x10^{16}$ O]	55					
	Fe,Cu,Ti	1.5-$4x10^{16}$	various	RT	K-15	1.1-1.4		(38)
	W,Mo							
	Ni	10^{17}	300	RT	K-25	1.3		(39)
	Ni	10^{15}	300	100°K	K-25	1.5		(39)
	Ni	10^{17}	300	100°K	K-25	0.6	amorphous	(39)
a-axis	Y	$3x10^{16}$	300	RT	K-25	1.57		(33)
	Y	$6x10^{17}$	300	RT	K-25	0.7	amorphous	(33)
	Ti	$3.4x10^{16}$	300	RT	K-25	1.3		(40)
	Cr	$3.15x10^{16}$	300	RT	K-25	1.11		(40)
MgO								
{100}	Ti	$2x10^{16}$	300	RT	K-10	2.3		(41)
	Ti	$3.5x10^{17}$	300	RT	K-10	0.8		(41)
	Cr	$6x10^{16}$	300	RT	K-10	2.0		(41)
ZrO_2								
Y-FSZ	Al	$1x10^{16}$	190	RT	K-50	1.28		(42)
	Al	$4x10^{17}$	190	RT	K-50	0.83	amorphous	(42)
	Ti	$3x10^{16}$	400	RT	K-10	1.6		(43)
	Ti	$1x10^{17}$	400	RT	K-10	0.9	amorphous	(43)
TiB_2								
Sintered	Ni	$1x10^{17}$	1000	RT	K-15	1.7-2.1		(44)
SiC								
c-axis	Cr	$4x10^{14}$	280	RT	K-15	1.2		(31)
	Cr	$2x10^{15}$	280	RT	K-15	0.55	amorphous	(35)
	N_2	$8x10^{17}$	80	RT	V-25	0.37	amorphous	(45)
Sintered	Ar	$1x10^{16}$	800	RT	V-100	0.5	amorphous	(46)

*K = Knoop; V = Vickers; ULL = Ultra-low load.
The number following K or V indicates the load used in the hardness test in values of grams (force).

**The value listed here is the maximum (or minimum) reported in the indicated reference.

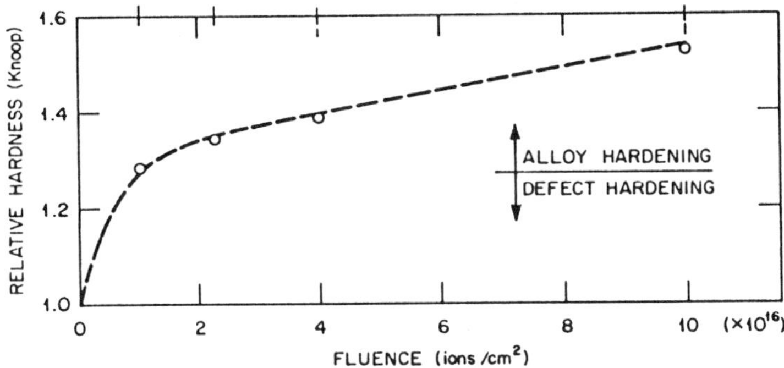

Figure 4. Relative hardness (implanted/unimplanted) for Al_2O_3 implanted with chromium (280 keV) at 300 K. Knoop indentations were made on the (0001) face with 15 g (f). (Ref. 32.)

Increases of 15 to 100% in the indentation fracture toughness have been reported for ion implanted Al_2O_3 (35)(39)(50)(51), SiC (51), MgO (41), and TiB_2 (35). Figure 5 illustrates the effects of fluence (of Ni^+ into Al_2O_3) and substrate temperature on the relative fracture toughness. Samples implanted at 100°K were amorphous, whereas those implanted at 300 and 533°K were crystalline. The increases in fracture toughness were attributed to the residual compressive stress induced by the implantation. Detailed examination of cracks around Vickers hardness indents made on the surface of Al_2O_3 implanted with 10^{17} Ni/cm^2 showed that implantation had little effect upon the incidence of radial cracking but inhibited the propagation of lateral cracks (51).

The changes in the mechanical properties of the ion implanted regions of ceramics summarized above indicate that increases in wear resistance should be expected for common ceramics. Most models for wear suggest that the wear rate is inversely related to hardness and fracture toughness. Even in instances where implantation produces a softer amorphous surface, improvements in wear resistance may be expected due to the change in deformation mechanisms and the inhibition of lateral cracking by the amorphous surface.

As noted above, friction is important in determining the manner in which a tangential force is transferred between contacting moving components. Ion implantation of single crystal sapphire to fluences less than those required for amorphization increases the coefficient of friction measured for metal pins under lubricated sliding (52) and metal, sapphire, and diamond

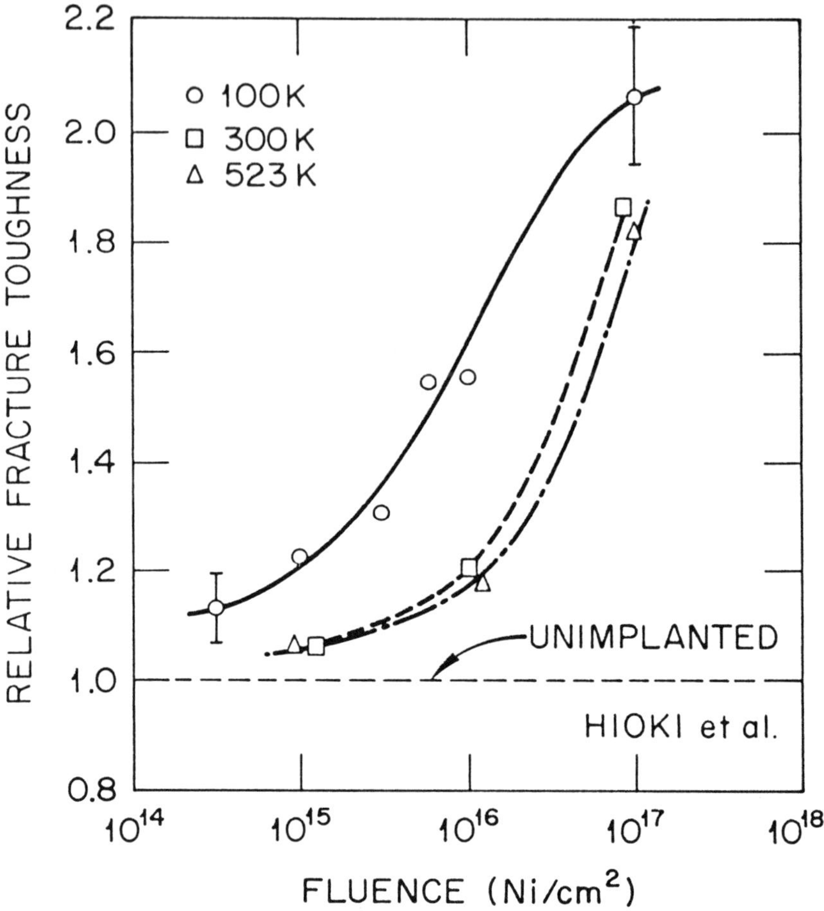

Figure 5. Indentation fracture toughness as function of fluence for Al_2O_3 implanted with 300 keV Ni ions at 100, 300, or 523 K. The K_c value was evaluated for Vickers indentations at a load of 0.49 N. After Hioki, et al. (Ref. 39.)

pins under dry sliding conditions (46)(53). The onset of amorphization is accompanied by a decrease in friction (31).

The scratch test corresponds to a single-pass pin-on-disk test and gives an indication of resistance to abrasive wear where gouging or plowing may be involved. The work of material removal may be calculated from the cross-sectional area of the groove and the measured tangential force.

Figure 6 shows a SEM photograph of scratches made by a loaded diamond stylus on a single crystal of Al_2O_3. The implanted (and crystalline) region is to the right-hand side and the unimplanted/implanted interface is indicated by the arrows. It is immediately obvious that the amount of lateral cracking and the resultant spalling of material is much less in the implanted region. The depth of each scratch extends beyond the implanted zone. As noted, the implanted surface inhibits the propagation of subsurface cracks.

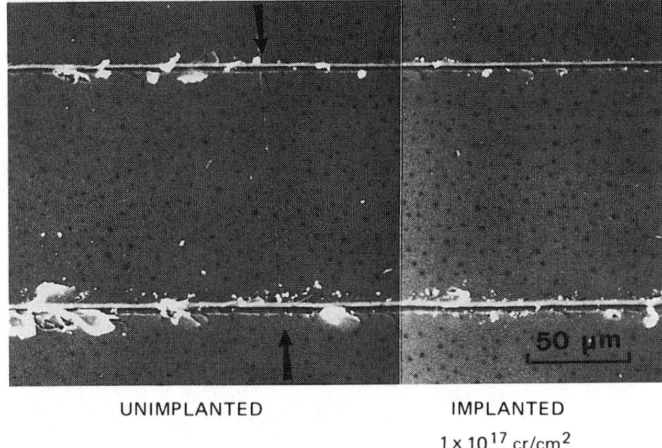

Figure 6. SEM photograph of scratch made by diamond stylus with normal forces of 0.29 N (upper) and 0.49 N (lower) in Al_2O_3. The interface between the implanted (1×10^{17} Cr/cm^2, 280 keV) and unimplanted regions is marked by the arrows (implanted region to the right). (Ref. 31.)

The specific work of material removal is 50 to 100 times greater for implanted regions of TiB_2 than for unimplanted regions (44). In these polycrystalline samples, material removal was primarily due to grain boundary cracking which was greatly suppressed by implantation. Transgranular cracks were found in the wear paths made in unimplanted regions but not in implanted areas. In this case, implantation produced residual compressive stresses at the surface which were in the range of 1 to 4 GPa. Such stresses apparently were very effective in preventing crack formation at the grain boundaries.

The production of an amorphous surface on ceramics causes a compressive stress due to the volume change accompanying the transformation. In addition, deformation of amorphous phases occurs by viscous flow, shear band propagation, or densification, rather than by dislocation slip and cleavage fracture typical of brittle solids. Both effects should affect the wear properties. Figure 7 is a SEM photograph of

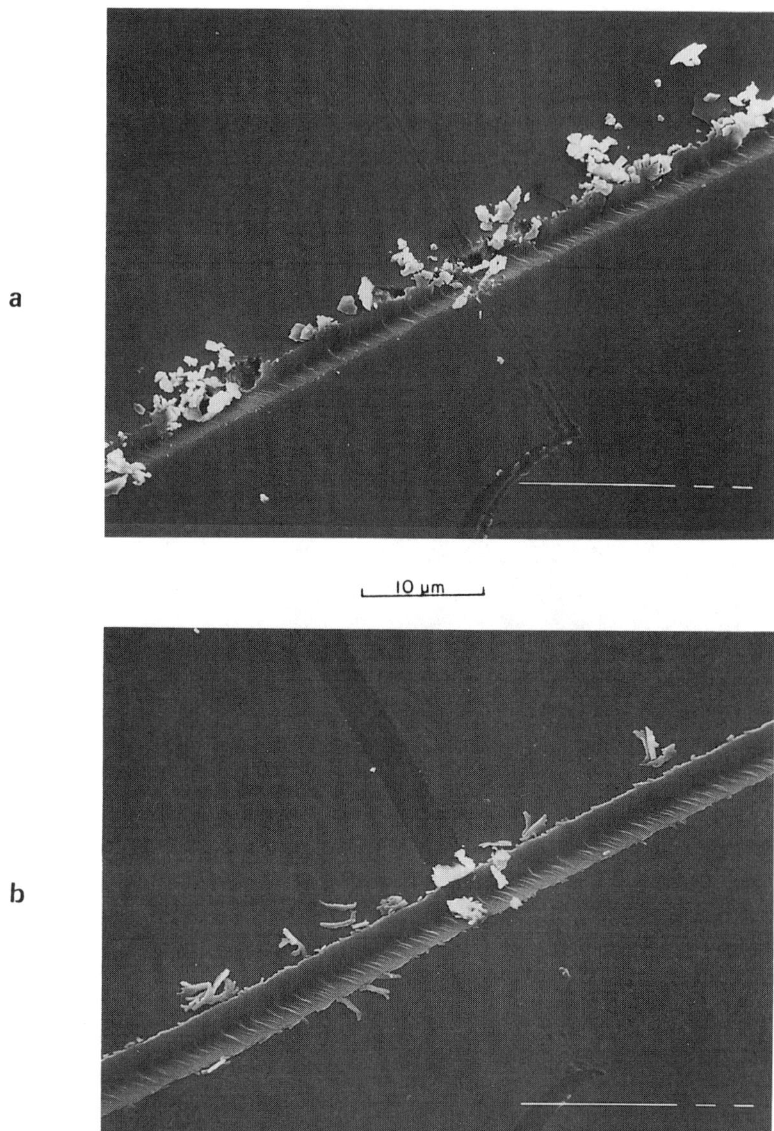

Figure 7. SEM photographs of scratch made by a diamond stylus in Al_2O_3. (a) Crystalline, unimplanted; (b) amorphous, implanted with 4×10^{16} Cr/cm^2 (150 keV) at 77 K. (Ref. 31.)

scratches made in Al_2O_3 by a diamond stylus. The groove in the unimplanted (crystalline) region is accompanied by cracks that extend into the surrounding matrix for distances comparable to the groove width. There is profuse cracking in the bottom of the groove, and material has spalled due to lateral cracking. Much of the wear debris on the surface consists of angular pieces fractured from the wear rack. On the other hand, the track in the implanted (amorphous) region is characterized by ductile-appearing chips and there are no visible cracks (either radial or lateral). Profilometer traces across the track in the amorphous region indicate that much of the material from the groove is piled up as ridges next to it. Even though the amorphous material is only 50% as hard as the crystalline material, the suppression of cracking has greatly increased its resistance to gouging or ploughing during sliding wear.

4.2 Compound Synthesis by Ion Implantation

Attempts to synthesize surface compounds by direct ion implantation of carbon, oxygen, or nitrogen into metal substrates usually result in precipitate formation and rarely in a continuous surface layer of a compound. Kelly (54) considered the factors that control the nature of the phases formed when metals are implanted with oxygen or nitrogen and concluded that phases form which are in local thermodynamic equilibrium.

Polycrystalline stoichiometric Al_2O_3 has been prepared by high fluence (10^{18} cm^{-2}) implantation of oxygen into aluminum films (55)(56). The compound formation was accompanied by large compressive stresses perpendicular to the surface. Lower fluence implantation resulted in polycrystalline Al_2O_3 plus aluminum metal surfaces.

The formation of a continuous, polycrystalline, 100-nm thick layer of TiC on iron and M2 tool steel has been reported by Singer et al. (57). Fluences of titanium greater than 5×10^{17} cm^{-2} and temperatures of implantation greater than 600°C were required. The carbon was contained in the M2 steel or was incorporated as a contaminant from the atmosphere in the implantation chamber. This TiC had a wear resistance ten times greater than the substrate material in abrasive wear tests. The coefficient of friction for dry sliding against hardened steel balls was 60% less than for the substrate.

Continuous surface layers of TiC were also produced by room temperature implantation of carbon (8×10^{17} cm^{-2}) into the alloy Ti-6 Al-4V. This layer exhibited an improvement in wear resistance of 70X in abrasive wear tests. Implantation of boron under similar conditions yielded a dispersion of TiB_2

in amorphous Ti. The wear resistance of this dispersion was 10X that of the substrate material (58).

Elevated temperature implantation of titanium into SiC and Si_3N_3 resulted in compound formation but continuous, single phased layers were not obtained. The data suggest that a layer of mostly TiC was obtained for implantation of SiC at 800 to 900°C. Three surface layers were identified for Ti-implanted silicon nitride. The outermost layer was comprised of Si_3N_4 + Ti; the center layer was TiN + Si, and the innermost layer was TiN plus a small amount of Si. Although no wear tests were reported, the mechanical properties of these surfaces were different from those of the substrates.

5.0 ION BEAM MIXING

Ion beam mixing uses energetic ions to cause the intermixing of layers during ion bombardment. Ion mixing has several advantages over ion implantation: *(i)* larger changes in concentration occur for the same irradiation fluence; *(ii)* the influence of sputtering is reduced; (iii) changes in composition and structure are insensitive to the species of the mixing ion; and *(iv)* new phases are more apt to be formed. Figure 8 illustrates the experimental arrangements often used. The mixing ion ("C") may be one of the target species or, usually, an inert gas ion.

In most ion mixing experiments the beam parameters are chosen to give a damage energy deposition profile as indicated by the dashed line in Fig. 8a, i.e., maximum energy deposition near the film/substrate interface. Many of the processes described in an earlier section of this chapter contribute to the intermixing. For convenience of discussion, these processes may be categorized as atomic mixing, cascade mixing and quenching, and defect interactions and enhanced diffusion. The relative importance of each mechanism of material transport depends upon many parameters, including substrate temperature, mass of the mixing ion, and chemical and thermodynamic properties of the particular system. The interplay of ballistic, cascade, and thermochemical factors complicates the task of developing detailed models. More than one mechanism is likely to be involved and analysis of experimental data shows that there are both temperature-dependent and temperature-independent components.

Atomic mixing due to primary or secondary recoils is ballistic in nature and is often called recoil implantation. The elastic collision of the incident ion causes long-range transport of the impurity (alloying) species across the interface. It is a result of the direct interaction with the incident ion (primary knock-on) or a previously displaced energetic target atom (secondary

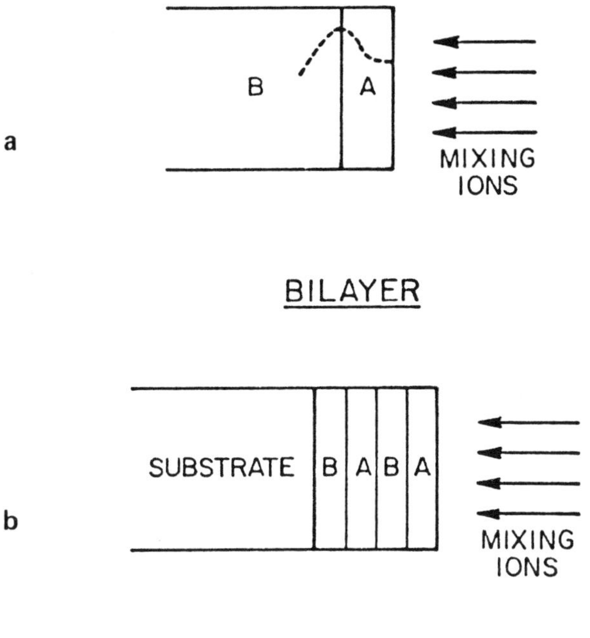

Figure 8. Schematic view of (a) bi-layer and (b) multi-layer configuration used in ion beam mixing studies.

knock-on). This process should be independent of temperature and have the largest relative effect at very low temperatures. A linear relationship is often found between the number of mixed atoms and the fluence of bombarding ions.

Cascade mixing refers to intermixing caused by displacements in the elastic collision cascade. The transport of the impurity is short-range in nature. Monte Carlo simulations and random-walk type of treatments have been used to model the intermixing. In this case, the mixed atom is displaced many times in small steps in successive collision cascades. Observations that mixing is not solely a function of deposited damage energy but of energy density in the cascade suggest that mixing may arise from cooperative phenomena within the cascade (59).

Differences in the amount of mixing induced in collisionally similar but chemically different metallic systems suggests that cascade mixing is strongly influenced by local chemical processes.

The third mechanism for mixing involves atomic transport by the large number of defects or impurity-defect complexes generated during the slowing down of the bombarding ion. Both radiation-enhanced diffusion (RED) and radiation-induced segregation (RIS) may contribute to the material transport across the interface. These processes will be temperature-dependent.

The initial interest in ion beam mixing was directed to silicide-forming systems due to applications for electrical contacts for silicon devices. Later research has been concerned with formation of amorphous and crystalline metastable phases. The reader is directed to Refs. 60 and 61 for recent reviews of the large number of systems studied to date.

The situation with regard to ion mixing when at least one of the components is a chemical compound (e.g., metal-insulator bilayer) is complex and not understood. An "enthalpy rule" has been proposed which states that mixing will occur if the sum of the standard enthalpies of formation of the products of any possible chemical reaction between the film and substrate is less than that of the reactants (62). This rule was tested in a study of thirty-three metal-compound substrate combinations (63). Mixing always occurred if the enthalpy rule predicted it, but also occurred for some material combinations which were not favorable according to this rule. It appears that chemical kinetic factors should also be considered.

Another indication of the importance of kinetic factors is given by the work of Banwell and co-workers (64)(65). The collisionally similar but chemically different systems of Ti-, Cr-, and Ni-SiO_2 were irradiated with 290 keV Xe in the temperature range of 77 to 750 K. At room temperature and below, the net transport of metal into the substrate was similar for all three metals. Thus, the amount of intermixing does not correlate with the thermodynamic predictions. This observation suggests that there are kinetic limitations associated with the complicated reactions in these ternary systems and the short duration of the cascade. These limitations were circumvented at elevated temperatures where the mixing did show a positive correlation with reactivity.

Because of the similarities between ion beam mixing and ion beam assisted deposition, many of the systems studied could be discussed under either heading.

Most of the interest in ion beam mixing of metal films on ceramic substrates has been directed to studies of the adhesion of the film to the substrate and the enhancement of this adhesion by ion beam treatments. Relatively little attention has been given to the structures so formed or to their mechanical properties.

Ion beam induced mixing has been observed for Ti films deposited on SiC substrates bombarded with 2 x 10^{17} N^+/cm^2 (100 keV) at 1000°C (66). The Auger results suggest that TiC formed under these conditions. Similar experiments at room temperature failed to produce any mixing or reaction.

Solnick-Legg and co-workers (67) formed a mixture of TiN and TiO_2 by bombarding a deposit of TiN on M2 and M43 tool steels with nitrogen ions. The oxygen apparently was incorporated from the atmosphere of the ion bombardment chamber. This ion beam mixed coating showed a lower coefficient of friction, better wear resistance, and the absence of stick-slip compared to the as-deposited TiN film.

Researchers at Southwest Research Institute (68) mixed Cr, Co, and TiNi films into PSZ and Si_3N_4 with 140 keV Ar at room temperature. Pin-on-disk tests at temperatures to 800°C showed significant improvements in wear resistance and lower coefficients of friction. The lower coefficient of friction was attributed to the formation of lubricating oxides in the ion mixed layer. After the 800°C test, the surface of the TiNi/Si_3N_4 specimen was found to be mostly NiO with some TiO_2 present. The lower wear was thought to be due to both the lower friction and to the increased fracture toughness of the ion beam treated ceramic substrate.

6.0 ION BEAM ASSISTED DEPOSITION

Most methods for preparing coatings and films have both strengths and limitations. Ion implantation is limited by the generally available ion energies to the formation of relatively thin coatings, generally less than 0.3 µm thick. The composition of the implanted species is limited to 30 at.% or less due to removal of the surface by sputtering. Ion beam mixing requires deposition of many layers by a conventional method before the ion beam treatment. There are also thermodynamic or kinetic factors that limit the mixing achieved in some material combinations, i.e., metal films/ceramic substrates. Conventional physical vapor deposition techniques can produce thick coatings rapidly, but such coatings are often characterized by poor adherence to the substrate and contain voids or other defects.

In recent years, there has been a concerted effort to combine vapor coating techniques with ion bombardment to use the strength of each to offset the limitation of the other. The large activity in this area has contributed to the proliferation of terminology: ion beam assisted deposition (IBAD), ion beam enhanced deposition (IBED), ion assisted coating (IAC), ion vapor deposition (IVD), dynamic recoil mixing (DRM), radiation enhanced deposition (RED), and reactive ion beam enhanced deposition (RIBED).

The field has developed along two contrasting but related lines: low energy ion bombardment (energies of 1 keV and less) and high energy ion bombardment (energies as high as a few MeV). This chapter deals mostly with the latter area since the low energy regime has been covered in a recent review by Harper et al. (69).

Concurrent ion bombardment during film deposition can modify the structural and chemical properties of the resultant film or coating. This procedure may produce films with properties entirely different from those made by the same deposition technique but without the ion bombardment. The bombarding ion beam is subject to a high degree of control with respect to the energy, particle flux, and ion species. Moreover, each of these parameters can be independently varied and they are not coupled to the process parameters of the deposited species.

During the ion bombardment a number of processes occur that may contribute to the structure and properties of the final film or coating. The incident ion contributes its energy to irreversible changes in the dynamics of film nucleation and growth. The ion may also be incorporated into the growing film and change the chemical nature of the system. As mentioned previously, the stopping of the energetic ion involves highly localized and extremely short time events such that the results are often not characteristic of equilibrium thermodynamics.

Ion beam induced sputtering can cause the formation or elimination of surface topographic features in addition to the constant removal of impurity atoms that may condense on the growing surface. The latter process provides a continuous cleaning. Preferential sputtering may be highly important in the case of chemical compounds. This process results in the enrichment of one component and depletion of the other (for a binary system) at the surface.

The deposition of the energy of the incident ion as phonons (heat), ionization, and defect formation (elastic collisions) gives additional mobility to the atoms in the growing film. The resultant structure may be altered and changes in the degree of crystallization, grain orientation, grain size, and density are often observed.

The adhesion between the film and substrate and between subsequently deposited layers of the film itself may be increased due to the ion bombardment. This enhancement may be due to cleaning of the surface just before arrival of the depositing atoms, to material transport across the interfaces due to ion beam mixing effects, or to ion beam induced chemical reactions at the interfaces.

In summary, the use of ion bombardment during the nucleation and growth of a film can affect both structure and composition and thus, properties. Among the changes reported are: *(a)* increased adhesion between the film and substrate; *(b)* reduced or change in sign of residual stresses; *(c)* increased nucleation density and hence effects on grain size and density; *(d)* changes in grain orientation; and *(e)* lower substrate temperatures for compound formation.

A number of systems have been developed for the simultaneous or sequential ion bombardment and deposition. A schematic view of such a system is given in Fig. 9. The ion beam is typically produced in a commercial ion implanter and may be ultra-pure if magnetic separation is used in the accelerating column. The vapor deposition may be provided by a conventional physical deposition source such as electron beam evaporation or sputtering. In a few instances, two accelerators have been used to provide two simultaneous ion beams.

Most of the studies on IBAD films applied for wear resistance or mechanical property enhancement have been concerned with preparation of metal nitrides. Selected IBAD experiments will be discussed in the remainder of this section.

Satou and Fujimoto (70) first reported the formation of cubic boron nitride during energetic N_2^+ bombardment of evaporated boron films. Further studies (71) used simultaneous evaporation of boron and nitrogen ion beams of 2 to 25 keV. X-ray diffraction studies showed a mixture of boron nitride and metallic boron with a ratio of cubic to hexagonal BN of 1:1 at the optimum deposition conditions. These BN + B films had Vickers hardness values of 3000 - 5000 which did not seem to depend upon the B:N ratio.

Films of BN 120 nm thick were made by Bricault et al. (72) by 120 keV N_2^+ bombardment of evaporated boron at a substrate temperature of about 300°C. The Knoop hardness values were greater than 2500 (kg/mm^2) and depended upon the B:N ratio. Ball-on-disk wear tests showed excellent film adhesion to the substrate, volumetric wear rates 1000 times less than for bare boron, and low coefficients of friction (e.g., 0.15 for B:N = 3:1). Likewise, Guzman et al. (73) prepared hard and adherent BN films by sequentially depositing 27 nm boron followed by 30 keV nitrogen implantation.

The formation of cubic (NaCl-structure) MoN has been reported for 5 and 25 keV N_2^+ implantation of evaporated molybdenum with substrate temperatures of 20 and 400 - 500°C (74). The films grown at room temperature had [110] perpendicular to the surface, whereas, those grown at the higher temperatures were oriented with [100] perpendicular to the surface.

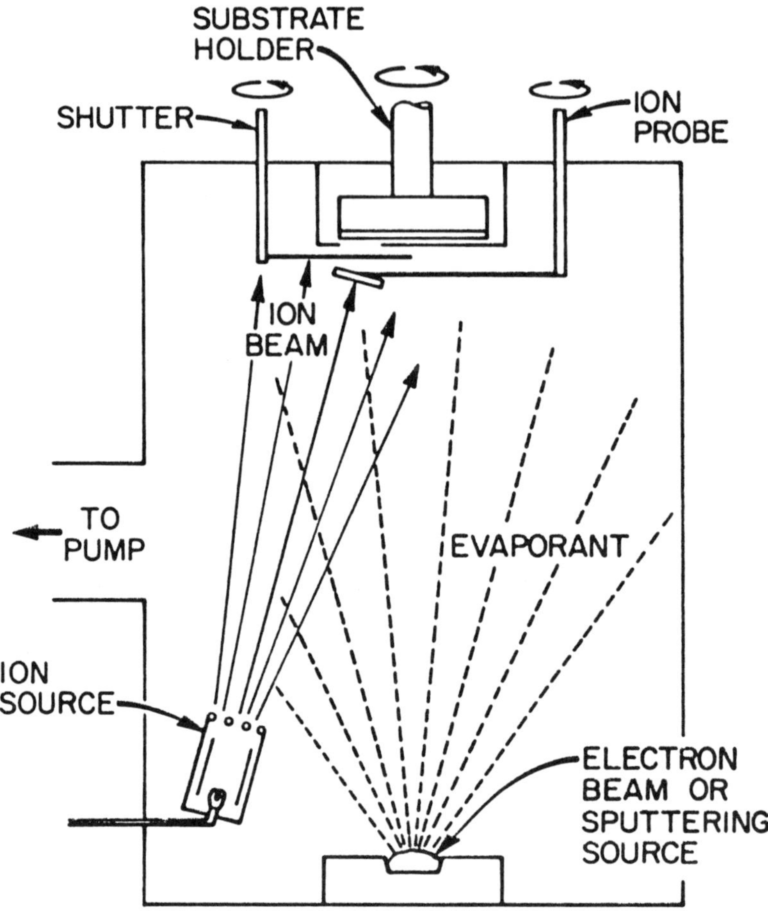

Figure 9. Experimental system for concurrent ion beam bombardment during physical vapor evaporation.

Simultaneous or sequential deposition of electron beam evaporated titanium and bombardment with 20 to 30 keV nitrogen ions produces very adherent TiN films on both metallic (75)-(78) and ceramic (79) substrates. These films generally contain less oxygen and carbon contamination than TiN films prepared by conventional PVD with the amount of contamination being less than the higher deposition rates. Transmission electron microscopy shows the films to have the TiN structure with a grain size in the range of 10 nm. The IBAD films were as much as 60% more dense than PVD films (77). In all instances, the IBAD TiN was softer than bulk TiN. The reasons for these films being softer has not yet been determined. The coefficient of friction measured in pin-on-disk tests was also lower than for conventionally prepared TiN (0.2 vs. 0.6) (Ref. 78). The wear resistance of the IBAD films was reported to be very good although no direct comparison to bulk TiN has been published. It has been reported that the IBAD films exhibit ductile behavior in scratch and pin-on-disk tests (78)(79).

Aluminum nitride (AlN) films have been grown by bombarding growing aluminum films with 5 keV (Ref. 80) and 20 keV (Ref. 76) nitrogen. In the latter instance, up to 50% of the nitrogen ions were incorporated into the growing film. Again, there was considerable oxygen contamination with the AlN containing 10 to 20% oxide.

The structure of chromium nitride films grown by simultaneous or sequential deposition and bombardment by 30 keV N^+ depends upon the substrate material. Studies by Guzman et al. (73) reported face-centered cubic CrN to be formed by sequentially depositing 10 nm of Cr and implanting with 8.4×10^{16} N_2^+/cm^2 if a copper substrate was used. If an iron substrate was used, they found β-Cr_2N at the outer surface and a mixture of CrN, Cr_2N, and Fe_2N in an inner layer. It appears that the ion beam caused mixing of the Cr film with the iron substrate and the nitrogen reacted with this "alloyed" surface layer. These films were reported to be "hard and adherent," but no data were given for these properties.

Films of $Si_{1-x}N_x$ made by 25 keV N_2^+ implantation of silicon contained lower internal stresses and lower void content than conventionally prepared films (74)(81). The as-deposited films were amorphous but crystallized in 2 hrs at 900°C to a fine grained polycrystalline structure. The hardness of the as-deposited films was reported to be greater than that of the silicon substrate (81).

Since the mass and charge of Si^+ and N_2^+ are the same, it is possible to accelerate a mixture of these ions in the same beam. Anttila and co-workers have deposited Si_xN_y by such a technique (82). The current of each species was chosen so as to yield Si_3N_4 but the deposited material was not analyzed. The film again exhibited qualitatively good wear resistance.

Pranevicius (83) has reported that implantation of 5 keV C^+, O^+, or P^+ into growing aluminum films results in the formation of Al_4C_3, Al_2O_3, or AlP. Films of these chemical compounds with thicknesses of 0.2 to 0.3 µm were prepared on substrates held at room temperature. The structure of the films depended upon the ion fluence. For example, polycrystalline Al_2O_3 was formed at low fluences but amorphous Al_2O_3 was formed at higher fluences of oxygen. Oxygen concentrations greater than $6 \times 10^{22}/cm^3$ caused bubble formation (80).

7.0 SUMMARY

The use of ion beam processes to prepare wear-resistant ceramic films and coatings is a promising field of research but has progressed to only a few instances of application. With further research and development, it may become an important addition to coating technology since these processes may produce structures and properties not attainable by other techniques.

Direct implantation can be used to alter the structure and properties of the near-surface region of a material and thus produce a modified surface having approximately the same composition as the substrate or it can be used for compound synthesis. Compounds generally form during ion beam processing at lower substrate temperatures than are used in conventional processing. The structure of ceramics subjected to ion implantation is a function of the implantation parameters (energy, fluence, ion species, substrate temperature) and material characteristics (chemical bonding). Both crystalline and amorphous structures exhibit improved wear resistance as measured by laboratory pin-on-disk tests.

Ion beam mixing to produce surface compounds has been less studied than direct implantation or ion beam assisted deposition. The processes involved are less understood because of the complexities of both the mixing mechanisms and the kinetics and thermodynamics of the multi-component systems.

Ion beam assisted deposition attempts to combine the best attributes of ion implantation with those of PVD. Results to date indicate that compounds can be synthesized which have higher densities, smaller grain sizes, and lower internal stresses than their counterparts grown by conventional techniques. Oxygen and carbon contamination is a problem unless extreme care is taken. The properties of IBAD films may differ from PVD films of similar composition. The causes for these differences have not been identified. Again, more research on the fundamentals of these non-equilibrium reactions will be required before the process is technologically accepted.

REFERENCES

1. Tabor, D., *J. Lubr. Technol.* 103:169-179 (1981)

2. McHargue, C. J., Yust, C. S., Angelini, P., Sklad, P. S. and Lewis, M. B., in: *Science of Hard Materials* (E. A. Almond, C. A. Brookes and R. Warren, eds.), No. 75, pp. 803-812, Adam Hilger, Ltd., Bristol (1986)

3. Archard, J. F., *J. Appl. Phys.* 24:981-988 (1953)

4. Bowden, F. P. and Tabor, D., in: *The Friction and Lubrication of Solids.* Vol. I (1950), Vol. II (1964), Oxford University Press, London

5. Rabinowicz, E., *Wear* 7:9-22 (1964)

6. Hornbogen, E., *Wear* 33:251-259 (1975)

7. Suh, N. P., *Wear* 44:1-16 (1977)

8. Mulhearn, T. O. and Samuels, L. E., *Wear* 33:478-498 (1962)

9. Kruschov, M. M., *Wear* 28:69-88 (1974)

10. Evans, A. G. and Wilshaw, T. R., *Acta Metall.* 24:939-956 (1976)

11. Yust, C. S. and McHargue, C. J., in: *Emergent Process Methods for High Technology Ceramics* (R. F. Davis, H. Palmour III and R. L. Porter, eds.), pp 533-547, Plenum Publishing Company, New York (1984)

12. Lindhard, J., Scharff, M. and Schiott, H. E., *Mat. Fip. Medd. Dan. Vid. Selsk.* 33(14): (1963)

13. Firsov, O. B., *Soviet Phys. JET* 5:1192 (1957); 7:308 (1958); and 9:1076 (1959)

14. Wilson, W. D., Haggmark, L. G. and Biersack, J. P., *Phys. Rev. B* 15:2458 (1977)

15. Biersack, J. P. and Haggmark, L. G., *Nucl. Instrum. Methods* 174:257 (1980)

16. Manning, I. and Mueller, G. P., *Computer Phys. Commun.* 7:85 (1974); 12:339 (1976)

17. Davisson, C. M. and Manning, I., *Naval Research Laboratory Report 8859* (1986)

18. Brice, D. K., in: *Ion Implantation Range and Energy Deposition Distributions,* Vol. 1, IFI, Plenum, New York (1975)

19. Winterbon, K. B. in: *Ion Implantation Range and Energy Deposition Distributions,* Vol. 2, IFI, Plenum, New York (1975)

20. Ryssel, H., in: "Ion Implantation Techniques." *Springer Series in Electrophysics,* Vol. 10 (H. Ryssel, ed.), Springer Verlag (1982)

21. Jahnel, F., Ryssel, H., Prinke, G., Hoffmann, K., Muller, K., Biersack, J. and Henkelmann, R., *Nucl. Instr. Methods* 182/183:223 (1981)

22. Biersack, J. P., in: *Ion Beam Modification of Insulators* (P. Mazzoldi and G. W. Arnold, eds.), pp 648-740, Elsevier, New York (1987)

23. Stathapoulous, A. Y. and Pells, G. P., *Phil. Mag.* 47:369 (1983)

24. Robinson, M. T. and Oen, O. S., *J. Nucl. Mater.* 110:147 (1982)

25. Chen, Y., Trueblood, D. L., Schow, O. E. and Tohver, H. T., *J. Phys. C.* 3:2501 (1970)

26. Crawford, J. H., Jr., Lee, K. H.and White, G. S., *Bull. Am. Phys. Soc.* 23:253 (1978)

27. Soullard, J. and Alamo, A., *Rad. Eff.* 38:133 (1978)

28. Locker, D. R. and Meese, J. M., *IEEE Trans. Nucl. Sci.* NS19:237 (1972)

29. Dell, G. F. and Goland A. N., *J. Nucl. Mater.* 102:246 (1981)

30. Perez, A. and Thevenard, P., in: *Ion Beam Modification of Insulators* (P. Mazzoldi and G. W. Arnold, eds.) Elsevier Publishing Company, New York (1987)

31. McHargue, C. J., *Nucl. Instr. Methods Phys. Res.* B, 19/20, 797 (1987)

32. McHargue, C. J., Farlow, G. C., White, G. W., Williams, J. M., Angelini, P. and Begun, G. M., *Mater. Sci. Engr.* 69:123 (1985)

33. Burnett, P. J. and Page, T. F., *J. Mater. Sci.* 19:845 (1984)

34. McHargue, C. J., in: *Ion Implantation* (F. H. Wohlbier, ed.), Defect and Diffusion Forum, 57-58:359, Trans. Tech. Pub, Switzerland (1988)

35. McHargue, C. J., Lewis, M. B., Appleton, B. R., Naramoto, H., White, C. W. and Williams, J. M., in: *Science of Hard Materials* (R. K. Viswanadham, D. J. Rowcliffe and J. Gurland, eds.), pp 451-465, Plenum Publishing Co., New York (1983)

36. McHargue, C. J., Farlow, G. C., White, C. W., Appleton, B. R., Williams, J. M., Sklad, P. S., Angelini, P. and Yust, C. S., in: *Application of Ion Plating and Ion Implantation to Materials* (R. Hochman, ed.), pp 255-266, American Society for Metals, Metals Park, OH (1986)

37. Oliver, W. C., McHargue, C. J., Farlow, G. C. and White, C. W., in: *Defect Properties and Processing of High-Technology Nonmetallic Materials* (Y. Chen, W. D. Kingerly and R. J. Stokes, eds.), pp 515-523, Materials Research Society, Pittsburgh, PA (1986)

38. McHargue, C. J., in: *Ion Beam Modification of Insulators* (P. Mazzoldi and G. W. Arnold, eds.), Chap. 6, Elsevier, Amsterdam (1987)

39. Hioki, T., Itoh, A., Ohkubo, M., Noda, S., Doi, H., Kawamoto, J. and Kamigaito, O., *J. Mater. Sci.* 21:1321 (1986)

40. Barnett, P. J. and Page, T. F., in: *Plastic Deformations of Ceramic Materials* (R. C. Bradt and R. E. Tressler, eds.), pp 669-680, Plenum Press, New York (1984)

41. Burnett, P. J. and Page, T. F., in: *Ion Implantation and Ion Beam Processing of Materials* (G. K. Hubler, O. W. Holland, C. R. Clayton and C. W. White, eds.), pp 401-406, North-Holland, Amsterdam (1984)

42. Legg, K. O., Cochran, J. K., Solnick-Legg, H. F. and Mann, X. L., *Nucl. Instr. Methods Phys. Res.* B7/8:535 (1985)

43. Burnett, P. J. and Page, T. F., *Ceram. Bull.* 65:1393 (1986)

44. McHargue, C. J., Sklad, P. S., Angelini, P. and Lewis, M. B., *Nucl. Instr. Methods Phys. Res.* B1:246 (1984)

45. Roberts, S. G. and Page, T. F., *J. Mater. Sci.* 21:457 (1986)

46. Doi, H., Private Communication (1987)

47. Bull, S. G., Cambridge University, Private Communication (1987)

48. Spitznagel, J. A., Wood, S., Choyke, W. J., Doyle, N. J., Bradshaw, J. and Fishman, S. G., *Nucl. Instr. Methods Phys. Res.* B16:797 (1987)

49. McHargue, C. J., Unpublished work at Oak Ridge National Laboratory

50. Burnett, P. J. and Page, T. F., J. *Mater. Sci.* 20:4624 (1985)

51. Burnett, P. J. and Page, T. F., *Proc. British Ceram. Soc.* 34:65-76 (1984)

52. Burnett, P. J. and Page, T. F., *Wear* 114:85 (1987)

53. Bull, S. J. and Page, T. F., *J. Mater. Sci.*33:4217 (1988)

54. Kelly, R., *Rad. Eff.* 64:205 (1982)

55. Chereckdjian, S. and Wilson, I. H., *Nucl. Instr. Methods Phys. Res.* B1:258 (1984)

56. Musket, R. G., Brown, D. W. and Hayden, H. C., *Nucl. Instr. Methods Phys. Rev.* B7/8:31 (1985)

57. Singer, I. L., Bolster, R. N., Sprague, J. A., Kim, K., Ramelingam, S., Jeffries, R. A. and Ramseyer, G. O., *J. Appl. Phys.* 58:1255 (1985)

58. Bolster, R. N., Singer, I. L. and Vardiman, R. G., *Surf. Coat. Technol.* 33:469 (1987)

59. Averback, R. S., in: *Ion Mixing and Surface Layer Alloying* (M-A. Nicolet and S. T. Picraux, eds.), pp 8-16d, Noyes Publications, Park Ridge, NJ (1984)

60. Matteson, S. and Nicolet, M-A., *Ann. Rev. Mater. Sci.* 13:339 (1983)

61. Paine, B. M. and Averback, R. S., *Nucl. Instr. Methods Phys. Rev.* 7/8:666 (1985)

62. Banwell, T., Liu, B. X., Golecki, I. and Nicolet, M-A., *Nucl. Instr. Methods Phys. Res.* 209/210:125 (1983)

63. Farlow, G. C., Appleton, B. R., Boatner, L. A., McHargue, C. J., White, C. W., Clark, G. J. and Baglin, J. E. E., in: *Ion Beam Processes in Advanced Electronic Materials and Device Technology* (B. R. Appleton, F. N. Eisen and T. W. Sigmon, eds.), pp 137-145, North Holland, New York (1985)

64. Banwell, T., Nicolet, M-A., Sands, T. and Grunthaner, P. *J., Appl. Phys. Lett.* 50:571 (1987)

65. Banwell, T., Liu, B-X., Golecki, I. and Nicolet, M-A., *Nucl. Instr. Methods Phys. Rev.* 209/210:125 (1983)

66. Miyagawa, Y. and Miyagawa, S., *Nucl. Instr. Methods Phys. Res.* B28:27 (1987)

67. Solnick-Legg, H., Legg, K. O., Rinker, J. G. and Freeman, G. B., *J. Vac. Sci. Technol.* A4:2844 (1986)

68. Wei, W., Lankford, J. L. and Kossowsky, R., *Mater. Sci. Eng.* 90:307 (1987)

69. Harper, J. M. E., Cuomo, J. J., Gambino, R. J. and Kaufman, H. R., in: *Ion Bombardment Modification of Surfaces* (O. Auciello and R. Kelly, eds.), pp 127-163, Elsevier, New York (1984)

70. Satou, M. and Fujimoto, F., Japan *J. Appl. Phys.* 22:L171 (1983)

71. Andoh, Y., Ogata, K., Suzuki, Y., Kamijo, E., Satou, M. and Fujimoto, F., *Nucl. Instr. Methods Phys. Res.* B19/20:791 (1987)

72. Bricault, R. J., Sioshansi, P. and Bunker, S. N., *Nucl. Instr. Methods Phys. Res.* B21:586 (1987)

73. Guzman, L., Giacomozzi, F., Margesin, B., Calliari, L., Fedrizzi, L., Ossi, P. and Scotoni, M., *Mater. Sci. Eng.* 40:349 (1987)

74. Fujimoto, F., Nakane, Y., Satou, M., Komori, F., Ogata, K. and Andoh, Y., *Nucl. Instr. Methods Phys. Res.* B19/20:791 (1987)

75. Sato, T., Ohata, K., Asahi, N., Ono, Y., Oka, Y., Mashimoto, I. and Arimatsu, K., *Nucl. Instr. Methods Phys. Res.* B19/20:644 (1987)

76. Fatkin, J., Kohno, A. and Kanekama, N., Japan *J. Appl. Phys.* 26:856 (1987)

77. Sartwell, B. D., *J. Mater. Energy Systems* 8:246 (1986)

78. Kant, R. A. and Sartwell, B. D., *Mater. Sci. Eng.* 90:357 (1987)

79. Legg, K. O., Nucl. *Instr. Methods Phys. Res.* B24/25:565 (1987)

80. Pranevicius, L., *Thin Solid Films* 63:77 (1979)

81. Donovan, E. P., Brighton, D. R., Huber, G. K. and Van Vechten, D., *Nucl. Instr. Methods. Phys. Res.* B19/20:983 (1987)

82. Anttila, A., University of Helsinki, Private Communication

83. Pranevicius, L., *Nucl. Instr. Methods Phys. Res.* 182/183:251 (1981)

4

Corrosion Resistant Thick Films by Enamelling

Frank A. Kuchinski

1.0 INTRODUCTION TO PORCELAIN ENAMELS

Porcelain enamel is an inorganic coating material which is applied to a metal substrate and fused at a high temperature to form a continuous, adherent and protective coating. Other terms which are used to describe porcelain enamels include glass coatings or linings, vitreous enamels or coatings, high temperature coatings, ceramic coatings or just enamels. Some of these terms may be easily confused with other materials, such as organic paints, or are too restrictive and include only amorphous coatings. Porcelain enamels usually include more than just an amorphous phase, hence they have been classed as solutions. Porcelain enamels are "super-cooled solutions or glasses holding certain materials in suspension" (1). These certain materials are usually colloidal in nature and include color oxides, opacifiers and gases (2). These materials may be added to the coating before it is applied or may be formed during the heating process when the coating is fused to the metal.

A *groundcoat* enamel is applied directly to the metal with the primary function of adhering to it. For steel substrates, these groundcoats are generally dark in color (nearly black) and contain smelted-in oxides of cobalt, nickel, and copper to aid adherence. The dark coating on the inside of an oven cavity is considered a groundcoat. *Covercoat* enamels are light in color and are applied over the groundcoat to provide the required surface properties. These required properties include color, gloss, texture, and corrosion and abrasion resistance. The white or almond colored porcelain enamel on a stove top or clothes washer lid is considered a covercoat.

The thickness of a porcelain enamel coating varies depending on its purpose but can range from 25 or 50 μm to several millimeters. The temperature at which porcelain enamel is fused to the metal is 750 - 870°C for steel (3)(4), 760 - 930°C for cast iron (5), and 500 - 600°C for aluminum (6). Other metals can also be coated; specific details are given later in this chapter.

This chapter is organized into three sections: an Introduction, Porcelain Enamelling Principles and Theories, and Applications and Improvement Methods for Protective Porcelain Enamel Coatings. It is intended to provide the reader with a fundamental understanding of porcelain enamel materials, processes, theories and applications. A large number of references are provided throughout the chapter for readers interested in further or specific details on any of these subjects.

1.1 History of Porcelain Enamelling

The complete history of porcelain enamelling has been covered extensively by several authors (7)-(10). The application of vitreous coatings to metals dates back to Egyptian times. The first products made were jewelry and the metals coated were gold, silver and copper. Although the enamel compositions and enamelling methods changed over 2000 years, the metals coated and product purpose did not change until the industrial revolution. The new metals coated in the eighteenth century were ferrous and included cast iron and sheet iron.

In 1761, J. Gottlieb Justi described a method for porcelain enamelling iron vessels, and by 1764 commercial trade had begun (7). The porcelain enamel coatings were formed by heating cast iron to red heat and applying the coating as a powder, then further heating to fuse the enamel. This process was repeated to increase the coating thickness (8). Other early commercial applications included cooking vessel production in Germany around 1840 (9), and sheet iron enamelling in Germany and Austria around 1850 (8). By 1890, the usefulness of cobalt and nickel oxides in the glass composition for improved adhesion was known (7). During the turn of the century, it became clear that porcelain enamelling had advanced from the art of the goldsmith to a new technology. During the twentieth century, it attracted scientists from all over the world. In an effort to bring these researchers together, the Porcelain Enamel Institute was founded in America on Nov. 6, 1930 (8). This was the first technical forum dedicated to porcelain enamels. The first European forum was founded in Britain in

1934 and called the Institute of Vitreous Enamellers (9). As a result of these institutes and other technical forums, a considerable amount of literature was generated during the 1930's through the present day regarding the principles of porcelain enamelling. The majority of this literature discusses the nature of porcelain enamels in relation to cast iron or sheet steel. Only a small portion is dedicated to other metals. The details of many of these publications are discussed later in this chapter.

1.2 Reasons for Porcelain Enamelling

It is rare that one material can provide the optimum bulk and surface properties for a given application (9). Through a combination of materials, or alteration of a surface, satisfactory bulk characteristics can be achieved with enhanced surface properties, provided that economic justification exists for such processing. Porcelain enamelling is one method of applying a hard, durable inorganic coating over a metal to provide a smooth, attractive finish. This coating can be nearly any color, glossy or matte, and is generally an easy-to-clean, non-stick surface (11). Most porcelain enamel coatings consist of a continuous glassy phase, with isolated crystalline phase(s) and isolated pores. The pores are usually spherical and are generated from the steel or the enamel itself. Other coatings, such as continuous clean porcelain enamels for oven cavities, include a large, continuous, open pore network.

The porcelain enamel/metal substrate system combines the bulk properties of the metal and some bulk, but mainly surface, properties of the porcelain enamel. The metal is a good conductor of heat, has a high thermal expansion coefficient and is soft and shock resistant, while the porcelain enamel is a thermal insulator with a lower thermal expansion coefficient and is relatively hard and brittle (12).

Porcelain enamels are used to impart protective and/or aesthetic qualities to metals. The aesthetic qualities are smoothness, luster, gloss, color, and color stability, uniformity and durability (7)(10). The protection is usually against mechanical or abrasive wear, chemical corrosion and high temperature oxidation (7)(13)-(15). In all cases, it is likely that the porcelain enamel coating enhances certain properties of the coated metal, e.g., abrasion resistance and corrosion resistance, but can detract from other properties, e.g., thermal shock or impact resistance. The cracking, or chipping, of a porcelain enamel is one of its most common drawbacks (13).

Porcelain enamels also offer a wide range of electrical properties. Some enamels are conductors at room temperature, but most act as insulators up to their glass transition temperature (16). Enamelled steels are used as electronic substrates, while modified glass compositions are used for other electronic applications, including hermetic feed-throughs for high vacuum systems, electrical leads in light bulbs and end-seals for high pressure sodium vapor lamps (17).

1.3 General Applications for Porcelain Enamels

Porcelain enamels were first applied to gold and silver to make jewelry or trinkets. Other decorative uses include lamp stands, ashtrays and snuff boxes (18). Household applications include stoves, grills, refrigerators, dishwashers, cooking containers and utensils, clothes washers and dryers, small appliances, cabinets, sinks and bathtubs, and hot water tanks (18)-(20).

Several architectural uses exist due to the excellent weather and abrasion resistance of porcelain enamels. These include the exterior finish of office buildings, store fronts and gasoline fill stations (21), tunnel walls (18), and porcelain enamel on aluminum, which can be bent, drilled and sawed (22). Street and commercial signs as well as interior wall panels and chalk boards are also porcelain enamelled.

The industrial uses are primarily chemical and foodstuffs storage and processing. Some of the applications are smokestacks (19)(23), vessels, pipes, valves and stirrers for processing or storing strong acids (18) and hot concentrated alkali solutions (23)(24). Other industrial uses are heat exchangers (13)(15)(25) and solar collectors (4)(26). One unusual application for porcelain enamel is its use as a vibration damping material (27).

Although this brief summary of applications is not exhaustive, it is intended to show the diversity and economy of porcelain enamels. Most of these applications require the protective nature of porcelain enamel coating in order to be successful. The protective nature of porcelain enamels will be discussed in Sec. 3 of this chapter.

2.0 PORCELAIN ENAMELLING PRINCIPLES AND THEORIES

The major steps in the manufacture of porcelain enamel products are shown in Fig. 1. The entire porcelain enamelling process consists of smelting and fritting, wet or dry milling of the materials, application to the metal substrate, and subsequent heat treatment. Each of these steps has a direct effect on the resultant properties of the finished product. These properties include the degree of bond, color, corrosion and abrasion resistance, reflectivity, gloss and porcelain enamel thickness to name a few. In addition, the resultant properties are affected by the type of metal, its fabrication method and its pre-treatment process.

The smelting and fritting operations are carried out by the frit supplier and this glass is provided to the enameller in flake or powder form. The milling process is performed in the enamelling plant to combine the frit with other components prior to application. Milling may be performed either wet or dry, depending on the application method. However, in the case of electrostatic powder application, the frit supplier performs the milling operation. Metal fabrication and cleaning is conducted in the enamel plant and followed by the application and subsequent heat treatment.

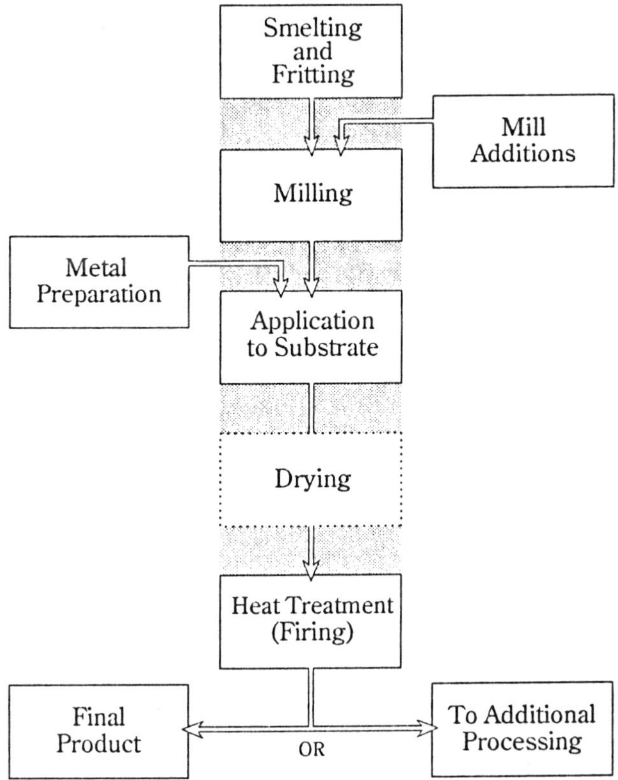

Figure 1. The major steps in the porcelain enamel manufacturing process.

In this section, each of the processing steps are described along with glass/metal considerations and the theories regarding porcelain enamel to steel adherence and covercoat opacity development.

2.1 Porcelain Enamel Smelting and Fritting

Andrews (28) defines smelting as the "melting together of the raw materials entering the enamel composition until a fairly uniform glass is formed." The attainment of a uniform glass requires proper weighing and mixing of selected mineral and chemical raw materials, and appropriate heating rates and heat distribution in the smelter. The smelting operation may be conducted in batch or continuous smelters, usually between 1150 and 1350°C, and is followed by a quenching operation. The molten glass is either water- or roller-quenched in order to facilitate grinding.

Porcelain enamel frits are primarily alkaliborosilicate glasses with other chemical oxides added to yield the desired properties (25). Groundcoat frits incorporate several percent of cobalt, nickel, copper, iron and molybdenum oxides in order to control fired color and improve the adherence to the steel. The metal oxides can be used alone or in combination. Covercoat frits include 12 - 25% TiO_2, ZrO_2, or TiO_2 and ZrO_2, which devitrify during firing to generate opacity. The oxides of calcium, magnesium, barium, zinc, phosphorus and aluminum are added to groundcoats and covercoats to further modify viscosity, surface tension, corrosion resistance and other properties. In addition, porcelain enamel frits containing 3% Sb_2O_3 have been shown to absorb hydrogen generated during firing (29). Also, fluorine may be added as a flux.

Both continuous and batch smelters are used in the production of porcelain enamel frit (30). Small crucible furnaces are used for special applications, such as laboratory developments, jewelry enamels and colors. The hearth furnace is a simple box design which is charged, plugged and heated, then tapped to remove the molten glass. The rotary smelter is a hollow cylinder which rotates about its axis during smelting. This provides better mixing versus a box type hearth furnace, however the mechanical rotation makes the manufacturing process more complex and therefore more costly. In all cases, the glass volume must be small in order to use any batch smelter efficiently.

For large quantities of material, continuous hearth-type furnaces are most economical. These furnaces may be gas fired or electrically heated. Although a much higher throughput is achieved in continuous furnaces, and less labor employed per pound of glass, special care must be taken during change-over from one composition to another. The frit produced during change-over is often rejected and must be reworked back into the smelter and compensated into the raw batch formula. Scheduling of similar compositions back-to-back in the smelter helps to minimize this quantity of change-over rejects.

2.2 Metals Selection and Preparation for Porcelain Enamelling

The metal composition, thermal and mechanical history, and surface preparation procedure are the most critical aspects in attainment of a high quality surface for porcelain enamels. For steel substrates, a high quality porcelain enamel surface requires a minimum of gas producing agents within the metal, such as carbon and hydrogen (31), and on the surface, such as grease, oil, detergents and dirt. Other porcelain enamelling requirements include the ability to withstand the high firing temperatures (6), and to develop adherence to the coating.

In addition to providing all of the required enamelling properties, the metal must meet all of the mechanical and physical requirements for the forming operations and end use application. The forming operations include drawing, welding, punching, drilling, blanking, spinning, and bending (32). Also, each of these forming operations have an impact on the porcelain enamel quality. For example, in the case of welding, the weld composition must be similar to the base steel to prevent thermal expansion and conductivity problems during porcelain enamel firing or subsequent usage. Also, low hydrogen welding electrodes must be used to prevent high weld porosity, and hydrogen gases from generating blisters or spall defects in the porcelain enamel coating (11). Two examples of end use applications are porcelain enamelled hot water heaters and kitchen cooking ovens. The steel used for hot water heaters must withstand high internal pressures from water vapor generated in the heating process. In the other example, the steel used for kitchen oven cavities must withstand repeated cooking (200 - 300°C) and cleaning cycles (500 - 600°C) with minimum deformation.

Many different metals have been porcelain enamelled, but the bulk of the industry uses sheet steels and gray or cast iron. Copper, silver, and gold are coated to make jewelry and trinkets. Aluminum and its alloys are generally used for architectural applications, particularly outdoors, due to the excellent corrosion resistance of aluminum metal. However, the aluminum purity must be controlled closely, since as little as 1.0% magnesium causes porcelain enamel spalling due to poor adhesion (33). Nickel, brass and brazing mixtures cannot be coated (34). Kyri (34) reports that brazing mixtures are not wet by the molten enamel. He also states that porcelain enamel cannot develop bond with a nickel substrate, and reacts strongly with zinc in brass to yield very poor surface quality.

Biswas et al. (14) have reported successful porcelain enamelling of mild steel, stainless steel and nimonic alloy. The firing temperatures required to develop adhesion for both steels was 800 - 900°C and 1160 - 1200°C for the nimonic alloy. Subsequent heat treatment was employed to further devitrify the porcelain enamel coatings. Gackenbach (11) has shown that many ferrous alloys can be porcelain enamelled for use as chemical processing equipment. These metals include cast iron, carbon steel, high tensile steel, select stainless steels and high chrome and nickel alloys.

Although most porcelain enamelled articles employ steel as the substrate, many variations exist. The steel may be ingot or continuous cast, rimmed or killed, and hot or cold rolled. All of these process steps, in addition to the steel composition and impurity levels, effect the porcelain enamelling process and final results. Also, the thickness of the steel varies from one quarter inch to 34 gauge (4). Five different types of flat rolled carbon steels are currently used in the porcelain enamel industry (35). These include enamelling iron, decarburized extra low carbon steels, common cold rolled steels, interstitial free steels and enamelling iron replacements. Bowley (35) and "Porcelain Enamelling" in the *Metals Handbook* (36) provide overviews of the various compositions, production methods, mechanical properties and basic pros and cons for these and other steels.

Ingot casting is accomplished by pouring molten steel into a mold and permitting it to solidify from the outside walls. The conversion of any iron oxide plus carbon to carbon dioxide and iron causes a boiling action which removes the impurities from the solidifying iron at the walls and concentrates them in the center (37). This results in a very pure ingot exterior, hence the term "rimmed steel". Ingot cast steels can be *killed* to remove iron oxide, instead of being rimmed (38). The killing operation is accomplished through additions of silicon, titanium, aluminum or calcium prior to casting the ingot. These additions are made directly to the molten steel in the ladle and convert the iron oxide to silicon dioxide, titanium dioxide, etc. Since little or no carbon dioxide is generated in the ingot, the iron is very still and is referred to as "killed" (37).

Over the last decade, continuous casting of steel has grown due to its improved economy over ingot casting. In continuous casting operations, the steel must be killed in the ladle. It is then transferred to the tundish which feeds the mold used for casting. A glass powder is used to prevent surface oxidation and extract impurities on the top of the mold. This glass powder is also a lubricant as the steel begins to solidify and pass through the stationary walls of the mold. Once through the mold, the glass spalls off the solid iron surface. At this stage, the inner portion of the steel is still molten. Further cooling results in solid continuous steel which must be cut into sections for further handling.

After either of the casting processes are completed, the steel must be hot rolled or cold rolled into a sheet prior to being supplied to the enameller. Cold rolling is accomplished by passing the steel through two rollers to reduce its thickness. This is accomplished at less than one-half of the melting temperature (38). The hot rolling process is similar, except that the work done to the steel is at an elevated temperature (nearer the melting point) which makes it a much more rapid and less expensive process than cold-rolling. The cold rolling process produces a better quality surface with different mechanical properties (39). Cold rolled steels result in better quality porcelain enamel surfaces, particularly in relation to hydrogen-generated defects (38) (also see Sec. 2.7 in this chapter), thus are preferred over hot rolled stock.

The steel composition is extremely important in controlling its final properties. Andrews (40) discusses the impact of carbon, silicon, manganese, phosphorus, sulfur and other impurities on the enamellability of cast iron and steels. Many authors discuss the minimization of carbon content as the most critical aspect in steel selection for high quality porcelain enamelling (6)(7)(13). Maskall and White (41) report a considerable improvement in surface quality as the percent carbon decreases from 0.2% in mild steel to 0.1% in enamelling steel to 0.005% in zero carbon steel. These decarburized steels are more costly to produce due to additional processing steps required of steel manufacturers. Low carbon steels in the 0.002 to 0.003% carbon range are commonly available to the enameller, but at a higher cost.

As the steel quality has improved over the years, so too has the need for uniform and thorough preparation of the surface prior to enamelling. Even a low carbon steel, if improperly prepared, will result in a very poor quality porcelain enamel surface. The surface preparation may consist of alkali cleaning, sand or shot blasting or acid etching with a subsequent nickel flash (25). The cleaning and etching steps are conducted using hot alkali detergents and hot sulfuric acid, respectively (4). Maskall and White (42) and Andrews (43) provide details regarding solution concentrations, temperatures and metal exposure times for all of these processes. Grease and oil residues are removed in alkali detergent solution or by annealing at 500 - 550°C for just two to three minutes (42). Andrews (43) notes that annealing also removes stresses induced into the steel during the forming operations. Rust or scale is removed in 10% hydrochloric acid for one-half hour with subsequent neutralization in 0.1% hydrated borax and soda ash (42). The acid treatment to remove scale is also referred to as *pickling*. The action of hydrogen at the iron surface loosens the scale and exposes clean metal surface (43). The steel surface is then flashed with nickel to improve surface uniformity and aid adherence. This step is accomplished through galvanic reduction of nickel using a nickel sulfate solution (43).

Cast iron samples are sand- or shot-blasted to remove scale and roughen the surface. This is critical for both surface quality and adherence (11). Annealing at 600 - 700°C prior to enamelling is necessary to remove most of the entrapped carbon, which can be as high as three or four percent (41). Failure to anneal properly results in blistering due to outgassing of carbon dioxide during firing. One major drawback to cast iron enamelling is that the required porcelain enamel thickness is very large to successfully coat the rough iron surface (44). The thickness may be three or four times that required for sheet steel, and multiple porcelain enamel coats and firings further increase the cost for cast iron enamelling. However, the use of a vacuum casting process (44) has helped to keep porcelain enamelled cast iron competitive.

Finally, two additional metal preparation techniques are worth mentioning. Aluminum substrates must be cleaned by chemical means or annealed at 540°C for 10 minutes to remove surface grease, oils and dirt (23). Also, to show the extent of pre-treatment methods considered, Nelson and Bacher (45) reported improved edge coverage for porcelain enamel on steel through use of a flame spraying procedure. The procedure involved flame spraying a metal and glass powder on the steel prior to the enamelling process.

Although the general steel preparation steps have been described, each porcelain enamelling plant develops its own procedures. The time, temperatures, concentrations, types of solutions and number of steps vary widely from plant to plant. Many modern plants have eliminated the pickle and nickel process due to tight EPA and OSHA restrictions. In most cases, the steel is cleaned with a series of alkaline solutions, often employing a spray wash, followed by water rinsing and anti-rusting solutions. This is referred to as the "cleaned only" process. This is followed by drying at 110 - 150°C and subsequent transfer to the porcelain enamelling process line for immediate coating application.

The requirements of porcelain enamel formulations are different for these "cleaned only" steels. The adherence of the porcelain enamel to the steel must be provided entirely by the porcelain enamel composition, since the pickle and nickel flash are no longer present to aid adherence. The newest developments have shown the feasibility of enamelling *uncleaned* steel. This process is being used commercially by at least one major European appliance manufacturer. This places further demands on the porcelain enamel composition. Also, certain oils and drawing compounds must be used to produce acceptable quality surfaces on uncleaned steel. The considerations given to development of porcelain enamel formulations for these new, as well as traditional, steel preparation techniques are discussed later in this section.

2.3 Porcelain Enamel Milling

Frit and other raw materials must be mixed and reduced in size to yield a slurry or powder suitable for the selected application process. This step is usually accomplished in a ball mill. The milling may be performed using either a wet or dry process, and is dependent on the subsequent application technique. Wet milling is used for dipping, slushing, flow coating, wet spraying, wet electrostatic spraying, and electrophoretic deposition application methods. The required slip properties are different for each of these methods. The control of these properties is accomplished in the milling operation. Dry milling is employed for dredging and sifting operations on cast iron and dry electrostatic spraying on steel. Again, careful control of the powder characteristics is crucial for proper application. Except for dry electrostatic powder milling, porcelain enamel milling is usually conducted in the enameller's plant. Electrostatic powder is manufactured by the frit producer and supplied to the enameller in a ready to use form.

The wet milling process employs water as a suspension medium and combines frit with various mill additions. Maskall and White (46) report that the mill lining is usually porcelain or steatite and the grinding media is alumina. The typical ball charge is 55% of the mill volume while the frit and mill additions are 22 - 25% of the mill volume. The mill additions are divided into five categories: suspending agents, electrolytes, bisque strengtheners, refractories, and opacifiers and colorants. These mill additions consist of beneficiated minerals, clays and rocks, and chemical processing by-products (47). Careful consideration must be given to particle size, shape and size distribution, composition, purity, hardness, solubility in water, coefficient of thermal expansion, melting temperature and other chemical, physical and thermal properties depending on the porcelain enamel coating purpose.

Suspending agents, or floating agents, are used to suspend frit particles in the slip (48). Clays are most commonly used for this purpose, although colloidal materials, such as silica, can be used in some cases. The particle size of the clay controls its suspension ability, while the impurity levels significantly effect the porcelain enamel fired properties. Organic impurities cause bubble formation during firing which is desirable in groundcoat enamels, but may create defects for covercoat enamels. Metallic or metal oxide impurities cause localized black specking, blistering or general discolorations in covercoat enamels.

Electrolytes are added to the mill to further control the slip rheology. These salts of sodium, potassium and magnesium dissolve in the aqueous medium and interact with the clay surface to aid in suspension of the frit particles (49). Similar to clay additions, improper use of electrolytes can also cause enamelling defects. Concentrated salt crusts at the porcelain enamel surface due to incorrect water evaporation rates can result in drying cracks, tears and blisters. Impurities can also cause discoloration. Various components of the porcelain enamel frit are leached out and influence the effect of the electrolytes (50). The alkali and boron have a significant effect, with dissolved boric acid leading to tearing of the dried bisque (51). A one to one molecular ratio of Na to B is required in order to eliminate the tearing (52)(53).

Bisque strengtheners are added to provide mechanical adhesion of the frit and other particles once the hydrostatic forces imparted by the water have been eliminated through drying. The improved strength of the dried enamel bisque helps to overcome handling defects prior to the firing operation. The materials added for this function are clay, bentonite, gums, carboxymethylcellulose and sodium alginates.

Refractories are high temperature melting materials added to the porcelain enamel system for reasons varying from reductions in costs to increases in molten viscosity. Silica is added to hot water heater porcelain enamels in quantities up to 40 wt% to reduce costs and improve aqueous corrosion resistance. Other silica containing mill additions have been evaluated by Svetlov et al. (54) for improved corrosion resistance. Alumina is added to certain porcelain enamels to increase the rheological set point of the slip and raise the molten viscosity (49).

Finally, opacifiers and colors are added to impart the desired aesthetic properties in the fired porcelain enamel. The theories of opacity and color are briefly discussed later in this chapter. It is the suspension of colloidal materials in the solid glass of the porcelain enamel after firing that yield the opacity as seen by the viewer. Opacity is provided by oxides of tin, cerium, titanium, antimony and zirconium (55). The added colorants are complex crystal structures formed through sintering processes. These colorants are finer than 10 micrometers and must be stable in the molten frit. Color can also be provided in a porcelain enamel through small additions of transition metal oxides smelted into the frit during its manufacture. Opacity is usually provided through devitrification of titania opacified covercoats. Careful control of the covercoat frit composition can result in nucleation and growth of anatase and/or rutile particles in the molten glass during firing. The homogeneous dispersion of these crystals provide opacity.

In a similar fashion to the metal preparation steps, the milling operation must be modified or adjusted to meet the specific needs of each porcelain enamelling plant. Within each plant, there may be several application methods, different color or end use porcelain enamel requirements, and different mills, raw materials and furnace conditions. All of these factors further complicate the milling procedure. In fact, Page (6) points out that a well-trained, highly skilled employee is required in the mill room in order to maintain high quality. This individual must maintain tight control over slip properties, such as slump, set, wet and dry pick-up, fineness, specific gravity and drain time, and fired enamel properties, such as adherence, color and surface quality. The required slip properties are discussed later in relation to their effect on the individual application methods.

As mentioned previously, dry grinding is employed for the dredging or sifting application methods. Very few, if any, mill additions are used and the only critical parameter for this dry grinding operation is the resulting particle size distribution (11). Mill additions which may be required include refractories to affect cost, corrosion or abrasion resistance, and opacifiers and colorants to yield desired aesthetic characteristics.

The dry milling of electrostatic porcelain enamel powders is the newest of all the milling methods. The process is carried out by the porcelain enamel frit producers and is supplied to the enameller in a powder form ready for application. Kuchinski and Labant (56) define the materials and process considerations for the milling of electrostatic porcelain enamels and discuss the effect of several variables on the resultant fired properties. The major considerations for the powder manufacturer are the electrical characteristics of the frit and its reactivity with the encapsulant. An organic silane encapsulant is milled with the frit and mill additions to provide the high surface resistivity required for this application method (57)(58). One advantage to the powder producer is that the solubility and rheology in water of the frit may be ignored since it is not made into a slip. The only other mill additions needed are refractories, colorants and opacifiers as discussed for the dredging or sifting powders.

During the milling of dry electrostatic powders, the manufacturer must carefully control particle size distribution, as the enameller does for wet enamels. However, in addition, the dry electrostatic powder must be manufactured, stored, transported and applied under certain environmental conditions. Of primary importance are temperature and humidity. These process variables, as well as materials properties, affect the resultant properties of the dry electrostatic powders and are described in the next sections.

2.4 Porcelain Enamel Application Methods

As mentioned previously, there are several methods for applying porcelain enamel to a metal substrate. These application methods fall into two basic categories, wet and dry methods. The wet application methods include dipping, slushing, flow coating, wet spraying, wet electrostatic spraying and electrophoretic deposition. The dry application methods include dredging and sifting on cast iron and dry electrostatic spraying on sheet steel. The following section describes each method, discusses important features or considerations and provides advantages and disadvantages for each method.

Dipping involves submerging an entire part into the porcelain enamel slip and removing the part to permit drainage (59). The rheology of the slip is extremely important for controlling the porcelain enamel coating thickness. A specific gravity of 1.63 to 1.75 provides a fired thickness of 75 to 125 μm (60). Slushing is similar to dipping. However, a thicker slip is used which must be shaken off the part (59). Also, the slip may be poured onto the surface, rather than submerging the part (11). The major disadvantage for dipping and slushing are resultant drain lines. This poorer quality surface appearance limits their usage to nonappearance parts or dark colored enamels.

Flow coating is the application of porcelain enamel slip to a metal substrate using a directed stream of material with a subsequent drain time. Since a directed stream of enamel is employed, only one side of a part may be coated if desired (60). This results in a considerable economy of enamel usage compared to dipping which coats both sides. Flow coating is particularly cost effective for internal porcelain enamel coatings, such as dishwasher or oven cavities. The drain lines encountered for dipping or slushing are not eliminated by conversion to flow coating, thus flow coating is not usually employed for high visibility parts.

Spraying is accomplished by atomizing a porcelain enamel slip with compressed air through a spray gun (59). The gun may be stationary or moving, and is most often hand-held. Spraying provides a smoother surface than dipping, slushing or flow coating, thus is often employed for high visibility parts. Also, enamel slip utilization is maximized since only the desired area is coated. However, since wet spraying is a directional application method, only relatively flat parts can be coated successfully. Deep recesses or crevices will not be coated sufficiently and will likely result in defective ware. Also, hand spraying is usually more labor intensive, thus it is more costly than dipping or flow-coating.

Wet electrostatic spraying is similar to wet spraying, except that an electric charge of 100 to 200 kV is imparted to each particle to attract it to the metal substrate (36). This technique is much more efficient in enamel usage due to the electrostatic attraction and produces an improved quality surface, but the added capital expense often outweighs the minor improvements.

Electrophoretic deposition (25), or electrodeposition (60)(61), consists of an electric cell using the part to be coated as the anode. Negatively charged frit particles are attracted to the substrate and form a dense coating of uniform thickness. This technique is the best method for coating edges and holes. Electrodeposition is efficient in enamel utilization, but requires high capital investment for plant equipment (25).

Cast iron substrates can be coated using dipping or spraying as discussed for sheet steels, but often employ dry application techniques. Usually, the part is heated to drive off most of the carbon and then coated with porcelain enamel powder while still hot. Maskall and White (60) and Andrews (62) discuss this technique. The powder is sifted, or dredged, through a screen so that a relatively uniform thickness can be developed. The part is then reheated to further melt the coating. The process of dredging and firing needs to be repeated several times to increase the coating thickness and achieve the desired properties.

The newest of the porcelain enamel application techniques is the dry electrostatic spray process. A dry porcelain enamel powder is fluidized and transported to the part using compressed air. As the powder passes through the spray gun, a voltage up to 100 kV is applied to the particles. This causes the powder to be attracted and adherent to the grounded metal substrate. The electrostatic spray process achieves nearly 100% materials utilization and eliminates the wet mill room in the enamel shop since the powder is provided in ready to use form by the frit supplier. Another advantage is the elimination of drain lines compared to several of the wet methods. The major drawbacks of this method include the high capital investment and the added care required during post-spray handling. Any sudden jar of the part causes some of the powder enamel to fall off, rendering a poorer quality fired part.

The dry electrostatic powder application process has provided the porcelain enameller with the ability to produce a covercoated part on cleaned-only steel with a single firing cycle. Nearly all wet application techniques require groundcoat application, drying and firing followed by covercoat application, drying and firing on cleaned-only steels. The groundcoat is required for adherence on these steels. This conventional process is referred to as "two coat/two fire." The dry electrostatic process is known as "two coat/one fire," and consists of a thin layer (25 μm) of powder groundcoat, referred to as "basecoat," and a thicker layer (100 - 150 μm) of covercoat. Both powders are applied to the substrate and the part is fired just once to provide adhesion through the basecoat and the finished surface properties by the covercoat. "Two coat/one fire" wet systems are presently being developed and tested in enamelling plants in Europe with promising results.

As mentioned previously, each of these methods has advantages and disadvantages. This causes most major porcelain enamel plants to employ more than one application method depending on the requirements for each of the individual parts being coated. For example, one stove plant may choose to flow coat oven cavities, apply electrostatic powder on oven flatware parts and spray wet enamel onto range tops and outside door panels. However, a different stove plant may employ the two coat/one fire process for all covercoats while applying all groundcoats using dipping and wet spray reinforcing. The methods selected depend on such factors as the number of parts to be coated, the available equipment, labor and capital, environmental conditions and restrictions, required enamel properties and quality, materials availability, size and shape of parts, and number of colors needed.

2.5 Porcelain Enamel Bond Theories

The theory of porcelain enamel adherence to steel has been the subject of intense investigations, and controversy, for over fifty years. The porcelain enamelling industry most commonly refers to this adherence as "bond". Pask (63) and others (17)(64)(65) indicate that the best bond is obtained through chemical bonding and minimal stress differentials. However, mechanical interlocking can play a role in the bond strength, but is not sufficient; some consideration of chemical nature and thermal expansion is required.

The term "bond" must be defined and quantified in order to understand the various theories, however Andrews (66) points out that no good definition or test exist for bond. He suggests that bond can be measured by the amount of damage sustained due to impact, torsion, bending or thermal shock. "Porcelain Enamelling" (67) defines adherence as "the degree of attachment of enamel to the metal substrate," but also states that none of the commonly used tests provide force per unit area values for detachment of the enamel from the steel. Rather, the common tests (Table 1) involve deformation of the steel and measurement of the amount of enamel removed (67). Oftentimes, the amount of enamel removed is estimated visually for comparative bond values only.

Table 1. Bond Tests (49)

Number	Name	Application
ASTM C313	Adherence of Porcelain Enamel and Ceramic Coatings to Sheet Metal	Steel substrate thickness from 0.4 mm to 2.0 mm
PEI Bulletin T-29	Test for Adherence of Porcelain Enamel Cover Coats Direct to Steel	For Direct-On Cover Coats with substrate thickness from 0.7 mm to 1.3 mm

Considering the complexity of porcelain enamel adherence to steel, the best approach to understanding the current theory is to review the previous theories in chronological order. Several authors (63)(68)-(72) provide excellent reviews of the various theories, in addition to promoting their own explanation. Therefore, only a basic overview of the major bond theories developed since the 1930's is presented here. Occasionally, apparently conflicting data was presented, but after decades of debate, most of the information was scientifically explainable. Portions of the earliest theories, and of the most clearly disproved theories and explanations are still valid today, and comprise the current theory. The following pages summarize this history of porcelain enamel bond theory.

During the 1930's, significant progress was made regarding bond theories. In 1933, King (73) promoted the *dendritic theory*. He found that dendrites of alpha-iron were present at the glass/metal interface whenever the bond was good (74). It was believed that the dendrites provided a mechanical interlocking network and they reduced strains at the glass/metal interface caused by differential thermal expansion, thus promoting bond (69).

In 1934, Staley (75) proposed the *electrolytic theory.* This theory exhibits similarities to King's. It includes the effects of mechanical interlocking and reduces strains caused by differential thermal expansion. The mechanical interlocking is still believed to result from dendrites, and perhaps surface irregularities. However, cobalt, nickel and antimony are shown to promote adhesion through a plating action on the steel surface and a tenacious adherence to it, thus providing bond. This plating process results from galvanic corrosion of the base metal by oxides more noble than iron that are dissolved in the molten glass. Also, the thermal expansion coefficients of the three metals mentioned are reportedly between that of iron and most enamels, thus reducing the stress between the iron and enamel caused by differential thermal expansions.

In 1935, Dietzel (76) furthered the electrolytic theory. He indicated that CoO and NiO in the glass are reduced by iron to form the metals of Co and Ni on the iron surface. This deposition set up local currents which caused selective corrosion of the iron substrate. The selective corrosion led to an irregular surface which provided a basis from improved mechanical gripping (69). Although Dietzel was able to prove this effect for Co and Ni, his theory was weakened by the fact that the more noble metals did not produce a similar or enhanced effect (72).

In 1936, Kautz (77) advanced the *oxide layer theory.* Again, an intermediate phase, in this case, iron oxide, was responsible for minimizing stresses caused by differential thermal expansions. Also, it was believed that this iron oxide layer provided a transition zone by adhering to the iron through metallic bonding and adhering to the enamel as an oxide. Kautz (77) showed that roughened surfaces of various metals and alloys produced no better adherence than smooth surfaces. Also, the required presence of oxygen was identified since iron enamelled in oxygen free atmospheres produced no bond. He explained the role of cobalt to be that of an oxygen carrier and a promoter of solution of the ferrous phase. Although significant data was provided to support his finding, Kautz (77)(78) was more successful in refuting previous theories than he was in promoting his own. It is well known that most oxides, including iron oxides, are not strongly adherent to their metals and cannot generate adequate bond (72). Therefore, his observations were correct, but bond resulting from a thick layer of iron oxide that remained between the glass and the metal was highly unlikely.

Even though the oxide layer did not provide bond directly, it became apparent that its presence was a prerequisite for bond. Andrews and Swift

(79) studied the solution of iron oxide in porcelain enamel and found the saturation point to coincide with the first signs of bond. Douglas and Zander (80) suggested that the dissolution of the surface iron oxide film resulted in a sharing of oxygen bonds at the iron/enamel interface. They also proposed that nickel coated pickled steel enhanced bond due to the smaller amount of iron oxide formed during firing as a result of the nickel layer, thus less oxide needed to be dissolved by the enamel coating. In addition, Douglas and Zander (80) observed dendrites, or metallic precipitates, in systems that produced no adherence. This refuted a portion of the previous theories.

Although the addition of Co and Ni oxides to porcelain enamel frit was known to enhance bond for many decades, Healy and Andrews (81) proposed the *cobalt-reduction theory* in 1951. They observed cobalt metal particles near the glass/metal interface and postulated that hydrogen, evolved from the steel during firing, reduced the cobalt oxide in the glass to form these particles. Verification that these cobalt particles in fact came from cobalt oxide in the porcelain enamel was provided by Harrison et al. (69) in 1952. Harrison et al. (69) used a radioisotope tracer of cobalt and suggested that the precipitated layer of cobalt metal was about 0.01 μm thick, if computed on a continuous layer basis across the interface. They also indicated that the optimum concentration of cobalt oxide in frit was 0.5 - 1.2%. In 1953, Richmond et al. (68) further studied the effects of cobalt and other metal oxide additions to porcelain enamel frits and observed optimum bond at 0.8 wt% CoO. Also, they indicated that "a positive correlation was found between adherence and roughness of interface" and went on further to state "that roughness of the interface was a necessary, but not sufficient condition for the development of bond" (68).

In 1954, Moore et al. (70) investigated the electrolytic theory and showed that although galvanic corrosion did in fact take place during normal porcelain enamel firing conditions, "mechanical anchoring was not the only important factor affecting bond strength." Moore et al. (70) questioned the overall validity of mechanical anchoring since the use of copper oxide in the frit and sandblasting of the iron both provided a roughened surface which permits many points for mechanical interlocking, but little or no bond was observed.

Eubanks and Moore (82) again showed the positive correlation between surface roughness and bond during an investigation of various cobalt oxide percentages in the frit and different concentrations of oxygen in the furnace atmosphere during firing. The cobalt oxide apparently supplied oxygen to the porcelain enamel interface (a necessity for adhesion) since a decrease in oxygen content in the furnace was compensated by a higher cobalt oxide concentration in the frit. Other important findings of Eubanks and Moore (82) were that no bond developed in frits free of cobalt oxide, even at 99+% oxygen, and that at levels of oxygen higher than 20% (about that of air), no additional surface roughness or bond increase was observed for frits containing cobalt oxide.

Although copper oxide was not observed to promote adhesion on normal enamelling iron, Moore and Eubanks (83) identified copper oxide frit additions to be effective on AISI type 321 stainless steel. They showed similarities of the copper oxide/stainless steel system to that of cobalt oxide/iron in regard to surface roughness, reduction of the oxide at the interface and the degree of bond. These two studies (82) (83) led Moore and Eubanks (83) to state that "the metal oxide formed in the plating-out reaction, and not the action of the plated-out metal, is largely responsible for the bond development."

The concept of the iron oxide layer being necessary continued to receive researchers attention. This seemed to be the only underlying factor consistent in all of the theories. In many cases, the promoters of the various theories did not discuss this point, but nevertheless, it was evident given a careful review of the data.

Researchers also investigated the wettability of several metallic substrates by various glasses (84)-(89), the gases evolved during firing (90)(91) and iron oxide solubilities in glasses (88)(89)(92)-(94). This vast amount of work, combined with further studies at the University of California, Berkeley, led Borom and Pask (95) to the proposal that chemical bonding is the principle mechanism for porcelain enamel bond and that the maintenance of equilibrium compositions across the interface is the key requirement. The balance of bond energies between the iron and enamel is accomplished via saturation of the enamel with iron oxide and the maintenance of a mono-layer of iron oxide at the iron surface (96). Borom, Pask, and others (97)-(100) continued to pursue the *chemical bond theory* and in 1973, Brennan and Pask (100) showed that a glass saturated with iron oxide (44.5% FeO) would bond to an iron substrate when fired in an oxygen free ($Po_2 \approx 10^{-10}$ atm) atmosphere. This seemed to confirm the chemical bond theory. However, it is important to note that other factors mentioned in previous theories, such as mechanical interlocking, dendrites and galvanic corrosion may still play a role and contribute to bond in addition to its chemical nature.

Brennan and Pask (100) also investigated Co and Ni oxide glass additions and adhesion characteristics on Co, Ni, Ni-Fe and Ni-Co alloy substrates. In all cases, saturation of the glass by oxides of the metal substrate provided adhesion. Even though chemical bonding was identified as the nature of porcelain enamel adhesion to steel, production facilities did not operate under ideal laboratory conditions, hence the effects of adhesion oxide additions, rates of iron oxide dissolution and effect of furnace atmosphere conditions still required investigation. Much work (101)-(104) has been devoted to these pursuits in the laboratories and in the plants. Some overall guidelines have been developed for good adhesion, but the relationship between the numerous materials and process variables are still not fully understood. This is even further complicated by the conditions necessary for the attainment of other porcelain enamel requirements, such as color, cost, production rates, etc., in addition to adhesion.

2.6 Covercoat Opacity Mechanisms

As mentioned earlier, the groundcoats are used to provide adherence to the steel and the covercoats provide the remaining aesthetic and other required properties. This section provides an overview of the development of devitrified titania-opacified porcelain enamels.

Kinzie and Plunkett (105) reported that titanium compounds were known fluxes for porcelain enamels in the nineteenth century and that crystallites of TiO_2 formed during firing which generated opacity. However, this opacity did not yield white covercoats. The colors varied from light tan to dark tan and from pea green to strong blue-greys. Yee and Andrews (106) indicated that impurities of iron, chrome and other transition metals present in various TiO_2 raw materials produced these non-white covercoats until purer forms became available in the 1940's.

The purer forms of TiO_2 raw materials led to the development of "superopaque" porcelain enamels in the late 1940's. In 1948, Friedberg et al. (107) reported on the relationship between the TiO_2 crystallite size and shape and the resulting reflectance and color. They identified rutile and anatase phases, with the acicular rutile particles occurring at higher temperatures or longer firing times and the rounded anatase particles forming earlier in the firing process. Also, they associated the blue color with the smaller and rounded anatase particles and suggested that the yellow colors formed at the later stages of firing were associated with the dissolution of the anatase particles at the expense of the larger rutile particles. This was confirmed through the observation of a sharp color change (blue to yellow) and a corresponding large increase in the rutile/anatase ratio (107).

During the 1970's, Engel, Eppler and Parsons conducted several studies using transmission electron microscopy (TEM) to investigate TiO_2-opacified porcelain enamels. Engel et al. (108) confirmed that rutile crystals were always present with an acicular morphology. However, anatase crystals were observed with cubic, rectangular and acicular morphologies. Therefore, all rutile was acicular, but not all acicular particles were rutile. Eppler (109) further reported that anatase nucleation was a bulk phenomena and the rutile nucleation occurred at the interface between adjacent frit particles. He also showed the depletion of anatase crystallites near rutile surface crystals which indicated that the observed anatase to rutile inversion was a ripening process as opposed to a phase transformation.

The nucleation and growth rates which ultimately control the resultant crystal phases and morphologies for the "superopaque" porcelain enamels were investigated by several authors. In 1953, Olympia (110) provided an interpretation of differential thermal analysis (DTA) data for several TiO_2-opacified porcelain enamels. He identified nucleation at 425 - 460°C, accompanied by the onset of melting, and the crystallization, or growth, from 620 - 720°C.

In 1956, Yee and Andrews (106) attempted to relate glass viscosity and nucleation and growth rates for TiO_2-opacified porcelain enamels to time and temperature. They found that temperature exhibited a greater effect than time, and observed the same anatase to rutile transformation as other researchers. However, their most significant, and somewhat startling discovery, was that the thermal history of the porcelain enamel had a major effect (2- or 3-fold) on the measured viscosity. Thus, the consistency of the smelting and quenching process for the manufacture of TiO_2-opacified frits was even more important than for many other porcelain enamels.

Eppler and McLeran, Jr. (111) developed a quantitative model in 1967 to predict the rutile and anatase concentrations in TiO_2-opacified covercoats as a function of time and temperature. Assuming three reactions, crystallization of anatase, crystallization of rutile, and conversion of anatase to rutile, the model predicted the crystal concentrations well under most conditions, except short time and low temperature conditions. However, the observed color data did not correlate well with the observed, or predicted, crystalline concentration data. In 1969, Eppler (112) modified the model to include solubility factors similar to the precipitation of salts from solutions. In comparison to experimental results, he showed that the solubility model was accurate from 660 - 940°C for 1 to 64 minutes (113). However, the correlation with the color data was not improved.

In addition to the studies relating porcelain enamel physical properties (e.g., viscosity) to resultant reflectance and color values, a considerable amount of literature deals with compositional influences (114)-(123). Andrews (123) defines reflectance as the "ratio of the amount of diffused light reflected from an enamel as compared to the amount from a freshly prepared MgO surface under similar conditions." Porcelain enamels produce reflectance via opacifiers, including crystallites (such as TiO_2) but may also include insoluble materials, phase separated glasses and gas bubbles. Andrews (123) also lists the items which control the opacity in porcelain enamels:

1. Index of refraction difference between the glass and opacifier
2. Absorption of the glass and opacifier
3. Sizes and shapes of the opacifier
4. Distribution of the opacifier
5. Number of particles
6. Wavelength of the incident light
7. Porcelain enamel thickness

Friedberg et al. (114) indicate that the high indices of refraction for TiO_2 (rutile: 2.76, and anatase: 2.52) compared to that of typical porcelain enamel glasses (1.50 to 1.55) make TiO_2 the best selection for opacification of porcelain enamels. Other typical opacifiers, SnO_2 (2.04), Sb_2O_3 (2.09) and ZrO_2 (2.17), are not as effective as TiO_2.

Friedberg et al. (114) and Beals et al. (115) reported on numerous compositional effects on TiO_2-opacified porcelain enamels. Na_2O was preferable to K_2O for flow, color and reflectance, with K_2O causing a decrease in viscosity, resulting in more anatase crystals and a bluish-white color. Cole (116) also points out that K^+ has been used to stabilize the anatase phase in pigment research. This supports the findings of Friedberg et al. (114).

Eppler and Spencer-Strong (122) showed P_2O_5 to enhance the blueness of porcelain enamels through an acceleration of anatase formation relative to rutile. Both crystals exhibited enhanced nucleation rates, but anatase was accelerated by twice that of rutile. This was again supported by pigment research in which PO_4^{3-} was used to stabilize the anatase structure (116). Blair and Beals (117) showed the feasibility of a silica-free composition using high concentrations of P_2O_5.

Cook and Essenpreis (118) investigated the effects of alkali on the resultant porcelain enamel properties. They found that a substitution of Li_2O for K_2O caused an increase in reflectance and viscosity, a color shift toward yellow, and no effect on gloss or acid resistance. Furthermore, a 50/50 mixture of K_2O and Na_2O (total R_2O = 9%) yielded the optimum reflectance. However, the R_2O/B_2O_3 ratio was not evaluated which severely limits the universal application of these results over all TiO_2-opacified porcelain enamel compositions.

Antimony and niobium additions were shown by Patrick (119) to enhance the blueness of titania-opacified enamels. The suggested reason for this effect was that these elements retarded crystal growth, hence produced a coating with smaller crystals and a larger number of anatase crystals.

A reduction in viscosity and an improvement in acid resistance are other beneficial effects of TiO_2 additions to porcelain enamels. After an extensive study of many glass compositions, TiO_2 was shown to decrease the viscosity of all porcelain enamels except high (> 20%) P_2O_5 compositions (120). Yee, et al. (121) confirmed these results and observed that only additions of < 11% TiO_2 resulted in a decrease in viscosity and that greater amounts caused an increase in viscosity.

In addition to time/temperature relationships and compositional effects, numerous other factors can effect the crystallization and resultant reflectance and color of porcelain enamel covercoats. The two other major influences are mill additions and furnace atmosphere conditions. Mill additions can directly influence the viscosity of the porcelain enamel as well as the oxidation/reduction conditions in the coating. Both effects will change the fired coating. The influence of electrolytes was studied by Marbaker et al. (124) and the effects of several coloring oxide additions were evaluated by Russel et al. (125).

The furnace atmosphere can also change the oxidation/reduction conditions, and can alter the glass viscosity through moisture variations. These process variables, as well as others not mentioned, all lead to tight controls on production conditions and specialized development of covercoat systems for most enamelling plants on an individual basis.

2.7 Drying, Firing and Defects in Porcelain Enamels

It is only appropriate to discuss the drying and firing processes after an understanding of groundcoat bond theory and covercoat opacity mechanisms. The firing process is the point at which nearly all defects become apparent, even though the cause may be earlier in the process. For example, a misloaded mill may lead to an incorrect color, lack of bond or poor surface quality, but these defects will usually go undetected until the porcelain enamel is fired. Even after firing, it is often difficult to identify the exact cause, and correct the problem to ensure that it will not occur again.

Drying is probably the simplest process of all of the porcelain enamelling steps. However, it must still be carried out correctly in order to achieve an acceptable coating. Andrews (126) indicates that dryers are employed because the iron would oxidize too heavily if one waited for nature to dry the enamel. Furthermore, a large number of parts would need to be stacked somewhere to facilitate adequate drying time which would not be economical. Andrews (126) also points out that the temperature of the firing furnace is too high to perform the drying step. This approach would create large local vapor pressure differentials which in turn would disrupt the porcelain enamel surface. In addition, the firing furnace could not tolerate a high concentration of water vapor because it would cause other enamelling defects which will be discussed later in this section.

Drying is also performed to "permit the application of additional porcelain enamel slip" and to "permit brushing of the coated parts" (127). The temperature and/or humidity of the dryer are varied in order to control the drying rate, which is generally on the order of 2 - 5 minutes in a continuous dryer (127). Most enamelled parts go from the steel pre-treatment process through the firing process in less than an hour, therefore, minimization of the drying time is important in achieving that objective. However, if a part is dried too rapidly, a hard surface film forms, trapping moisture which leads to *tearing* of the surface (127).

In contrast to drying, the firing process is perhaps the most critical of all the steps. Since all furnaces have some inherent degree of variability, including temperature gradients and atmosphere compositions, these variations accentuate any potential flaws introduced by the previous processing steps or the materials employed.

The control of furnace conditions has become much more sophisticated and reliable during the past ten years. However, proper firing of porcelain enamel remains an economical, as well as a technical challenge. Most new furnace designs are very well insulated and air-tight, which makes them energy efficient, but this traps and accumulates moisture, which leads to enamel defects and overall process variability. The concentration of water vapor in the furnace will actually vary depending on the rate and number of parts being run through the furnace, and the amount of water being generated from each part. For example, the concentration of water vapor at the end of the first shift will be higher than at the start, since it will have accumulated during the shift. This condition of a high and variable concentration of moisture in the furnace is worse than just a constant high concentration of moisture in the furnace.

The logical solution to the control of moisture in the furnace is to purposely vent it to remove the excess moisture. This is generally not done with sophisticated controls, but is based on trial and error, periodic furnace moisture measurements, and manual operation of the vents. The interrelationship of ambient humidity, furnace chain load and rate, type of parts and amount of enamel on each part make this method of controlling furnace moisture extremely difficult, but manual control is performed successfully in many plants. Generally, the observance of high water vapor induced defects at the inspection line is the best indicator of when to open the vents further.

Opening and closing of the vents may be an excellent method for controlling the furnace moisture, but it creates large and variable temperature gradients within the furnace. These variations can be up to 100°F from the furnace set-point, with actual side by side, or top to bottom part temperatures varying by a similar amount. A 100°F temperature gradient is often more than the selected materials and previous processing steps can tolerate for one finished enamel property or another. This temperature gradient can produce an unacceptable color variation, extremely poor bond at the low end, or unacceptable surface quality.

The firing process is performed with the intent of fusing the particles of the applied porcelain enamel slip (or powder) into a continuous layer of glass (128). The top outer enamel surface generally fuses first and proceeds inward, rather than from the metal outward (52). The firing cycle is considered complete for groundcoats once sufficient bond is achieved and the proper *bubble structure* is developed. A good bubble structure contains a rather uniform size distribution of spherical pores with an average size less than half the thickness of the coating (129). Andrews (130) indicates that excessive firing results in oversized bubbles which lead to pinhole type surface defects. Covercoat systems are properly fired once the surface has fused and the titania has devitrified to yield the desired color and opacity. Overfiring causes further crystal growth, and additional conversion from anatase to rutile. These changes alter the color (more yellow and green) and reflectance (darker). It is also typical for covercoats to fuse at lower temperatures than groundcoats (131).

"Porcelain Enamelling" (132) points out that time and temperature can be varied to compensate for one another. They suggest that certain enamels can be fired for longer times at lower temperatures to achieve similar results. This may be acceptable for many groundcoat systems but for covercoats that exhibit titania re-crystallization, the nucleation and growth behavior is extremely sensitive to the time/temperature relationship. Although many enamels have different firing requirements, Andrews (133) best summarizes the subject by stating, "The ultimate objective is to get heat into the ware being fired as quickly, uniformly and economically as possible."

Many studies of the various gases generated during the firing of porcelain enamels have been carried out over the past 40 years (29) (52)(90)(91)(134)-(142). This subject has received so much attention because most defects are believed to result from these gases, with hydrogen causing the greatest number of gas-induced defects. Using deuterium as a tracer, Moore et al. (91) showed that dissolved water in the porcelain enamel frit was the principal source of defect-producing hydrogen. This study also included investigation of water introduced during the frit quenching operation, water from the pickling of the steel, mill added water, and chemically combined water in the clay.

In 1953, Moore, Mason and Harrison (134) found the principle gases evolved during the firing of porcelain enamels to be carbon monoxide, carbon dioxide and hydrogen. They indicated that the carbon gases were formed through oxidation of carbon in the steel, and from organic matter associated with the mill added clay. The hydrogen was most likely generated from the reaction between the iron and water according to the following:

Eq. (1) $$Fe + H_2O \rightarrow FeO + H_2\uparrow$$

Moore, Mason and Harrison (134) also pointed out that in addition to causing surface defects, hydrogen and carbon dioxide could effect the oxidation state of metallic oxides in the porcelain enamel which in turn could affect the bond. During the same year, Chu et al. (29) confirmed that hydrogen gas was the principal cause of delayed enamel defects. In addition, he showed that higher contents of structural water in clays and higher decomposition temperatures led to a greater number of gas-induced defects.

Hydrogen-induced defects include *blisters, fishscale* and *reboil.* The blisters are the result of trapped hydrogen bubbles which raise the glass surface. This occurs as hydrogen passes through the glass while it is still viscous (135). Since the solubility of hydrogen in steel increases with temperature, hydrogen is continuously expelled from the steel during the cooling process. After the glass has become extremely viscous, the hydrogen collects and forms pockets at the steel/enamel interface. This hydrogen can build up sufficient pressure to actually spall off a small piece of the porcelain enamel. This defect is called "fishscale" and may occur as much as several weeks after firing. "Reboil" occurs on a second firing of a part due to the escape of hydrogen gas which draws groundcoat to the outer surface of the covercoat resulting in a black speck defect. Benzel et al. (90) confirmed the positive relationship between the tendency to reboil and the amount of trapped hydrogen during the first firing.

In 1962, Chu presented a series of articles to address the issue of hydrogen in porcelain enamels (136)-(141). He indicated that molecular hydrogen could be trapped in steel voids or imperfections, but only atomic hydrogen could diffuse through steel (136). Therefore, he stated, the hydrogen must move through the glass in the molecular form (141). Chu also showed that fishscaling tendencies were directly related to the amount of combined water in mill added clays (138).

Through the use of deuterium, and furnace dew points from -90°F to +120°F, Sullivan et al. (142) showed in 1962 that water vapor in the furnace atmosphere was the principle source of defect-producing hydrogen. This was not necessarily contradictory to the work of Moore et al. (91), since they had kept the furnace atmosphere constant in their study. Other findings made by Sullivan et al. (142) include the decrease of hydrogen absorption by the steel due to surface iron oxide, and an increase in fishscaling tendency for groundcoats exhibiting a high iron oxide solubility.

Most other surface defects result from an improper balance of mill added material, large surface tension differences between two frits in a multi-frit porcelain enamel, or contamination. Contamination can be introduced nearly anywhere in the process. Iron, or other metals, cause black specking. Large pieces of refractory, from the mills or the furnace lining, result in a protrusion from the surface, or *sticker,* and possible discoloration. Sulfur-containing gases, from the dryer or the furnace combustion process, alters the glass surface tension. As little as 0.002% sulfur gas, or 20 ppm, in the furnace atmosphere causes a surface scum on the enamel (143). Localized areas of sulfates, from solution or airborne contaminants, create large local surface tension variations. These variations generate small depressions, or *pits* in the surface of the fired enamel. This can also occur when two dissimilar surface tension frits are used in combination. An example of improperly selected mill additions is the required ratio of sodium and boron in solutions. Excess boron, whether introduced as a soluble mill addition or leached from the frit, leads to cracks in the enamel surface, or "tearing" (50)-(53)(144).

The effect of furnace moisture is not only limited to the generation of hydrogen. Parikh (145) showed that water vapor decreases the surface tension of commercial soda-lime-silica glasses. Cutler (146) proposed a decrease in viscosity for similar glasses proportional to the square root of the water vapor partial pressure. If this also occurs for porcelain enamel systems, which seems likely, the fusion process will change considerably. The change in surface tension and viscosity will alter sintering rates, wettability behavior, surface texture and quality, and the color or bubble structure. Andrews (143) points out that a high water vapor content in the furnace will cause a reduction in surface gloss of the finished enamel. This is most likely due to the altered surface tension.

2.8 Other Comments on Materials and Processing

After completion of the firing step, on-line inspectors send acceptable parts directly to assembly or to the second coat operation for "two coat, two fire" systems. Rejected parts are usually salvaged and returned to the porcelain enamel application area for reprocessing. The principal limiting factor in the number of recoats is the final thickness, which could result in excessive enamel chipping or warpage of the part due to the thermal expansion mismatch.

Nearly all porcelain enamel coatings are formulated with thermal expansion coefficients lower than that of the substrate. This is intended to place the coating in compression after firing, which will reduce its tendency to failure. Like other brittle ceramic materials, porcelain enamels are stronger in compression than under tension.

Other glass and metal characteristics must be considered for specific porcelain enamel applications, and the interrelationship of these various properties are also important. The requirements for thermal expansion differences, iron oxide solubilities, steel yield strengths, glass surface tensions or viscosities, and other properties will be discussed in the next section in relation to some specific protective applications.

3.0 APPLICATIONS AND IMPROVEMENT METHODS FOR PROTECTIVE PORCELAIN ENAMEL COATINGS

Earlier in this chapter, a list of applications for porcelain enamels was provided. This section focuses on those applications where the protective nature of porcelain enamels is the principal reason for their use. Aesthetic and other qualities are given less attention, except in their relation to characterizing the protective nature of these coatings. The test methods used by the porcelain enamel industry to evaluate finished coatings are also presented. Some of the pros and cons and limitations of these various test are discussed. Finally, a series of examples for improvement of the protective qualities of porcelain enamels is provided.

3.1 Applications and Competitive Coatings

Many types of coatings are available in addition to those discussed in this book. Since most porcelain enamels are applied to steel or iron, the major competitors are organic based coatings (147) such as powder paints, electroplated, or otherwise applied, metal coatings (13), and flame or arc spraying of oxides, carbides, nitrides or other materials (148). Porcelain enamel offers use of higher temperatures, and better corrosion and abrasion resistance than the organic based systems (147). Its color retention is also superior to paint (6). From an economic standpoint, the actual cost per square meter of finished coating is nearly equal, therefore, porcelain enamel is often selected, except for applications where its major drawback, chipping, or brittleness, cannot be tolerated. In comparison to aluminizing or galvanizing, porcelain enamelling is more costly, but generally offers a lifetime ten times greater than these metal coatings (15).

Zinc, tin, nickel and chromium are the most commonly applied metals by electroplating (13). Archer and Archibald (13) indicate that electroplated metals provide a decorative effect and corrosion resistance and can be either dull or bright. Porcelain enamels provide a much broader range of colors, excellent acid and alkali resistance and freedom from toxicity problems, but have poorer thermal shock resistance which can lead to cracking and chipping (13).

Another advantage in many applications is the mechanism of protection of steel by porcelain enamels. The glass coating is actually corrosion resistant, and the steel only corrodes at its normal rate if it becomes exposed. Since metallic coatings protect steels either anodically or cathodically, one of two scenarios takes place once the steel is exposed. Zinc, aluminum and cadmium provide anodic protection and are sacrificially corroded to protect the steel, while nickel, copper and chromium are cathodic to steel and actually accelerate the steel corrosion rate (23). For any type of metallic coating, either the coating or the substrate corrodes.

Many applications exist that specifically require the corrosion resistance of porcelain enamel. These include conventional household appliances, outdoor architectural panels and signs, water heaters, chemical processing tanks and heat exchangers. Household appliances must withstand food acid attack and detergent attack (25). Alikina and Sirotinskii (149) evaluated the rates of attack for several mineral and organic soluble compounds and found that tartaric acid exhibited the highest rate and acetic acid exhibited the lowest rate. In general, porcelain enamels exist that are highly resistant to any acid except hydrofluoric and concentrated phosphorics (36).

The principal requirement for corrosion resistant porcelain enamels (assuming they are corrosion resistant) is continuity of the coating. Any pits, pinholes, cracks, chips or fishscales which expose the metal (or a less corrosion resistance groundcoat) nearly always result in failure. Also, improper firing or contamination can cause a poor bubble structure. These large bubbles leave a very thin layer of glass to act as a barrier, which can easily be broken or quickly corroded. This further reduces the local corrosion resistance and exposes the metal and groundcoat more quickly. A method for testing the continuity of the porcelain enamel coating is described later in this section.

Exterior applications for porcelain led to the conductance of numerous outdoor exposure tests by the National Bureau of Standards and the Porcelain Enamel Institute beginning in 1939 (21)(150)-(152). The weathering studies were performed in seven states, Washington, D.C. and Canada, and lasted as long as thirty years. The principal finding of these extensive tests was that the weather resistance, based on gloss loss and color change, was related directly to the acid resistance of the porcelain enamel (21).

One potential application which takes advantage of porcelain enamel's excellent weatherability is that of solar panels. Smith and Eppler (4) justified this use based on the weatherability, the smooth surface which remains clean, and the high solar absorbance value (0.935 in the visible and near IR region) of enamels. Simonis et al. (26) described the use of a spectrally selective tin oxide coating over a dark porcelain enamel to enhance the absorbance and decrease the emittance. They also pointed out that since only 40% of the solar energy is radiated in the visible region, dark blue, brown and green enamels are often as effective as black enamels (26).

Water heater enamels must be resistant to hot water and steam which are always present (25). Since the water quality varies considerably in different areas of the world, water heater enamels must provide resistance to a broad range of hot aqueous solutions and condensates. In addition to the protection provided by the porcelain enamel coating, a sacrificial magnesium anode is included in the tank and will corrode to protect the iron if the iron becomes exposed (20). Many competitive materials have been tested for water heaters, however not until recently has a cost effective and technically satisfactory alternative to porcelain enamelling been found. The use of a polymer, Nylon II, to form a lining was begun around 1980 by several European manufacturers (20). Its use continues to increase as technical problems are overcome and capital is invested to provide the manufacturing facilities.

The chemical and foodstuffs industries have been using porcelain enamels as coatings for its production, transport and storage equipment for many years. Gackenbach (11) refers to these thick porcelain enamel coatings (375 - 625 μm) as "glass linings" and indicates that they can be used for acid solutions up to 175°C and alkaline solutions up to 135°C. He also points out that the common applications in the chemical industry are reactors and polymerizers with operating pressures to 600 psi and capacities of 25,000 gallons (11). Partridge (24) compares a number of vitreous and devitrified coatings for the chemical industry. Porcelain enamel coatings are used on silos due to their low coefficient of friction, good weatherability, and ability to withstand temperatures in the range of 400 - 500°C (87). Karyuk et al. (153) have shown the feasibility of porcelain enamel coatings for "the production of acetic acid, by the method of direct synthesis of carbon monoxide and methanol in the presence of triodothiocyanate and methyl iodide." Alikina and Sirotinskii (154) justified the use of porcelain enamels on tubes for drainage, power engineering and the oil drilling industries. They pointed out that the porcelain enamels provide adequate resistance to "earth corrosion" and salt-water environments (154).

Another industrial application, having some requirements similar to the chemical industry applications, is the use of porcelain enamels for heat exchangers. Warren (155) provides an excellent review of heat exchangers, from the invention by Ljungstrom in 1922 through the designs and concepts employed in 1982. Several authors (25)(156)(160) address the advantages of using porcelain enamels as coatings on heat exchanger components and discuss the various requirements for the special coatings. Nadyrov et al. (157) indicate that porcelain enamel extends the life of carbon steels by five to eight times and no metals can match the cost and corrosion resistance of porcelain enamels. Hackler and Dinulescu (25) confirm that porcelain enamels even exceed the corrosion resistance of most stainless steel materials.

The Battelle Columbus labs conducted an extensive study of various metals, plastics, ceramics and coating materials for use in heat exchangers from 1979 to 1985. Specifically, Sekercioglu et al. (160) reported on the corrosion results of monolithic cordierite, mullite, RBSC, RBSN, sintered α-SiC and a porcelain enamel coating on steel. After cyclic exposure up to 290°C in an oil-fired furnace for 900 hours, resultant projected corrosion rates varied from 1.5 to 58.0 µm/year (160). The porcelain enamel sample exhibited a projected corrosion rate of just 13.3 µm/year, making it technically acceptable, and a much more cost effective solution than the more corrosion resistant silicon nitrides and silicon carbides tested.

Dobrunova et al. (161) investigated the effect of γ-radiation on porcelain enamel heat exchangers. They reported that increases in TiO_2 and PbO levels were required in the glass "to increase the chemical stability and resistance to the action of γ-radiation" and that chromium and zinc oxide additions enhanced bond with the stainless steel substrate (161).

Bazayants et al. (158) proposed an equation to estimate the lifetime of porcelain enamelled heat exchangers based on thickness, corrosion resistance and porosity. The equation is

Eq. (2) $$\tau = \frac{0.9h}{K}(1 - 0.44 \log P)$$

where τ = service life (years), h = coating thickness (mm), K = corrosion rate (mm/year), and P = total content of chips, open pores, large bubbles, etc. (units/cm^2). They point out that this model generally overestimates the lifetime since it does not address failure due to mechanical or thermal factors. However thermal cycling is not necessarily a large problem if the system is properly designed and the materials carefully selected. Porcelain enamels have withstood cycling in household cooking ovens up to 485°C for many years (25).

The critical factors relating to thermal shock resistance are the porcelain enamel thickness, the glass-to-metal bond, the resultant stresses at room temperature, and the tensile strength and modulus of elasticity of the glass (11)(162). Thin coatings are best for high thermal shock resistance (162). Porcelain enamel coatings with strong glass-to-metal bonding and high glass tensile strength are also less susceptible to thermal shock failure (11).

To understand how the resultant stresses at room temperature affects thermal shock resistance, the three major factors which alter these stresses must be explained. These three factors are the curvature (or flatness) of the steel, the difference in the coefficients of thermal expansion between the steel and enamel, and the glass transition temperature (Tg). Highly curved areas exhibit the maximum strain and usually result in failure on convex surfaces (11). Convex surfaces, or edges, may be further susceptible due to the commonly thicker enamel layers in these locations. The actual stress perpendicular to the interface is derived from the differential shrinkage between the two materials over the temperature range below Tg. Above Tg, the glass can relax to relieve these thermally induced stresses.

The resultant stress in a porcelain enamel coating at room temperature is represented by the shaded area shown in Fig. 2(A). The upper region for Glass X (Area X1) places the glass in tension and the lower portion (Area X2) produces compression. Maximum tension is reached at temperature Ttx and zero stress occurs at temperature T1x. The change in stress with temperature is plotted in Fig. 2(B). The resultant stress at room temperature is the total compressive stress represented by area X2 minus the tensile

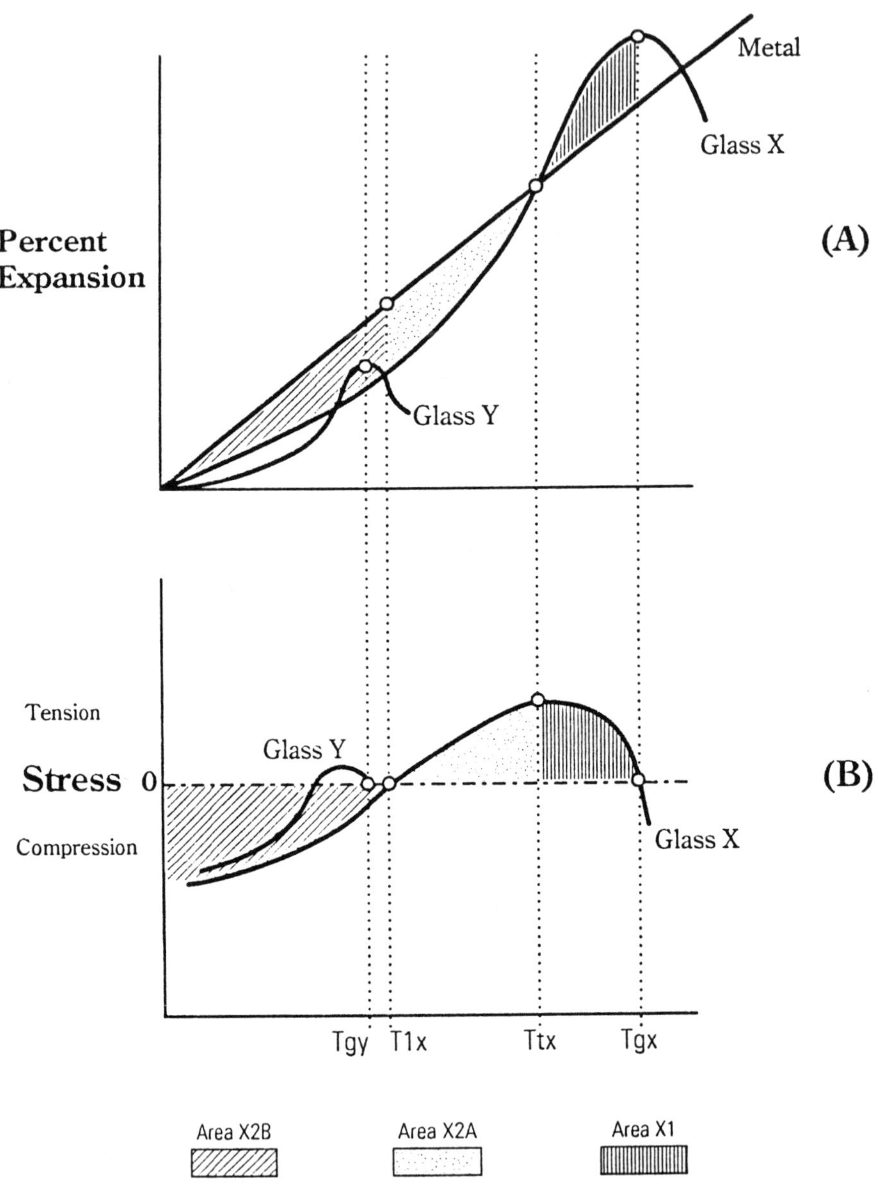

Figure 2. Percent expansion (A) and resultant stress (B) for porcelain enamel systems as a function of temperature

stress represented by Area X1. In this example Area X2 is larger than X1, therefore the resultant stress is compressive. A larger difference in CTE between the substrate and coating does not always generate a larger stress. Glass Y is shown with a lower CTE than Glass X, but due to a much lower Tg, Glass Y does not begin to develop stresses until a lower temperature, even though at room temperature the cumulative stress for both glasses is nearly identical.

Partridge (24) provides an in-depth review of the *glass bead theory* for glass to metal seals and how it relates to porcelain enamel shock resistance. Through experimental work, he justifies the use of this theory and shows that glass-ceramics provide improved resistance to thermal shock (24). Biswas et al. (14) confirms the enhancement of thermal shock resistance by glass-ceramic coatings and further shows improvements in impact, abrasion and corrosion resistance.

The abrasion resistance for porcelain enamels is dependent on the abrasion resistance of the various phases comprising the coating. In the cases mentioned above, crystalline phases provide an overall enhancement of the coating's abrasion resistance once the surface glass layer has been abraded away. However, the bubble structure of the porcelain enamel can often be detrimental to the abrasion resistance. Large bubbles, an increase in the total volume of porosity and their proximity to the surface cause dramatic decreases in abrasion resistance (11).

Although the next application described does provide protection of a metal substrate, it is a rather unique application that requires many properties opposite to those discussed to this point. Trubnikov et al. (163) showed the effective use of porcelain enamel coatings for the protection of metals and alloys from high temperature oxidation, decarburization and loss of alloying elements. However, these coatings were designed with very low coefficients of thermal expansion to promote spalling of the glass off of the metals in the temperature range of 25 - 300°C (163). These enamels also required good wettability characteristics, but could not exhibit bond to the substrate!

3.2 Porcelain Enamel Properties Testing

The porcelain enamel field is like any other engineering design or materials discipline: Tests are developed to best simulate actual use conditions and then standardized to permit comparison of different materials and processes over time and space. Furthermore, consideration is given to the length of time required for these standard tests in relation to the potential liability associated with inaccurate predictions of life expectancy or the application limits. The most commonly used standardized test for the porcelain enamel field in the U.S. are those set up by the American Society for Testing and Materials (ASTM) and the Porcelain Enamel Institute (PEI). During the past several years, the PEI has been updating and rewriting its procedures in an attempt to convert them to the ASTM designation.

The standardized testing of porcelain enamels used specifically for the protection of metal substrates can best be broken into three categories:

1. Corrosion or Chemical Resistance
2. Mechanical and Physical Characteristics
3. High Temperature Properties

The standardized tests are listed in Table 2 for these categories. Due to the numerous amount of tests, only some of them are described. Several non-standardized tests are also described in the case of new or special applications.

As seen in Table 2, many of the standardized porcelain enamel tests have been developed to evaluate the corrosion resistance of the coatings. The most commonly used tests are the Citric Acid Spot Test (ASTM 282) to classify acid resistant (AR) groundcoats and covercoats, and the Continuity of Porcelain Enamel Coatings (ASTM C743) to verify complete coverage on the interior of water heaters. The Citric Acid Spot Test is a 15 minute exposure to a few drops of a 10% citric acid solution covered by a one inch watch glass (164). The affected area is then classified by eye according to the visibility of the stain and its ability to reflect an image, with class AA being the best ranking, A, B, and C being intermediate, and D being the worst (164).

The continuity test involves the application of a high voltage between the enamel surface and the base metal which will discharge across any areas where it exceeds the dielectric strength of the coating (165). The discharge occurs at exposed metal locations, such as cracks, or at locations of thin glass coating, such as pits, pinholes, blisters and large bubbles (165). This test is helpful in determining the number and location of defects, which are detrimental to the corrosion resistance of porcelain enamel coatings but would go undetected in most of the other laboratory standard test procedures. The other commonly used technique for locating defects, particularly hairline cracks, is the Electrified Particle Inspection Method (166). Calcium carbonate, with an electrostatic positive charge is directed at the surface to be tested, where it preferentially collects on negative potential sites, or defects, and helps to identify their location (166). However, Baker (165)

Table 2. Selected Standardized Tests for Protective Porcelain Enamel Coatings (36)

Designation	Application*	Title
ASTM C346	1, 2	Gloss of Ceramic Materials, 45° Specular
ASTM C538	1	Color Retention of Red, Orange, and Yellow Porcelain Enamels
ASTM C703	1, 2, 3	Spalling Resistance of Porcelain Enamelled Aluminum
ASTM C282	1	Citric Acid Spot Test
ASTM C283	1	Boiling Acid Test
ASTM C614	1	Alkali Resistance of Porcelain Enamels
ASTM D2244	1	Instrumental Evaluation of Color Differences of Opaque Materials
ASTM C756	1, 2	Cleanability of Surface Finishes
ASTM D1567	1, 2	Testing of Detergent Cleaners for Evaluation of Corrosive Effects on Certain Porcelain Enamels
ASTM C872	1	Lead and Cadmium Releases from Porcelain Enamel Surfaces
ASTM C664	1	Thickness of Diffusion Coatings
ASTM D1186	1	Measurement of Dry Film Thickness of Non-Magnetic Organic Coatings Applied on a Magnetic Base
ASTM E376	1, 2, 3	Practice for Measuring Coating Thickness by Magnetic Field or Eddy-Current Test Methods
ASTM B117-571	1	Salt Spray Test
ASTM C409	2	Torsion Resistance of Laboratory Specimens of Porcelain Enamelled Iron and Steel
ASTM C448	2	Abrasion Resistance of Porcelain Enamels
ASTM C385	2, 3	Thermal Shock Resistance of Porcelain Enamelled Utensils
ASTM C536	1, 2	Continuity of Coatings in Glassed Steel Equipment by Electrical Testing
ASTM C743	1, 2, 3	Continuity of Porcelain Enamel Coatings

*1. Corrosion or Chemical Resistance
2. Mechanical and Physical Properties
3. High Temperature Properties

points out that small defects, such as pinholes and blisters are difficult to see using the electrified particle test.

The weatherability of porcelain enamel coatings is measured using ASTM C 346, C 538 and D 2244 to evaluate changes in surface color and gloss and ASTM C 703 to evaluate spalling tendency (167). The scanning electron microscope has also been used to assess the effects of weathering (168). Other standardized tests exist for alkali, water and salt spray resistance (Table 2), with nonstandardized tests being performed for water, soil corrosion, and other chemical resistance applications. Eppler et al. (169) describes a series of test procedures for pressurized water and sulfuric acid, citric acid, distilled water and caustic solution, all at elevated temperatures. The Battelle study for heat exchangers devised a test to best simulate actual production conditions, but in a controlled environment (160).

Numerous Soviet researchers developed special tests around their particular applications for porcelain enamels. Karyuk et al. (170) evaluated weight loss every 20 hours for specimens exposed to 20, 50 and 85% phosphoric acid and 20% hydrochloric acid solutions for 400 hours at temperatures over 100°C. Karyuk et al. (171) evaluated weight loss every 10 hours for specimens exposed to hydrochloric acid and sodium hydroxide solutions at their boiling points for 100 hours. Karyuk et al. (153) also evaluated weight loss every 10 hours for specimens exposed to various concentrations of acetic and hydroiodic acids at their boiling points for 100 hours. In all three test procedures, the solutions were changed after ten hours of testing. Mozhaeva and Golovko (172) evaluated weight loss for an enamel in 15 different acids and 6 different alkaline solutions in concentrations ranging from 5% to concentrated for 100 hours at their respective boiling points. Vargin et al. (173) evaluated the weight loss of an experimental iron-containing enamel after cyclic exposure and drying to 20.2% HCl for 50 hours. A seemingly endless number of non-standardized corrosion tests have been performed in the development and evaluation of porcelain enamel systems.

Table 2 also shows the standardized tests for mechanical and physical characteristics of porcelain enamel coatings. Bond mechanisms and test procedures were discussed earlier in this chapter. Another property of porcelain enamel coatings is tested using the Abrasion Resistance Test ASTM C448. In ASTM C448, six 4 3/8" square flat specimens are subject to the abrasive action of 5/32" diameter alloy ball bearing with "either -70 + 100 Pennsylvania glass sand, or No. 80 electric corundum" using a horizontal circular motion (approximately 300 RPM) for various times, and measuring the change or loss of gloss for each sample and the loss in weight

(174). The hardness of porcelain enamels is measured using the same methods as for other ceramics. The Mohs hardness of porcelain enamels ranges from 3.5 to 6, with most between 4 and 5.5 (175). No standardized tests exist for the evaluation of porcelain enamel lubricity, but inclined plane tests are the most commonly used (175). This involves raising one end of a panel and measuring the angle when a given load just begins to slide.

The coefficient of thermal expansion (CTE) is often difficult to determine exactly for porcelain enamels under actual conditions. This occurs because the CTE for glasses is affected by thermal history and the preparation of samples for typical CTE determinations requires a significantly different heat treatment schedule than the typical porcelain enamel firing cycle. Even if the thermal histories were identical, the incorporation of iron oxides, gases and other components from the steel substrate into the porcelain enamel coating changes the resultant CTE value. In spite of these problems, CTE measurements are made and usable information is provided given an understanding of these limitations.

The thermal shock resistance of porcelain enamels depends on the CTE of the coating as mentioned earlier and is evaluated using ASTM C385. A specimen is heated to successively higher temperatures, starting at about 250°C, and repeatedly quenched in a 20 - 22°C water bath until spalling of the coating occurs (176). Since thermal shock resistance also depends on coating thickness, ASTM methods C664, D1186 and E376 are applicable.

3.3 Enhancement of Porcelain Enamel Protective Properties

This final section provides examples of how special materials or processing changes can be made to enhance particular characteristics for specific applications. Similar to the non-standardized test methods, a seemingly endless list of property enhancement methods are available. However, each is specific to a particular application and composition, and often some other property is worsened, or compromised, in the process. Still, some general rules apply. The intention of this section is to provide a series of examples for enhancing the corrosion, abrasion and thermal shock resistance of porcelain enamel coatings.

The two approaches to enhancing porcelain enamel corrosion resistance are improvement of the glass (and other phases) durability and control of the bubble structure to minimize the exposed surface area. Ideally, a system with zero porosity would expose the least surface area, hence corrosion weight loss would be lowest with no bubble structure. However, several other porcelain enamel properties would be sacrificed in the process of trying to "boil" out the entire bubble structure. Essentially, the coating would be overfired causing poor color, susceptibility to chipping and thermal shock and excess substrate attack (which could produce poor adhesion). These mechanical defects in the coating would lead to localized increased corrosion rates and a poor quality surface. The corrosion rate for an underfired porcelain enamel would also be high. This would occur due to the high surface area throughout the coating resulting from incomplete fusion of the glass. Chernyavsky et al. (177) show the importance of minimizing surface area by showing a significant reduction in corrosion resistance after producing a smooth porcelain enamel surface via diamond polishing.

The bubble structure is controlled through the glass physical properties, the firing conditions (time, temperature and atmosphere) and the outgassing or decomposition nature of the substrate and mill additions. The most common approach used to control the bubble structure is through the choice of clay additions. The amount of organic material (loss on ignition) and the temperature of the decomposition affects the quantity and the size distribution of the resultant bubble structure. Murdoch (178) points out that several methods can be used to quantify the porcelain enamel corrosion rate. However, gloss loss is the best method, since the consumer will notice a small gloss loss much more readily than even a large weight loss.

The glass corrosion resistance is controlled by the frit composition and the dissolution of mill additions into the glass upon firing. The mill additions which have been shown to enhance the corrosion resistance of porcelain enamels are SiO_2, TiO_2, ZrO_2, certain clays and mullite (24)(179)-(184). Other aluminosilicates, zirconates and titanates can also be used in certain applications. The major drawbacks for excessive additions of all of these materials are the higher required firing temperatures, alteration of coefficient of thermal expansion and changes in surface appearance. The surface loses gloss, changes color and may become very rough.

Similar additions of SiO_2, TiO_2, ZrO_2, etc. smelted into to the frit composition also enhance corrosion durability. The adverse effects mentioned above are not as great in the frit. However, a significant increase in viscosity, except for TiO_2 ($<12\%$), is the limiting factor. Eppler and others (169)(185)-(186), Clark and Ethridge (188) and numerous other authors (189)-(194) address the relationship between glass chemistry and corrosion resistance for porcelain enamels. The nature of corrosion is rather similar to that of conventional glasses. However, the multi-phase nature of porcelain enamels usually leads to difficulty in modelling.

In spite of the limitations, some general rules exist for enhancing corrosion resistance of porcelain enamels through frit compositions:

1. Small additions (1 - 4%) of TiO_2 are the most effective for improving acid resistance without increasing the glass viscosity. The major drawback is a significant decrease in bond for groundcoats.
2. ZrO_2 (>10%) is most effective for improving high pH corrosion resistance. It, too, detracts from bond, and increases viscosity considerably.
3. Li_2O for Na_2O (on a molar basis) improves corrosion resistance and reduces viscosity. K_2O for Na_2O (on a molar basis) improves corrosion resistance, but increases viscosity. However, certain concentrations of Na_2O are required to produce stable TiO_2-opacified covercoats and to control costs.
4. The transition metals usually increase viscosity (except Fe_2O_3) and decrease corrosion resistance.
5. MgO provides better corrosion resistance than CaO, but produces a much higher viscosity.
6. Al_2O_3 has limited value in frit compositions. It detracts from acid and alkali resistance, increases viscosity and reduces bond in groundcoats. Only small percentages (1 - 5%) are used and usually enter the batch as contaminants of inexpensive raw materials.
7. Fluorine additions usually decrease corrosion resistance. However, its strong fluxing ability may permit higher concentrations of SiO_2 or ZrO_2, and lower concentrations of Na_2O, which could lead to an overall improvement in durability.

The abrasion resistance of porcelain enamels is improved through increased hardness of the surface. Again, the incorporation of mill additions, the frit composition and the bubble structure are the principal methods of controlling the hardness of the coating. In addition, it is fortunate that most conditions which favor improved corrosion resistance also yield increased abrasion resistance. The major difference is that abrasion resistant coatings can be produced from devitrified porcelain enamels since only the exposed portion of the coating is attacked.

Biswas et al. (14) produced high abrasion resistant coatings containing devitrified crystals of spinel phases, magnesium-alumino-titanates, rutile, anatase, lithium-alumino-silicates, and others. Berretz (182) patented a series of coatings containing numerous mill additions, including the carbides nitrides, borides, and silicides of titanium, zirconium, chromium, vanadium, molybdenum, tungsten, boron and silicon. Several of these were actually shown to enhance the abrasion resistance of a porcelain enamel coating.

Thermal shock resistance requires control of the thermal expansion coefficients (CTE) and the introduction of energy absorbing mechanisms. Andrews (195) indicates that the CTE must be continuous from the steel to the outer surface of the coating. Hence, the CTE of the groundcoat must be between that of the steel and the covercoat. Again, improvements in thermal shock resistance are made through mill additions (24)(182)(183) and glass chemistry changes (177)(196). However, the most effective methods to improve thermal shock resistance are to apply as thin a coating as possible and generate a large amount of very fine, homogeneously distributed bubbles.

Again, it is important to note that these recommendations are simply general guidelines that can be used as a starting point for porcelain enamel coatings development and applications. Some degree of experimentation will be required in most cases where stringent requirements must be met. Also, the references provided are excellent sources of more detailed information.

4.0 SUMMARY

Porcelain enamels are versatile, cost effective coatings for several applications. These appplications include large and small household appliances, industrial storage and processing vessels, heat exchangers and architectural panels. The high corrosion and abrasion resistance of porcelain enamels offer a significant advantage over many conventional uncoated steels and other metals. For many of these applications, porcelain enamels are superior to organic coatings because of their ability to withstand elevated temperatures.

The most promising technology within porcelain enamelling is the continued improvement of the electrostatic powder manufacturing method. Since this method recycles the product, nearly 100% of the coating powder is utilized. This will become increasingly important as tighter disposal regulations are implemented throughout the world. Therefore, porcelain enamel coatings will most likely continue to be utilized well into the foreseeable future.

REFERENCES

1. Andrews, A. I., *Porcelain Enamels,* p. 23, The Garrard Press, Champaign, ILL. (1961)

2. Burns, R. M. and Bradley, W. W., *Protective Coatings for Metals,* Ch. 17, p. 584, Reinhold Publishing Corp., New York (1955)

3. Maskall, K. A. and White, D., *Vitreous Enamelling: A Guide to Modern Enamelling Practice,* p. 1, Pergamon Press, Oxford (1986)

4. Smith, H. J. and Eppler, R. A., *AES Coatings for Solar Collectors,* pp. 45-50, American Electroplaters' Society, Inc., FL (1976)

5. Andrews, p. 423

6. Page, M. L., *Met. Mater.,* 34:40-41 (May 1979)

7. Kyri, H., *Handbook for Bayer Enamels,* Bayer AG, Leverkusen, Germany (1976)

8. Andrews, Ch. 1

9. Maskall & White, Ch. 1

10. Burns & Bradley, Ch. 17

11. Gackenbach, R. E., *Chem. Eng,* 85(26):132-137 (20 Nov. 1978)

12. *A Manual of Porcelain Enamelling,* (J. E. Hansen, ed.), The Enamelist Publishing Co., OH (1937)

13. Archer, N. J. and Archibald, L. C., *Chartered Mechanical Engineer,* 24(2):59-63 (Feb 1977)

14. Biswas, K. K., Datta, S., Das, S. K., Ghose, M. C., Mazumdar, A. and Roy, N., *Transactions of the Indian Ceramic Society,* 45(2):43-45 (Mar-Apr 1986)

15. Douglass, D., *Products Finishing,* 48(9):50-52, Cincinnati (June 1984)

16. "Properties of Porcelain Enamel, Electrical Properties", *Data Bulletin PEI 505*

17. Loehman, R. E. and Tomsia, A. P., *Cer. Bull.,* 67(2):375-380, (Feb 1988)

18. Maskall & White, pp. 4, 5

19. Andrews, p. 10

20. Chater, G. D. and Rowlands, H. A., *Australian Corrosion Association,* 1:C-3-1 - C-3-7, Australian Corros. Assoc., Parkville, Victoria, Aust. (1980)

21. Baker, M. A., *NBS Building Science Series 38,* (Aug 1971)

22. Hubbell, D. S. *Mat. Res. Std.,* 7(7): 291-294 (July 1967)

23. Henthorn, M., *Chem. Eng.,* pp. 103-108 (Jan 1972)

24. Partridge, G., *G.E.C. J. Sci. Tech.* 47(2):87-94 (1981)

25. Hackler, C. L. and Dinulescu, M., *Industrial Heat Exchangers,* (A. J. Hayes, W. W. Liang, S. L. Richlen and E. S. Tabb, eds.), American Society for Metals, Metals Park, OH (1985)

26. Simonis, F., Faber, A. J., and Hoogendoorn, C. J., *J. Solar Energy Eng.,* 109(1):22-25 (Feb. 1987)

27. Kumar, B. and Graves, G. A., Jr., *Cer. Bull.,* 61(4):480-483 (Apr. 1982)

28. Andrews, p. 309

29. Chu, P. K., Keeler, J. H., and Davis, H. M., *J. Am. Ceram. Soc.,* 36(2):48-59 (Feb. 1953)

30. Andrews, p. 324

31. Burns and Bradley, p. 587

32. Kyri, p. 130

33. *Engineering,* 206:171 (26 July 1968)

34. Kyri, p. 1

35. Bowley, D.L., *ASTM Standardization News,* pp. 50-52 (March 1990)

36. *Metals Handbook, Ninth Ed.,* 5:509-531 Amer. Soc. for Metals, (1982)

37. Andrews, p. 124

38. Maskall and White, pp. 55-58

39. *The Making, Shaping and Treating of Steel,* (Harold E. McGannon, ed.), Ch. 33, U. S. Steel Corp., Herbick and Held, Pittsburgh, PA (1971)

40. Andrews, p. 113

41. Maskall and White, p. 53

42. Ibid, p. 62

43. Andrews, p. 140-141

44. Zybell, M. M., Rocchetti, E. and Wagner, G., *Email Met.* 55:20-21 (Jan-Mar 1983)

45. Nelson, W. F. and Bacher, J. F., *Cer. Bull.,* 47(2):167-69 (1968)

46. Maskall and White, p. 37.

47. Andrews, p. 11.

48. Ibid, p. 361.

49. Labant, C. J. and Hackler, C. L., *Proc. PEI Tech. Forum,* Vol. 50 (1988)

50. McIntyre, G. H. and Bevis, R. E., *J. Am. Ceram. Soc.,* 21(5):184-88 (May 1938)

51. Blanchard, M. K. and Andrews, A. I., *J. Am. Ceram. Soc.,* 27(1):25-31 (1944)

52. Hurst, .L. and Andrews, A. I., *J. Am. Ceram. Soc.,* 24(5):171-78 (May 1945)

53. King, B. W., *J. Am. Ceram. Soc.,* 37(5):238-42 (May 1954)

54. Svetlov, V. A., Pervinov, A. A., Khodchenkov, V. L., *J. Applied Chemistry, USSR,* 57-2(7):1512-13 (Jul 1984)

55. Andrews, p. 23.

56. Kuchinski, F. A. and Labant, C. J., *Proc. PEI Tech. Forum,* 50:470-79 (1988)

57. Maskall and White, p. 83

58. Snow, J. D., US Patent 3,928,668 (Dec. 23, 1975)

59. Andrews, p. 391

60. Maskall and White, p. 82

61. Myasoedov, V. E., Kharitonov, E. B., and Belova, T. V., *Prot. Met. (USSR),* 21(6):800-802 (Nov-Dec 1985)

62. Andrews, p. 293

63. Pask, J. A., *Ceramic Bull.,* 11:1587-92 (Nov. 1987)

64. Tomsia, A. P. and Pask, J. A., *J. Am. Ceram. Soc.,* 64(9):523-28 (Sept 1981).

65. Kim, Y. W., *Proc. PEI Tech. Forum,* pp. 214-227, (1981)

66. Andrews, p. 517

67. *Metals Handbook,* p. 526

68. Richmond, J. C., Moore, D. G., Kirkpatrick, H. B., and Harrison, W. N., *J. Am. Ceram. Soc.,* 36(12):410-16 (Dec 1953)

69. Harrison, W. N., Richmond, J. C., Pitts, J. W. and Benner, S. G., *J. Am. Ceram. Soc.,* 35(5):113-120 (May 1952)

70. Moore, D. G., Pitts, J. W., Richmond, J. C. and Harrison, W. N., *J. Am. Ceram. Soc.,* 37(1):1-6 (Jan. 1954)

71. King, B. W., Tripp, H. P., and Duckworth, W. H., *J. Am. Ceram. Soc.,* 42(11):504-25 (Nov. 1959)

72. Cevales, M., *I.V.E. Bull.,*19(3):19-34 (Mar 1968)

73. King, R. M., *J. Am. Ceram. Soc.,* 16(5):232-38 (May 1933)

74. Spencer-Strong, G. H., Lord, J. O. and King, R. M., *J. Am. Ceram. Soc.,* 15(9):486-490 (Sept 1932)

75. Staley, H., *J. Am. Ceram. Soc.,* 17(3):163-67 (1934)

76. Dietzel, A. and Meures, K., (translated by R. M. King), *J. Am. Ceram. Soc.,* 18(2):35-37 (Feb. 1935)

77. Kautz, K. *J. Am. Ceram. Soc.,* 19(4):93-108 (Apr 1936)

78. Kautz, K. *J. Am. Ceram. Soc.,* 20(4):115-20 (Apr. 1937)

79. Andrews, A. I. and Swift, H. R., *J. Am. Ceram. Soc.,* 25(9):217-222 (May 1942)

80. Douglas, G. S. and Zander, J. M., *J. Am. Ceram. Soc.,* 34(2):52-59 (Feb. 1951)

81. Healy, J. H. and Andrews, A. I., *J. Am. Ceram. Soc.,* 34(7):207-213 (July 1951)

82. Eubanks, A. G. and Moore, D. G., *J. Am. Ceram. Soc.,* 38(7):226-230 (July 1955)

83. Moore, D. G. and Eubanks, A. G., *J. Am. Ceram. Soc.,* 39(10):357-61 (Oct. 1956)

84. Zackay, V. F., Mitchell, D. W., Mitoff, S. P., Pask, J. A., *J. Am. Ceram. Soc.,* 36(3):84-89 (Mar 1953)

85. Mitoff, S. P., *J. Am. Ceram. Soc.,* 40(4):118-20 (Apr. 1957)

86. Fulrath, R. M., Mitoff, S. P. and Pask, J. A., *J. Am. Ceram. Soc.,* 40(8):269-274 (Aug. 1957)

87. Volpe, M. L., Fulrath, R. M. and Pask, J. A., *J. Am. Ceram. Soc.,* 42(2):102-106 (Feb. 1959)

88. Hagan, L. G. and Ravitz, S. F., *J. Am. Ceram. Soc.,* 44(9):428-29 (Sept 1961).

89. Adams, R. B. and Pask, J. A., *J. Am. Ceram. Soc.,* 44(9):430-33 (Sept 1961)

90. Benzel, J. F., Uher, J. F., Allenbaugh, F. G. and Sweo, B. J., *J. Am. Ceram. Soc.,* 44(1):1-6 (Jan 1961)

91. Moore, D. G. and Mason, M. A., *J. Am. Ceram. Soc.,* 36(8):241-49 (Aug. 1953)

92. Johnston, W. D., *J. Am. Ceram. Soc.,* 47(4):198-201 (Apr 1964)

93. Baak, T. and Hornyak, E. J., *J. Am. Ceram. Soc.,* 44(11):541-44 (Nov. 1961)

94. Cline, R. W., Fulrath, R. M. and Pask, J. A., *J. Am. Ceram. Soc.,* 44(9):423-28 (Sept. 1961)

95. Borom, M. P. and Pask, J. A., *J. Am. Ceram. Soc.,* 49(1):1-6 (Jan 1966)

96. Pask, J. A. and Fulrath, R. M., *J. Am. Ceram. Soc.,* 45(12):592-96 (Dec. 1962)

97. Borom, M. P., Longwell, J. A., and Pask, J. A., *J. Am. Ceram. Soc.,* 50(2):61-66 (Feb. 1967)

98. Borom, M. P. and Pask, J. A., *J. Am. Ceram. Soc.,* 51(9):490-98 (Sept 1968)

99. Pask, J. A., *Proc. PEI Tech. Forum,* pp. 1-16

100. Brennan, J. J. and Pask, J. A., *J. Am. Ceram. Soc.,* 56(2):58-62 (Feb 1973)

101. Shook, W. B., *Proc. PEI Tech. Forum,* pp. 1-12 (1979)

102. Sullivan, J. D., *Proc. PEI Tech. Forum,* pp. 143-159 (1981)

103. Ohmura, A. and Nakano, T., U.S. Patent 4,361,654, (Nov. 30, 1982)

104. Sturgeon, A. J., Holland, D., Partridge, G. and Elyard, C. A., *Glass Technol.* 27(3):102-107 (June 1986)

105. Kinzie, C. J. and Plunkett, J. A., *J. Am. Ceram. Soc.,* 17(9):117-122 (Sept. 1948)

106. Yee, T. B. and Andrews, A. I., *J. Am. Ceram. Soc.,* 39(5):188-195 (May 1956)

107. Friedberg, A. L., Fischer, R. B. and Petersen, F. A., *J. Am. Ceram. Soc.,* 31(9):246-253 (Sept. 1948)

108. Engel, W. H., Eppler, R. A. and Parsons, D. W., *ACerS. Bull.,* 49(2):175-179 (Feb. 1970)

109. Eppler, R. A. *J. Am. Ceram. Soc.,* 54(12):595-600 (DEC. 1971)

110. Olympia, F. D., *Ceramic Bull.,* 32(12):412-414 (Dec. 1953)

111. Eppler, R. A. and McLeran, W. A. Jr., *J. Am. Ceram. Soc.,* 50(3):152-156 (Mar. 1976)

112. Eppler, R. A., *J. Am. Ceram. Soc.,* 52(2):89-94 (Feb. 1969)

113. Eppler, R. A., *J. Am. Ceram. Soc.,* 52(2):94-99 (Feb. 1969)

114. Friedberg, A. L., Petersen, F. A. and Andrews, A. I., *J. Am. Ceram. Soc.,* 31(9):246-253 (Sept. 1948)

115. Beals, M. D., Blair, L. R., Foraker, R. W. and Lasko, W. R., *J. Am. Ceram. Soc.,* 34(10):291-297 (Nov. 1951)

116. Cole, S. S., *J. Am. Ceram. Soc.,* 35(7):181-188 (July 1952)

117. Blair, L. R. and Beals, M. D., *J. Am. Ceram. Soc.,* 34(4,):110-115 (April 1951)

118. Cook, R. L. and Essenpreis, J. F., *J. Am. Ceram. Soc.,* 32(3):114-120 (Mar. 1949)

119. Patrick, R. F. *J. Am. Ceram. Soc.,* 34(3):96-102 (Mar. 1951)

120. Heimsoeth, W. and Meyer, F. R. *J. Am. Ceram. Soc.,* 34(12):366-370 (Dec. 1951)

121. Yee, T. B., Machin, J. S. and Andrews, A. I., *J. Am. Ceram. Soc.,* 38(10):378-381 (Oct. 1955)

122. Eppler, R. A. and Spencer-Strong, G. H., *J. Am. Ceram. Soc.,* 52(5):263-266 (May 1969)

123. Andrews, *Porcelain Enamels,* pp. 78, 81-83

124. Marbaker, E. E., Saunders, H. S. and Baumer, L. N., *J. Am. Ceram. Soc.,* 32(9):297-304 (Sept. 1949)

125. Russel, N. K., Friedberg, A. L. and Petersen, F. A., *J. Am. Ceram. Soc.,* 34(1):28-31 (Jan. 1951)

126. Andrews, p. 404

127. "Porcelain Enamelling", *Metals Handbook,* p. 518

128. Maskall and White, p. 86

129. Burns and Bradley, p. 591

130. Andrews, p. 508

131. Ibid, p. 263

132. "Porcelain Enamelling", *Metals Handbook,* p. 521

133. Andrews, p. 449

134. Moore, D. G., Mason, M. A. and Harrison, W. N., *J. Am. Ceram. Soc.,* 35(2):33-41 (Feb. 1952)

135. Andrews, p. 42

136. Chu, G. P. K., *Ceramic Industry,* pp. 98-101, 115-117 (Sept. 1961)

137. Chu, G. P. K., *Ceramic Industry,* pp. 60-62, 104 (Oct. 1961)

138. Chu, G. P. K., *Ceramic Industry,* pp. 62-63, 97 (Nov. 1961)

139. Chu, G. P. K., *Ceramic Industry,* pp. 39-41 (Feb. 1962)

140. Chu. G. P. K., *Ceramic Industry,* pp. 72-73 (July 1962)

141. Chu, G. P. K., *Ceramic Industry,* pp. 72-73 (Mar. 1962)

142. Sullivan, J. D., Nelson, D. H. and Nelson, F. W., *J. Am. Ceram. Soc.,* 45(11):509-512 (Nov. 1962)

143. Andrews, p. 418

144. King, B. W. Jr., Carter, H. D. and Draker, H. C., *J. Am. Ceram. Soc.,* 30(1):22-26 (Jan. 1947)

145. Parikh, N. M., *J. Am. Ceram. Soc.,* 41(1):18-22 (Jan. 1958)

146. Culter, I. B., *J. Am. Ceram. Soc.,* 52(1):11-13 (Jan. 1969)

147. Maskall and White, p. 1

148. Andrews, p. 452

149. Alikina, I. B. and Sirotinskii, A. A., *Glass and Ceramics,* 42(12):545-547 (Dec. 1985)

150. Baker, M. A., *NBS Building Science Series 50,* pp. 1-12 (July 1974)

151. Baker, M. A., *NBS Tech. Note 707,* pp. 1-14 (Dec. 1971)

152. Baker, M. A., *NBS Building Science Series 29,* pp. 1-11 (April 1970)

153. Karyuk, A. A., Shkolyar, P. S., Manzhelii, A. P. and Borodai, T. P., *Chemical and Petroleum Engineering,* 21(7-8):359-360 (Aug. 1985)

154. Alikina, I. B. and Sirotinski, A. A., *Glass and Ceramics,* 41(7-8):227-340 (Jul.-Aug. 1984)

155. Warren, I., *Heat Recovery Systems,* 2(3):257-271 (1982)

156. McRae, T. F., *Mater. Protect.,* pp. 41-42 (Dec. 1968)

157. Nadyrov, I. I., Tsirul'Nikov, L. M. and Rashkovan A. V., *Thermal Engineering,* 16(10):32-37 (Oct. 1969)

158. Bazayants, G. V., Svetlichnyi, V. A., Oleinik, M. I., Demchuk, V. V., Ryzhikov, V. A. and Sirotinskii, A. A., *Thermal Engineering,* 28(12):727-729 (Dec. 1981)

159. Bazayants, G. V., Svetlichnyi, V A., Demchuk, V. V. and Ryzhikov, V. A., *Glass and Ceramics,* 40(5-6):295-296 (May-June 1983)

160. Sekercioglu, I., Raxgaitis, R. and Lux, J., *Advances in Ceramics,* 14:359-370, Amer. Ceram. Soc., Columbus, OH (1985)

161. Dobrunova, V. M., Bakalin, Yu. I., Nesterenko, V. B., Doroshkevich, V. N. and Trubnikov, V. P., *Thermal Eng.,* 33(10):577-578 (Oct. 1986)

162. *Metals Handbook,* p. 529

163. Trubnikov, I. L., Korchagin, V. S. and Zusman, S. D., *Sov. Energy Technol.* (1):79-82 (1984)

164. Andrews, p. 584

165. Baker, M. A., *High Voltage Tests Porcelain Enamel Coatings*, 27(2):74-77 (Feb. 1970)

166. Staats, H. N., *Ceramic Bull.,* 31(2):33-38 (1952)

167. "Properties of Porcelain Enamel, Resistance to Corrosion", *Data Bulletin PEI 503*

168. Baker, M. A., *Proc. PEI Tech. Forum,* 33:84-90 (1971)

169. Eppler, R. A., Hyde, R. L. and Smalley, H. F., *Ceramic Bull.,* 56(12):1064-1067 (1977)

170. Karyuk, A. A., Stekhina, E. R., Bobovich, O. V., Borushko, O. I., Volkava, I. S. and Bulavkina, I. M., *Chemical and Petroleum Engineering,* 20(9-10):519-520 (Sept.-Oct. 1984)

171. Karyuk, A. A., Stekhina, E. R., Borushko, O. I., Borodai, T. P., Bulavkina, I. M., Litvinenko, L. I. and Fostova, V. V., *Chemical and Petroleum Engineering,* 21(11-12):602-605 (Nov.-Dec. 1985)

172. Mozhaeva, A. A. and Golovko, I. F., *Chemical and Petroleum Engineering No.* 9-10, pp. 697-700 (Sept.-Oct. 1969)

173. Vargin, V. V., Grachev, V. V., Zorina, M. L. and Ushakov, D. F., *Glass Ceramics,* 28(9-10):613-615 (Sept.-Oct. 1971)

174. Andrews, p. 553

175. "Properties of Porcelain Enamel, Mechanical and Physical Properties", *Data Bulletin PEI 502*

176. Andrews, p. 471

177. Chernyavsky, A. N., Preis, G. A., Smirnov, N. S. and Svidenyuk, T. A., *Soviet Mat. Sci.,* 11(1):61-63 (Jan.-Feb. 1975)

178. Murdoch, A. W., *I.V.E. Bull.,* 16(10):99-110 (Oct. 1965)

179. Svetlov, V. A., Pervinov, A. A. and Bovkun, N. P., *Glass and Ceramics,* 39(7-8):356-357 (Jul.-Aug. 1982)

180. Lorentz, R., *Werkstoffe und Korrosion,* 33(5):247-253 (May 1982)

181. Lorentz, R., *Werlstoffe und Korrosion,* 33(4):194-203 (April 4, 1982)

182. Berretz, M., US Patent 4,196,004, (April 1, 1980)

183. Barinov, Yu.D., Smakota, N. F., Ivanov, N. V. and Shatalova, L. G., *Glass and Ceramics,* 42(12):548-550 (Dec. 1985)

184. Viquesnel, A., *Email Metal,* No.7 pp. 63-69 (1970-1971)

185. Eppler, R. A., *ACerS. Bull.,* 56(12):1068-1070 (Dec. 1977)

186. Eppler, R. A., *ACerS. Bull.,* 60(6):618-622 (1981)

187. Eppler, R. A., *ACerS. Bull.,* 61(9):989-991 (1982)

188. Clark, D. E. and Ethridge, E. C., *ACerS. Bull.* 60(6):646-649 (1981)

189. Svetlov, V. A., Pavlichenko, T. I., Khodchenkov, V. L., *J. Applied Chemistry, USSR,* 57, Pt. 2(7):1512-1513 (July 1984)

190. Mozhaeva, A. A., Pilipenko, M. F., Shabrova, E. A., Stekhina, E. R., Tatarchenko, L. I. and Shabrov, B. M., *Chemical and Petroleum Engineering,* 18(9-10):423-425 (Sept.-Oct. 1982)

191. Volkov, S. I., Mizonov, V. M., Onishchenko, E. A., *J. Applied Chemistry, USSR,* 57(3-2):645-647 (March 1984)

192. Volkov. S. I., Mizonov, V. M., Shabrova, E. A. and Filippov, Yu. V., *J. Applied Chemistry, USSR,* 58(9):1985-1988 (Sept. 1985)

193. Volkov, S. I., Mizonov, V. M., Smirnov, N. S. and Shabrov, V M., *Protection of Metals,* 15(6):614-616 (Nov.-Dec. 1979)

194. Ivanov, I. V., Barinov, Yu. D. and Ivanova, L. N., *Glass and Ceramics,* 41(9-10):446-447 (Sept.-Oct. 1984)

195. Andrews, p. 472

196. Harrison, W. N., Moore, D. G. and Richmond, J. C., *J. Res. NBS,* 38:293-307 (March 1947)

5

Plasma Sprayed Ceramic Coatings

Herbert Herman, Christopher C. Berndt, and Hougong Wang

1.0 INTRODUCTION

Ceramic coatings are readily formed using a wide variety of methods (e.g., sputtering, electron-beam deposition). Plasma spraying has been used since the mid-1950's to form metal alloy, ceramic and cermet coatings on a range of metallic substrates. These coatings, usually greater than 50 micrometers in thickness, are used for a remarkable number of applications: wear/erosion and corrosion resistance; thermal barriers; electrical and magnetic components, etc. Plasma spraying has played a particularly effective role in depositing a variety of ceramic materials. These coatings are used in the aircraft industry for wear resistance and abradable seals and for thermal barriers, permitting hot sections of gas turbine engines to function at increased operating temperatures. Plasma sprayed refractory ceramic coatings are used for handling liquid metals (e.g., spinel, cordierite) and, increasingly, for electrically insulated metal substrates in the automotive electronics industry.

The variety of applications of plasma sprayed coatings is impressive and their use is evolving rapidly. This article reviews the fundamentals of plasma spraying with a special emphasis on why this process is so well suited for the processing of ceramics, particularly oxides. An overview is then given of the field of plasma sprayed ceramics, with special emphasis on two important fields: thermal barrier coatings and high temperature ceramic superconductors. These two subjects, though apparently very different, have in common the need to exercise great care and control in the production of high performance ceramic deposits. These subjects exemplify the great versatility and the exciting possibilities inherent in plasma spray technology.

2.0 PLASMA SPRAYING

Plasma is a gaseous collection of electrons, ions and neutral molecules. If the density of the plasma is sufficiently high (of the order of 10^6 electrons/m^3), the electrons of the plasma readily exchange energy with the ions and neutral species to yield a kinetic energy so high that any known material can be melted. This so-called "thermal plasma" can be contained and controlled within a small space so that high enthalpy densities can yield intense, high temperature flames at ambient pressure. In fact, such thermal plasmas can be operated at reduced pressures (VPS for "vacuum plasma spray"), resulting in plasma plumes exiting from the plasma torch at supersonic velocities. The plasma used for ceramic processing are generally operated in air and the flames rarely exceed sonic velocities. Overviews of dense plasmas as used in thermal spray technology can be found in the literature (1)(2).

A typical plasma gun is indicated schematically in Fig. 1. This kind of gun is commonly used to spray ceramics and operates with direct current. The heart of the gun is the cathode and the anode, both of which are internally water cooled. Plasma gas, usually an inert gas such as argon, enters the region between the thoriated tungsten cathode and a cylindrical water-cooled anode. The gas, in most renditions of this type of gun, enters as a vortex. To initiate the plasma, an arc is struck between the tip region of the cathode and the internal surface of the anode. The plasma is sustained by the continuous in-flow of the plasma gas. The plasma flame exits to the right, as indicated, through the circular orifice of diameter A. The ionized gas recombines and becomes neutral in the vicinity of the exit opening, yielding a high level of enthalpy. It is into this recombination region (beyond the plasma core) that the feedstock powder is introduced, carried out by the flame, melted, and accelerated to the workpiece, where it impacts and undergoes rapid solidification. Figure 2 is the schematic of an industrial level plasma gun. In this case, the powder is injected into the plasma flame internally as indicated by the "powder tube".

Of great importance is the power level and the kinds of plasma gases that are employed. Argon, for example, is commonly used as a plasma gas because it is non-reactive and has the properties needed to both initiate and

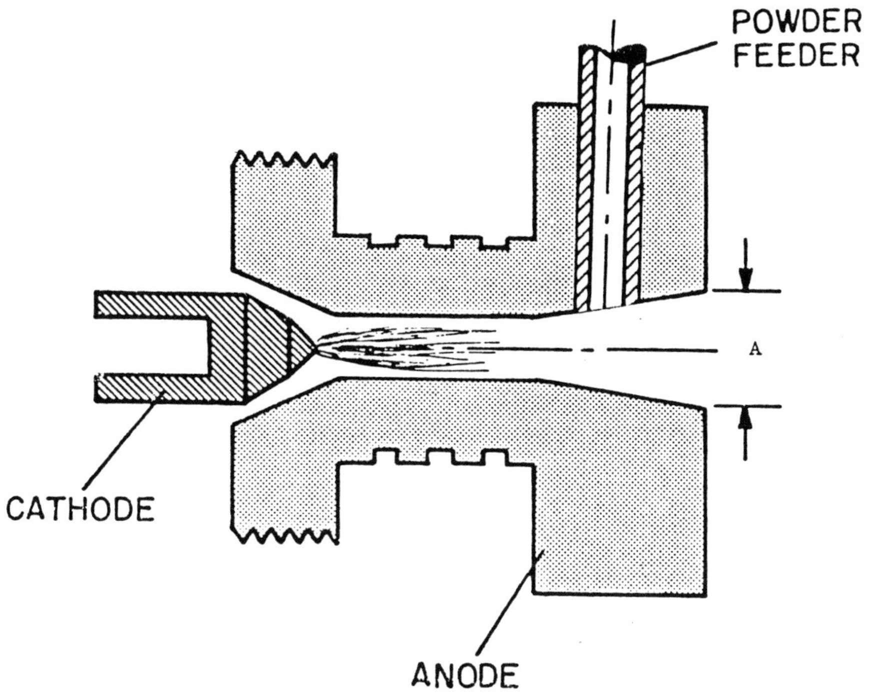

Figure 1. Schematic cross-section of non-transferred DC arc plasma spray gun with internal particle feed injector. The electrodes are water-cooled. The exit opening has a diameter *A* which is a design parameter.

to sustain the plasma. Again, energy needed to melt the powder particles results from the neutralization through recombination of the ionic argon and electrons. This energy can be enhanced through the use of bimolecular species such as hydrogen or nitrogen, which can be added as a secondary plasma gas. The resulting gas mixture yields much hotter flames, which are generally required to melt refractory materials. This is depicted in Fig. 3, where energy content is plotted versus effective gas temperature for a variety of plasma gases.

The rate at which, for example, alumina is fed into a typical 40 kW plasma gun of the kind depicted in Fig. 1 ranges from 2 to 5 kg/hr and higher. Higher power plasma spray guns, which can operate up to 250 kW, yield throughputs of up to 50 kg/hr (3). But the vast majority of the plasma spray guns are rated at between 35 to 80 kW. Details of plasma gun operation and the complexities inherent in their operation have been discussed in a number of review articles and books (e.g., Refs. 2, 4).

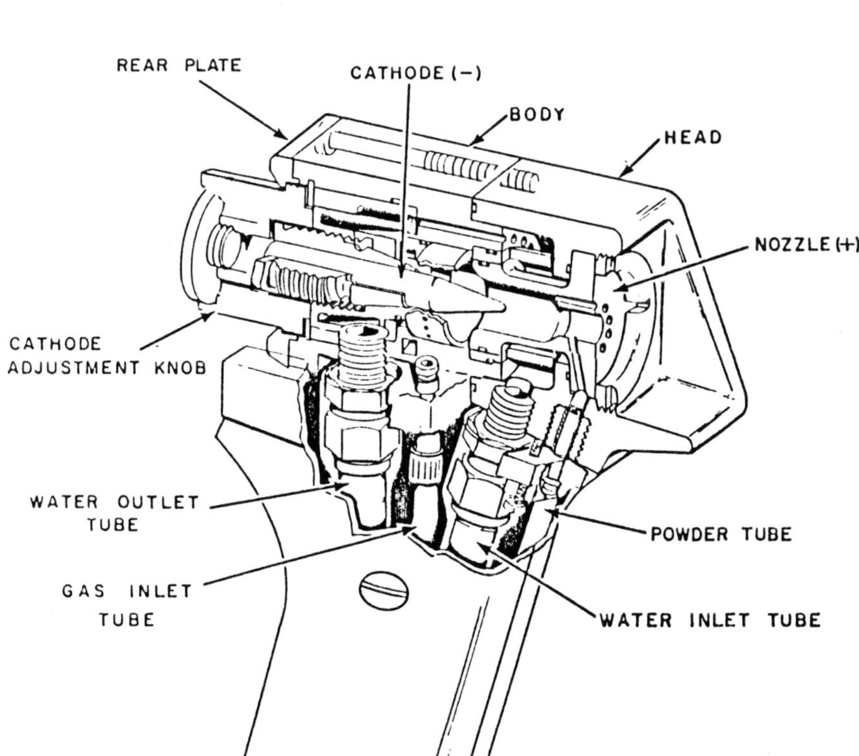

Figure 2. A view of the internal workings of a DC arc plasma gun. (After Bay State Abrasives).

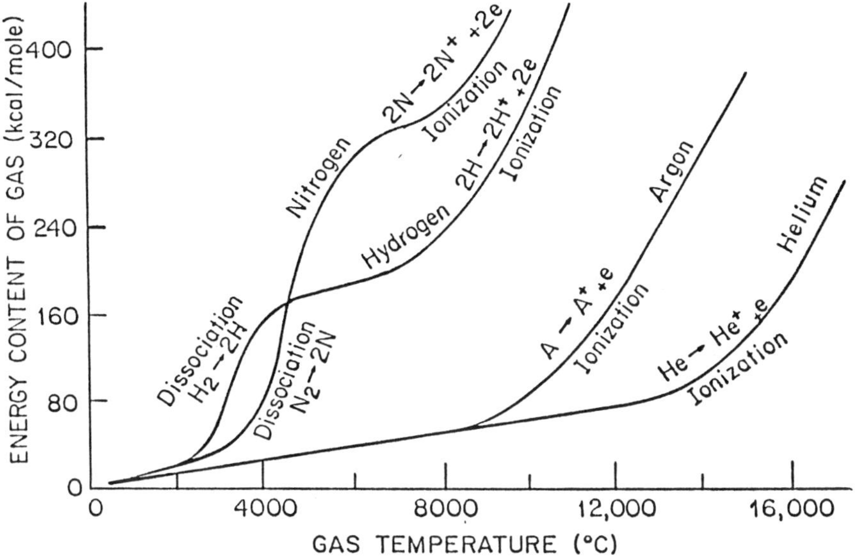

Figure 3. Energy content vs. gas temperature for a DC thermal plasma operating with different plasma gases (as quoted in Ref. 2).

2.1 Feedstock Powders

There are a number of interrelated parameters which determine the characteristics of the resulting plasma sprayed ceramic coating. These include: gas type, pressure and flow rate; power; spray distance; etc. Of great importance, and too frequently overlooked, are physical features of the feedstock powder. These include: particle size and size distribution, particle shape; and the level of chemical uniformity of the constituents in a mixed oxide. Size and shape are particularly significant because the former is important for meltability considerations, while the shape of the particles will determine the flowability of the powder into the flame. For example, a flake-shaped powder commonly will not display smooth flow, resulting in a discontinuous, pulsing feed of powder into the flame, leading to a non-uniform stream of molten particles and, thus, a poor coating. Spherical particles, on the other hand, enable smooth, uniform feeding, leading to a deposit with fewer discontinuities.

Much has been written on powders for plasma spraying. A number of issues are important and manifest themselves in the ceramic coating cross-sections, which are discussed later in this review.

2.2 The Ceramic Coating

Ceramics are generally brittle. As coatings on relatively ductile metal alloy substrates, ceramics behave in complex ways under a mechanical load or when temperature changes occur. Of special importance is the interface between the ceramic coating and the metallic substrate. Limited work has been carried out to characterize this interface, but, based on cross-sectional optical or electron metallography, it appears that the first ceramic layer to form is amorphous or composed of ultrafine sized grains (5). Little is understood of this first layer, but it is certain that proper surface preparation must be achieved to obtain a good adhesive bond. Surface preparation generally involves grit blasting of the substrate prior to plasma spraying. The blasting achieves a surface roughening, which leads to an essentially mechanical bond. A poorly prepared substrate surface (i.e., too smooth) will lack sufficient "tooth" to anchor the coating. The details of coating formation are relevant to bonding and are discussed below.

When a molten particle impacts the substrate (or previously deposited material), it spreads and solidifies rapidly. This is modelled in Fig. 4a, where it is seen that heat is immediately removed from the solidifying particle at the impact point. Concurrently, the remaining molten material is spread over the solidifying core. This process of coating build-up yields rapid solidification and the development of a highly defective microstructure, as pictured schematically in Fig. 4b. Observed are cracks and voids, which will, in general, contribute to poor mechanical properties.

If the deposit is a ceramic (e.g., alumina), the process-formed porosity and even micro-cracking may contribute to some apparent toughness and strain tolerance. Since the ceramic coating has no inherent ductility, apparent strain comes from variations of crack propagation, the extent of which can be limited by the occurrence of prior-formed micro-cracks and crack-blunting pores. In fact, it is commonly observed that highly dense ceramics cannot be sprayed to a great thickness (i.e., >0.5 mm) without delamination occurring during spraying. This results when large stresses cannot be relieved by limited crack motion. These issues of imperfections in ceramic coatings and how they can be controlled to yield strain-tolerant coatings are further discussed in Sec. 4.0 on thermal barrier coatings.

a

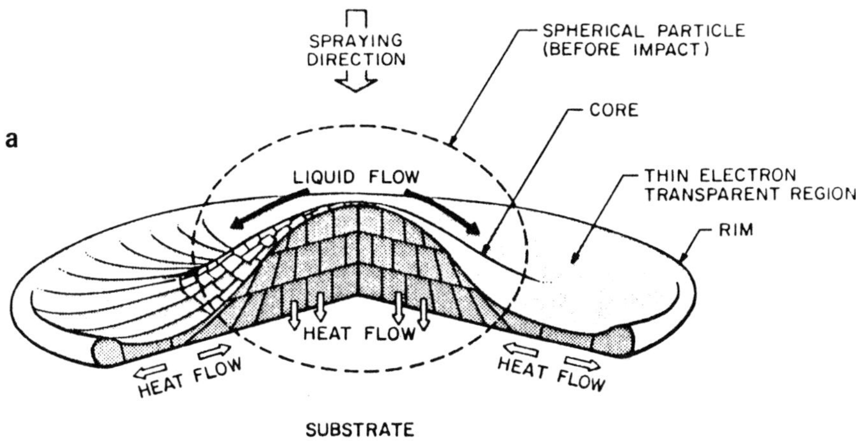

b

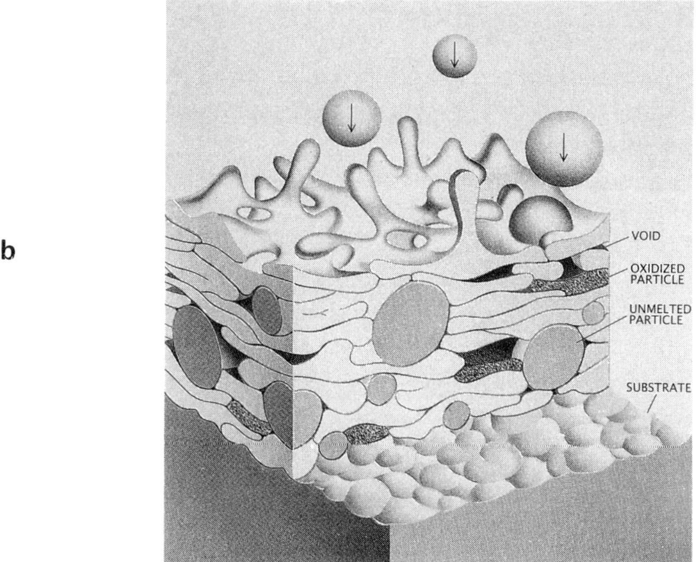

Figure 4. *a.* Model of solidifying splat showing dynamics of solidification (After Ref. 1). *b.* Model of the build-up of a defected coated by deposition of splats. Shown are voids, oxidized particles (for metals), unmelted particles associated with the cross-section of a plasma sprayed deposit (After Ref. 1).

Another important aspect of plasma spraying of ceramic coatings involves residual stresses which are created within both the coating and the substrate during spraying and, for elevated temperature applications (e.g., thermal barrier coatings), during use. The process-related stresses arise on cooling due to the differential thermal expansion coefficient between the ceramic coating and metallic substrates. In the most commonly quoted cases, the thermal expansion coefficient for a ceramic is about one-half that of the metallic substrate (i.e., partially-stabilized zirconia versus nickel-based alloy). These stresses can result in delamination of the coating and can lead to coating thickness limitations. These issues are further discussed in Sec. 6.0.

2.3 Special Features of Plasma-Sprayed Coatings

A brief review of the important and sometimes unique features that characterize plasma spray processing are listed below.

1. Plasma spraying can be used to deposit a wide range of ceramics and metals, and any combinations of these.
2. It is possible to deposit alloys and mixed ceramics (e.g., oxides) with components of widely differing vapor pressures without significant changes in composition.
3. Very homogeneous coatings can be formed that display no significant change of composition with thickness (that is, during the duration of deposition).
4. Microstructures can be formed having fine, equiaxed grains, without columnar defects (that might, for example, occur with electron-beam deposition).
5. It is possible to change from depositing a pure metal, to a continuously varying mixture of metal and ceramic, to a pure ceramic (for example, an oxide) using the same automated equipment and set-up without intermediate part-handling or readjustments. These are so-called "graded coatings," and variations of them are widely employed with thermal barrier coatings.
6. High deposition rates are possible without extreme investments in capital equipment.
7. Free-standing bulk forms can be plasma sprayed of virtually any ceramic or combinations of ceramics (e.g. mixed ceramics, composites, cermets); i.e., near-net shapes.
8. Plasma spraying can be carried out in virtually any environment: air, enclosed inert low and high pressure environments (relative to ambient), or underwater.

The versatility of plasma spraying has been widely recognized and is entering an increasing number of industries, but it is important to note that the apparent simplicity of the operation of a plasma spray torch can be deceptive. In fact, achieving a coating with the desired microstructure and properties is a rather complex exercise requiring great care and control. In the past, plasma spraying was carried out by specialists with considerable experience at torch handling. Similar to manual welding of the 1950's, plasma spraying was, and in some cases, still is, an art dependent on skill more than a technology. Current and future directions in industrial plasma spraying employ automated technology such as robotics and adaptive process controls. Of special significance has been the recognition of the inherent complexities of these processes, operating with numerous interactive and independent parameters. Thus, statistical process control is now becoming the norm for commercial plasma spray operations involving both large throughput and small-number high-value parts, i.e., ranging from millions of alumina-sprayed insulated metal substrates to thermal barrier coated gas turbine blades, produced by the dozens at a time.

The industry faces many challenges in achieving reproducibility and process control. It appears now that the science of the plasma is converging with the technology of process control, to the extent that on-line spray feedback control is emerging as a reality. This evolving philosophy of the control of plasma spraying is depicted in the schematic shown in Fig. 5, which represents the control that is sought in industrial-level spraying of high performance coatings.

The following reviews various salient features of some oxide ceramics in the form of plasma sprayed coatings. As an example of a widely used material, we shall explore plasma sprayed alumina-based ceramics.

3.0 ALUMINA-BASED CERAMICS

Alumina and mixed aluminas (e.g., with titania, magnesia, zirconia) are widely plasma sprayed as coatings and free-standing forms. Alumina has two important industrial properties: hardness and electrical insulation. The hardness qualities of alumina contribute to this material being employed as a wear resistant coating. Electrical insulation properties of alumina are widely appreciated; and it is that property, in addition to its reasonably high thermal conductivity, that creates a highly significant role as a plasma sprayed top coat for insulated metal substrates, for example, in automotive applications, where about 50 micrometers of alumina are plasma sprayed

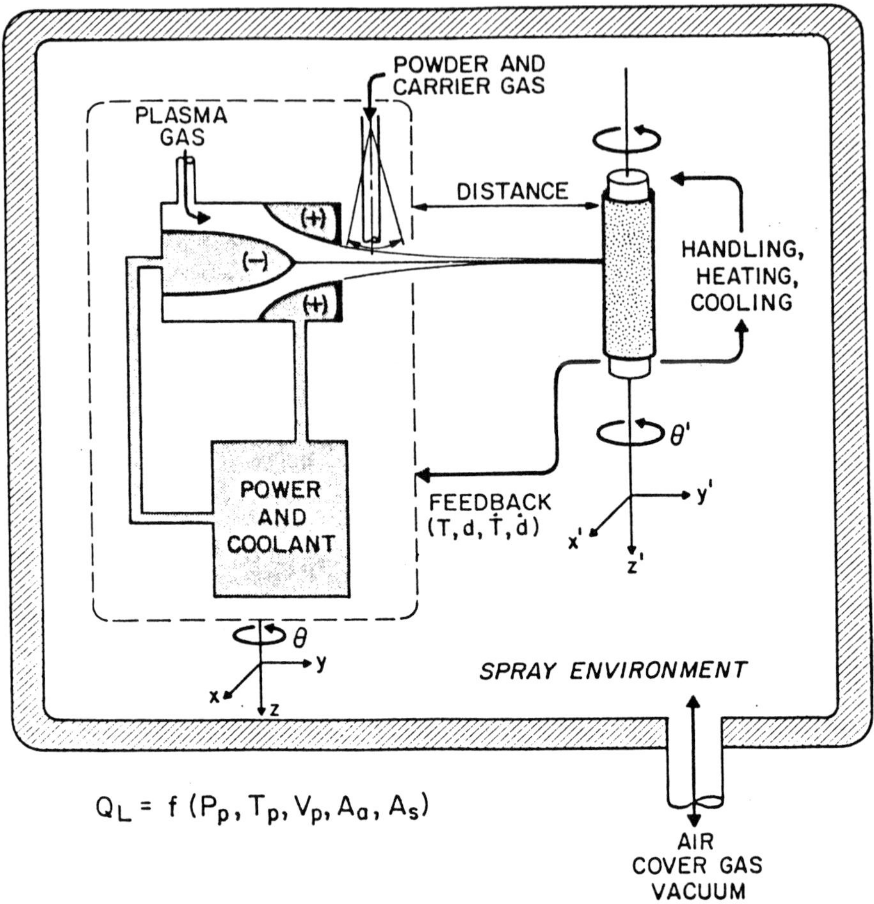

Figure 5. Automated plasma spray system showing the operational features that yield the best coatings by optimized spray parameters. The gun and the injector are controlled by robotic handling, and coating quality is a function of a number of independent parameters. Feedback, involving adaptive control, yields real-time adjustments of the process.

onto aluminum heat-sink substrates. Electronic circuits are then built up on the alumina top coat. This application, of course, requires that the sprayed alumina deposit has a high dielectric breakdown strength (in the range of several hundred volts for 25 micrometers in thickness). This requirement, while readily achievable, can be limited by the fact that plasma sprayed alumina has two crystal structures, alpha and gamma, the latter being metastable at room temperature. It is also important to note that gamma alumina absorbs water (6). The alpha phase is stable at room temperature and is chemically inert over a large temperature range in many aggressive environments. Therefore, it is important to plasma spray so that only alpha alumina is formed; a goal not readily realizable. It is also important to note that impurities in alumina will have significant effects on all of the dielectric properties. Thus, plasma spraying of insulated metal substrates requires extraordinary process control as well as great care in materials purity and feedstock powder specification.

Alumina with approximately 3 wt% titania is referred to as "grey alumina" and is used extensively as a wear-resistant coating. The hardness and friction coefficients decrease with greater levels of titania, 13 and 40 wt%, leading to a superior and increasingly used plasma sprayed coating. The means of preparation of the alumina-titania feedstock powder has been related to wear behavior (7). The fused-and-crushed powders enable considerably more latitude in spray parameters (e.g., spray distance) than do composite powders. This effect clearly resides in the superior chemical uniformity of the fused (electric-arc-melted and cast)-and-crushed materials vis-a-vis a composited powder. In the case of the alumina-13 wt% titania the chemical uniformity is very low since the submicron titania pigment is adhered to the large alumina particles. It should be noted that the positive effect of chemical uniformity within powder particles is displayed in other ceramic plasma spray systems, such as partially-stabilized zirconia and in ternary oxides such as cordierite (see below).

Kingswell et al. have vacuum plasma sprayed nominally pure alumina onto a variety of metallic substrates and have achieved a very dense well-bonded coating (8). Hard particle erosion experiments showed that vacuum sprayed alumina withstood erosion as well as sintered bulk products. Similarly, Chon et al. vacuum plasma sprayed various ratios of alumina-to-NiCrAlY alloy blends and obtained dense, well-bonded cermets (9). These workers observed that the slurry-wear-resistance increased with loading of ceramic. Recently, alumina-matrix composites have been studied for high temperature wear resistance.

Cordierite is a ternary oxide having the stoichiometry $2MgO \cdot 2Al_2O_3 \cdot 5SiO_2$. The only practical way to produce powders with cordierite's chemistry is through the fuse-and-crush technique. Cordierite has extremely good thermal shock resistance in the bulk form due to its very low thermal expansion coefficient.

Cordierite has been plasma sprayed, yielding an amorphous structure, which, on annealing, converts to a quartz-like metastable at 910°C and then to orthorhombic stable cordierite at temperatures above 1140°C (10). Of particular interest in the above study, as detected by transmission electron microscopy, is the occurrence of a pre-crystallization transition on annealing for 8 hrs at 870°C, which has been tentatively identified as spinodal decomposition. It is further interesting to note that substrate-free plasma sprayed cordierite demonstrates the same excellent thermomechanical properties as observed for normally prepared bulk cordierite (11).

A wide range of aluminas are plasma sprayed. Insufficient fundamental work has been done on these materials, but this has not limited their extensive utilization. There is a great need to examine the physical properties and the mechanical behavior of alumina, as well as the large numbers of oxides that are currently being plasma sprayed. Excellent sources for information on both the fundamental and applied aspects of these plasma sprayed coatings are the numerous proceedings which have been published by ASM International (Materials Park, Ohio, 44073) on the National Thermal Spray Conferences and by various publishers on the International Thermal Spray Conferences, which are convened at different venues every three or four years. A number of these proceedings are cited in this review.

Commercial ceramic powders that are currently plasma sprayed are listed below. Other powdered ceramics, both oxide and non-oxides, are under development.

Al_2O_3	Cr_2O_3
Al_2O_3-TiO_2	Cr_2O_3-TiO_2
Al_2O_3-Cr_2O_3	Cr_2O_3-SiO_2

Partially-stabilized zirconia (PSZ) is the oxide ceramic that has received the most attention from the plasma spray community. The interest in PSZ resides in its use as a thermal barrier coating for aircraft applications. In addition, bulk forms of PSZ's, produced using a variety of traditional ceramics processing approaches, have been comprehensively researched in the past 20 years as a toughenable structural ceramic. Thus, much is understood of the plasma sprayed PSZ systems. It should be emphasized, however, that strengthening mechanisms that apply to these bulk materials may not be directly applicable to plasma sprayed coatings.

The next section gives an overview of plasma sprayed thermal barrier PSZ coatings. These high performance coatings are reasonably well understood, and they point to the important potential of plasma spray methods as a processing technique for ceramics.

4.0 THERMAL BARRIER COATINGS

Plasma thermal spray coatings are used in numerous applications which take advantage of their excellent wear, corrosion, high temperature and thermal shock resistant characteristics. Thermal barrier coatings (TBC's) are a specific classification of plasma sprayed coatings and usually consist of ceramic alloys of zirconia with stabilizing oxides such as yttria, magnesia, calcia or ceria (12)-(19). These coatings experience severe thermal flux and high temperature environments in critical applications.

A variety of thermal spray techniques are used and these include flame spraying and atmospheric plasma spraying (APS) for the non-metallic and non-oxide materials which are intended for use in low temperature (<800°C) environments, vacuum plasma spraying (VPS) for the metallic bond coat materials and, generally, APS for all ceramic materials. Several articles (20)-(26) have addressed the application of TBC's in the turbine environment. The aerospace and utility turbine applications are specifically addressed in this chapter and an outline of the materials used in these applications is shown in Table 1. Similar coatings are being proposed for use in the automotive industry on the crowns of pistons and these are known as thick TBC's (or TTBC's). This section also be discusses the engineering science of these coatings, presenting a focused survey of experimental methods used to characterize the coatings.

The coating microstructure and phase distribution of plasma sprayed coatings are dissimilar to those of the bulk constituents, therefore, material properties of the coating such as thermal diffusivity, mechanical strength in tension and shear, and wear characteristics, will be different from the properties of the bulk material. The microstructure and thus the basic science of coatings determine their utility and this recognition enables the user to take full advantage of coating properties.

4.1 Applications

Plasma sprayed coatings are used on the compressor, combustion chamber, fuel vaporizers, nozzle guide vane platforms and turbine aerofoil components of aero-engines. These particular TBC's and other high temperature applications are discussed and two points emphasized. The first is that most ceramic coatings, and in particular TBC's and TTBC's, are used in conjunction with a metallic bond coat which is usually based on a Ni-Cr-Al-Y composition. The second clarification, as discussed above in Sec. 2.0, is that the quality and performance of the so-formed coating system is quite variable and depends on, among other factors, the thermal spray equipment, the skill of the applicator (whether a robot or a technician), and the quality of the spray powder.

Table 1. Use of Coatings in High Temperature Applications

Application Limitation	Temperature °C	Materials
Wear resistance		
	500	WC-Co
	800	Cr-C-Ni-Cr
	800	Cobalt materials
	800	Cermets
Abradable		
Minimize gas leakage	450	75/25 Ni-graphite
Clearance control	650	Ni-Cr-Al-bentonite
Rotating parts	325	60/40 Al-Si-polyester
Compressor seals	475	70/30 Al-graphite
Thermal Barrier		
Turbine blades	1050	Ni-Cr-Al-Y
Coatings and Combustor		stabilized zirconia
Salvage and Repair		
Hard surfaces for pump seals		Ni-Al base materials
Prevent adhesive wear on piston guides		
Hard bronze and babbit bearings for fuel pump rotors, impeller shafts, journals etc.		

Aero-Engines. The coatings, for example, for gas turbine aircraft engines are deposited to 0.38 mm thickness over the airfoil surface of turbine blades (27)-(32). It has been calculated that the incorporation of a thin coating decreases the metal/substrate temperature at the leading edge of vanes by 190°C (20). The metal temperatures of the turbine vane could be reduced by as much as 390°C when coated with 0.5 mm of zirconia. It has been predicted that engines with coated blades operate with a 40% reduction in the coolant-to-gas flow, and this calculates to a 1.3% improvement in the specific fuel consumption. Thus, the driving force for TBC development is either an economic advantage through increased efficiency in the domestic marketplace (with either increased fuel efficiency or increased component life) or improved performance for military applications.

Coatings which are based on WC-Co powder composites are applied to the compressor fan and disc mid-span stiffeners to prevent wear, and to the compressor airfoils to control particulate erosion (33). The powder and processing technology for the powders and the plasma spraying procedures have been intensely developed. The powders can be produced by a micropelletization process, by agglomeration and sintering or by a powder blending process. The material properties and performance of the resultant coating are markedly influenced by the quality and characteristics of the initial feedstock powder.

Some aircraft parts require good sealing between the rotating and stationary components to maintain high compression of gases. In these applications, a coating which can be abraded is sprayed onto the stationary component, such as the compressor, and an abrasive material is coated onto the tip of the turbine blade or onto the disc spacer (33)(34). The harder phase of the engineering assembly preferentially abrades the softer phase. The turbine will maintain the minimum clearance between the rotating components and, thereby, achieve optimum engine efficiency.

The abradable coatings are manufactured as composites of graphite with either nickel or aluminum. They may also consist of nichrome with a polyester or polyurethane. Two applications of abradable and abrasive coatings to seal airfoil systems which have rotating components are shown in Fig. 6 (33). The coating system may be multi-layered, and Fig. 7 shows an example where a combination of four coatings provides a thermal barrier and an abradable coating (33).

Diesel Engine Applications. Thermal barrier coatings have been used in diesel engines rather extensively (35)-(40). Zirconia-yttria alloy coatings may improve fuel efficiency by insulating the combustion chamber area of the engine, thereby recovering the 8 to 15% of the energy that is attributed to heat losses. The coatings have been applied to the cylinder head, the valves, the piston, and the liner top (to 1.5 mm thickness) (Fig. 8). A plasma spray molybdenum coating can be applied to piston rings to ensure long term sealing of the combustion chamber.

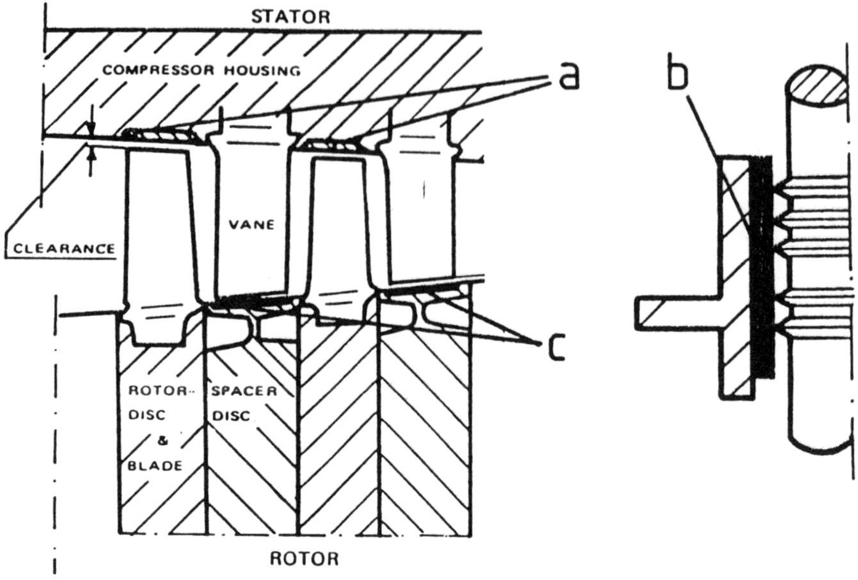

Figure 6. Applications for abradable coatings (After Ref. 33): *a.* Compressor housings; *b.* Labyrinth fins; *c.* Disc spacers

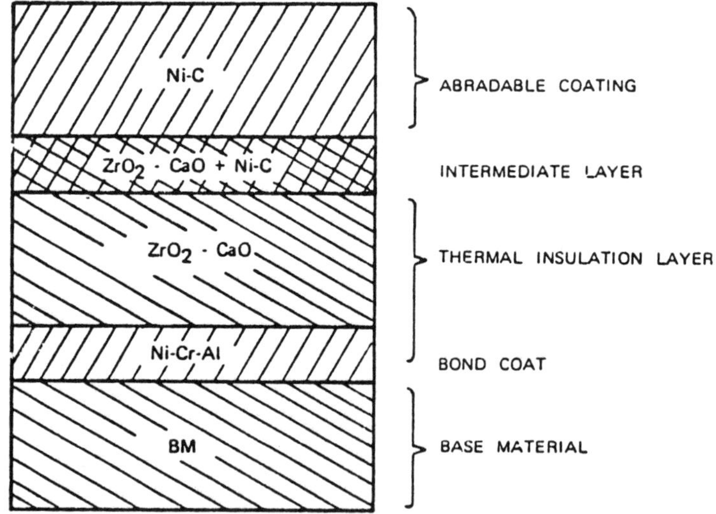

Figure 7. Multilayered system of thermal barrier and abradable coatings (After Ref. 33).

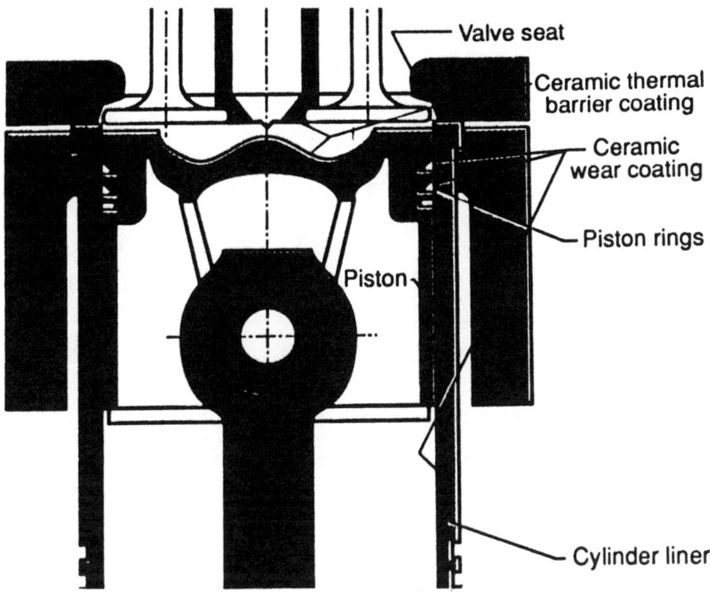

Figure 8. Ceramic coatings in auto-engines (After Ref. 39).

The problems associated with TTBC's are complex due to the layered nature of the deposits. The composite nature of the coating causes many interfaces that are subjected to process-induced strains, radical temperature cycles, and diffusion gradients. The lifetimes of TTBC's are quite modest compared to those of TBC's used by gas turbine manufacturers; this is in part due to the relatively low 0.5 mm thickness for gas turbines compared to 2.5 mm for diesel engines. Thus, TTBC's require a careful examination of the distribution of stresses, temperatures and chemistries and how these change with time, during processing and with use, both short and long term.

It is useful to relate the behavior of TBC's in gas turbine engines to that of TTBC's in diesel engines. There are two prime differences in the characteristics of coatings that operate under these environments. The TTBC will have a quite different stress distribution and microcrack network than the TBC, since the service thermal gradients are less severe than in the case of TBC's. In addition, the diesel engine operates at peak temperatures of some 200°C less than that of the gas turbine engine; therefore, oxidation of the bond coat and substrate is less severe. Nonetheless, the significant coating thickness presents special problems to the interfacial region between the ceramic and substrate.

The degree of insulation to the substrate and the mechanical toughness of the TTBC's can be enhanced by incorporating porosity into the coating, as discussed earlier. This is achieved through process control during the deposition or by including pore-forming components into the zirconia feedstock powder; for example, microballoons and evaporable plastics (e.g., polystyrene). Such TTBC's have porosity levels from 5 - 30 volume percent and allow control over the required thermal properties; although, the longevity of such low density coatings is a matter of current research efforts. Porosity levels can also be graded from the substrate to the outer levels of the TTBC such that the outer layers are smooth, while the inner layers are very porous. This is achieved through the plasma spraying of a finer particle ceramic or through modifying the gun-to-substrate distance.

The design specifications of an ideal TTBC are listed in Table 2 (in part from Ref. 40). In addition, such thermal barrier systems should have well-characterized thermal and mechanical behavior so that engineering reliability can be designed into their manufacture. Thus, it is important that both thermal and mechanical models be constructed that will permit an assessment of the temperature and stress distribution in the barrier system during operation and serve as a means of predicting and optimizing system performance.

Table 2. Requirements for Thick Thermal Barrier Coatings

- Low thermal diffusivity for maximum thermal protection
- Good thermal shock resistance over the temperature range of interest
- Elevated temperature stability during thermal exposure
- High surface emissivity for maximum heat rejection
- Relatively high surface finish for minimum friction
- Low cost of raw materials
- Ease of application
- Mechanical and metallurgical compatibility with the substrate and intermediate bonding material
- Strong bond with bond coat material
- Inspectable for flaws and defects in manufacture
- Durability and handlability during installation
- In service damage tolerant
- Resistant to environmental degradation

Power Generation Plant Applications. Other applications which can be considered, in the most broad sense, as TBC's are those that combat high temperature corrosion in power generation plants (41)-(43). The flame, plasma and arc metallization processes are well suited for the spraying of metal layers in these applications. Ceramics and cermets are also sprayed by the D-gun process (41). One coating that was successful was a composite of chromium carbide and nichrome applied 0.3 mm thick. This composite coating behaved more favorably than coatings of alumina, zirconia and tungsten carbide-cobalt under service conditions which control erosion due to fly ash.

4.2 Materials Properties

The materials property measurements from cyclic furnace, thermal rig, thermal expansion, acoustic emission and tensile adhesion test methods have been extensively examined. Although there is a wealth of engineering data on coatings, these measurements are not without ambiguity.

Routine Quality Control Tests. Metallographic examination of coatings allows qualitative assessment of the degree of porosity and oxide particles (for metallic coatings) at the substrate/coating or bond coat/ ceramic overlay interfaces. The microstructural quality of any coating or substrate interface must be uniform and exhibit a high degree of particle melting since these locations are critical for the overall integrity of the coating. At the same time, the number of unmelted particles within the coating system may be examined.

Tensile tests are used to assess the strength of the coating (i.e., the cohesive strength) and the strength of adhesion to the substrate (the adhesive strength). The strain tolerance of the coating is ascertained by indenting a sprayed panel of material with a 25.4 mm diameter ball. The coating in this case is subjected to a tensile strain. The observation of cracking which leads to spallation of the coating can preclude its acceptance for some applications within a turbine. All of the above routine quality control tests are destructive and, thus, are carried out on test panels which are sprayed at the same time as the engineering component.

The acceptance standard for a coating relies on a qualitative assessment; for example, a minimum adhesion strength must be attained; a maximum degree of cracking must not be exceeded for the ball penetration test; and the cleanliness (i.e., the amount and distribution of any porosity or unwanted inclusions) of the coating must meet certain visual standards. A further requirement for some coatings is that the dimensional accuracy of the component and, indeed, individual layers within the coating system and at precise locations on the component, such as over the leading and trailing edges of a turbine blade, must be ensured.

These quality control tests are far removed from the physical and chemical interactions that TBC's experience in service, especially since some tests are carried out under ambient conditions, although the coating application calls for high temperatures and pressures under corrosive atmospheres. However, the fact remains that the basis for choosing a specific application of a coating is drawn from low temperature tests, from the considerable practical experience of engine manufacturers and thermal spray contractors, and from some of the limited engine tests that have been performed. Some of the specialized testing methods are detailed below.

Mechanical Properties. The adhesion property of thermally sprayed coatings to the substrate may limit the utility of the TBC. The standard methods (44)-(47) of determining this adhesion are performed by first adhering a plug (or pull-off bar) to the coating and then using this mechanical attachment to pull off the coating in tension at ambient temperature. This method can be criticized (48)-(49); for example, a major shortcoming is the high variability in strength values which are obtained (50)-(51). Figure 9 (original data from Ref. 50) illustrates the large spread in the adhesion strength of yttria-stabilized zirconia (YSZ) coatings. The requirement that a batch of 5 samples lie above a certain tensile adhesion strength (44) to assess the suitability of the coating process is brought into question; and other methods which examine the statistical nature of results may be more appropriate (50)(52)(53).

Mechanical property measurements have been reported for YSZ TBC's (54)-(60) of similar compositions. The present discussion uses the term *Young's modulus* in the most general sense as a means to describe the almost linear relationship of the load-elongation curve. The Young's modulus has been measured as 735 GPa (55), 462 GPa (43); 47 GPa at low stress levels (57); 0.032 - 0.115 GPa over the entire stress range prior to failure (57); and approximately 0.125 - 1.56 GPa (calculated from the data in Ref. 59). Although there is disagreement with regard to the precise value of E, there is a consensus that TBC's exhibit pseudo-ductility when lamellae slide over each other. This ductility of the coating leads to a net extension of tensile adhesion specimens which are being loaded to high stress values (56)-(59). This strain may also account for a high temperature phenomenon which has been described as *creep* (60) of the YSZ coating. The term creep is used to describe the physical characteristics of the deformation process and should not be used to explain any phenomenological processes that occur within the coating.

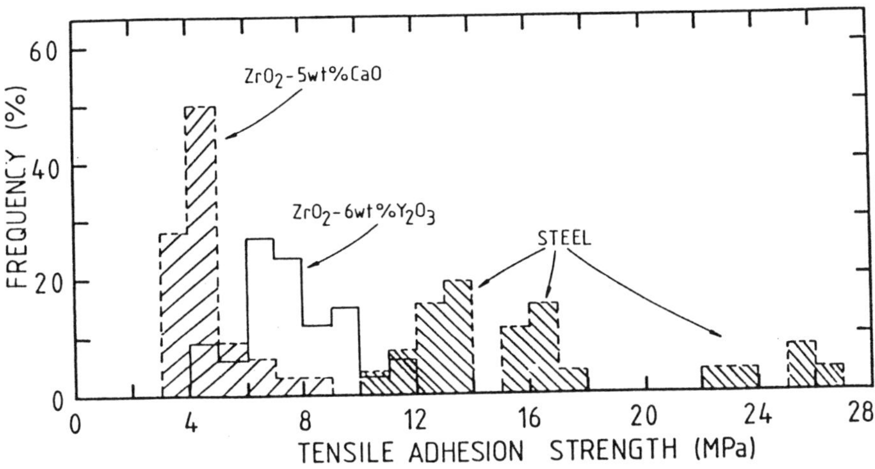

Figure 9. Tensile adhesion strength vs. frequency for ZrO_2 based coatings. The data for mild steel coatings is also indicated for comparison purposes.

A major concern of any tensile adhesion test data, especially with respect to a TBC application, is that any tests which require the adhesion of a test fixture are limited to near ambient conditions. These measurements may, therefore, have no relevance to the eventual high temperature application of TBC's. It can also be seen from Fig. 10 that the nature of the stress-extension plot of the coating alone (at room temperature) is dependent on the eventual failure morphology, and it is nonlinear. Thus "E" is highly variable, depending on both the failure mode and the stress level. These are difficult considerations to take into account during modelling studies (61)-(62).

It should be pointed out that the coating strength measurements referred to above are similar to the standard tests (44)(47) which are carried out as quality control procedures. However, the basic difference is that the scientific studies have been performed with a view to understanding the failure mechanisms of TBC's. Another point is that quality control tests are based on subjective assessments and do not allow any fundamental understanding of the TBC utility.

Another approach to establishing a measurement for the adhesion of the coatings is to perform fracture mechanics tests (55)(63)-(65). Such tests are not routinely used since specialized equipment and operators are

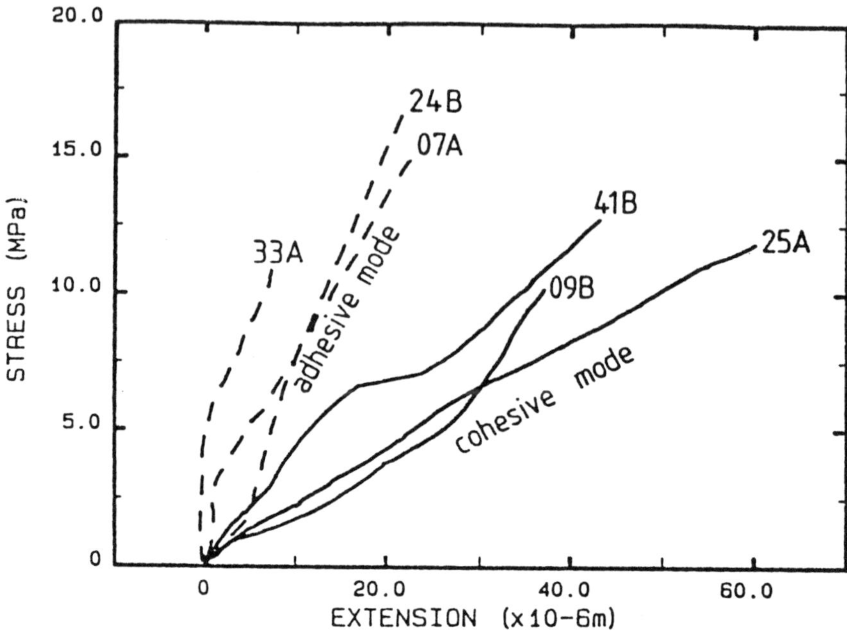

Figure 10. Stress/extension plots of tensile adhesion tests illustrate adhesive and cohesive modes of failure.

required, and such tests are more suited as a tool for coating development. Hardness tests (either macro-hardness or micro-hardness) are not suited to finding the fracture toughness of coatings because the required Boussinesq stress field is not attained (66). In fact, the results of hardness tests are probably determined by the laminar and defective structure of the coating rather than any true material property of the coating.

Measurement and understanding of the microhardness of coatings have implications with regard to both basic science and technological applications of coatings. The effective hardness of a microvolume of material depends on the cooling rate, phase structure, crack size and distribution, and residual stress and strain of the local environment. Thus, examining the variation of microhardness within coatings enables direct insights into the processing/structure/property relationships of coatings. Hardness tests may be related to the tensile adhesion tests (e.g., ASTM C633-69, Ref. 44) since both of these measurements rely on deformation under stress. Moreover, microhardness studies can give the variation of the strength and the flaw distribution throughout the specimen, whereas the strength tests yield the strength of the weakest link of the system.

Hardness tests have recently been used to study the anisotropic nature of coatings (67) and this continuing work shows promise in examining material property fluctuations across lamella-to-lamella or substrate-to-coating interfaces.

Cyclic Thermal Testing. The term *material property* has been used to include some of the qualitative measurements which will be mentioned below. Cyclic furnace tests were extensively used during the early development of TBC's (68). Small coupons with dimensions of typically 25 x 13 x 0.25 cm and having rounded edges (to about 0.16 cm radius) were coated and then subjected to cyclic furnace tests. The coupons had the same composition as the intended blade; such as B1900+Hf, MAR-M200+Hf, MAR-M509+Hf or Rene 41, among others.

In one type of furnace test, the specimens were heated to either 990 or about 1100°C in 6 min, held at that temperature for 60 min and then cooled over a period of 60 min to about 280°C. The samples were removed periodically (about every 12 cycles) and after cooling to room temperature were visually inspected (with the unaided eye) for failure (69). Failure occurred within the oxide layer but very close to the bond coating; and it always began with the formation of a small crack at one of the corners which eventually propagated along the edge (13).

An alternative heating schedule consisted of heating to 1100°C in 13 mins, holding at-temperature for periods of 1 to 20 hrs, then forced-air cooling to near ambient temperatures. Failure in this case was determined by examination under a 10x magnification; and cracking was usually observed to emit from a corner (70).

Another high temperature test (71) uses a natural gas oxygen torch which is directed at either the leading edge of a coated turbine blade or a flat coupon. The thermal cycle consisted of a 3 min heat-up, 60 min at-temperature (1185 - 1205°C) followed by a forced air cool over 5 mins to about 100°C. The thermal conditions for the above tests are contrasted to those experienced during burner rig, plasma torch and engine testing in Table 3. This table also includes details concerning the specimen size which was used for each test.

It is difficult to compare the results among different workers because of the various specimen geometries, different substrate materials, different coating compositions and spray deposition parameters, as well as the different experimental procedures and failure criteria that were employed.

Figure 11 summarizes the results of the thermal tests for the well-documented Ni-17Cr-54Al-0.35Y / ZrO_2-Y_2O_3 system. The intention is to illustrate the large differences in lifetimes that are observed among the

Table 3. Comparison of Thermal Test Conditions

Method	Test No.	Specimen size (cm)	Heating time (min)[1]	Maximum temp (°C)	Hold time (min)[1]	Cooling time (min)[1]	Minimum temp °C	References
Cyclic furnace	1	2.5x1.3x0.25	6	990-1095	60	60	280	70, 71, 72
	2	2.5x2.5x0.5	6	990-1095	60	60	280	70, 71, 72
	3	?	4	975	60	60	280	70, 71, 72
	4	2.5x2.5x0.5	13	1100	60, 360, 1200	90	?[2]	70, 73
Natural gas-oxygen torch	5	7.5x1.3x0.5[3]	3	1185-1250	60	5	100	71, 72
	6	J-75 blades[4]	3	1185-1250	60	5	100	71, 72
Burner rig	7[5]	1.3cm diameter	4	1040	0, 57	3	?	74
	8[6]	J-75 blades	0.5	1450-1570[7]	60	0.33	<75	71
Plasma	9	1.3cm diameter cylinders	0.5s	1100[8]	0.5s, 2.5s, 5.0s	30s 75s 120s	?[2]	75
Engine tests[9]	10	JT9D-7F first stage blades	2s 14s	870 1095		3s 12s	760 650	76

1. Times are expressed in minutes, unless specified otherwise.
2. Probably less than 50°C.
3. A hot zone of 2.5 x 1.3 x 0.6 cm is mentioned by this thickness is greater than the original specimen (10). Forced air cooling at a rate of 14 - 17 gs^{-1} air flow.
4. 6.5 cm^2 test area on the leading edge.
5. Test 7 was carried out at 0.3 Mach.
6. Test 8 was carried out at 1.0 Mach.
7. The surface temperature range of the blade is given. The corresponding substrate temperatures are 1450 - 1570°C.
8. Surface temperature.
9. The times to reach specific temperatures are given.

testing methods. The most attention has been paid to the ZrO_2-8 wt% Y_2O_3 ceramic composition, where results from cyclic furnace, natural gas oxygen rig, and burner rig tests are available. Turbine blades have been used as specimens for the two rig tests and these generally exhibit greater lifetimes than the rig tests on cylindrical specimens or the cyclic furnace tests, which have been carried out at equivalent temperatures (i.e., 1100°C)

The lines in Fig. 11 indicate the trend of cyclic lifetime with respect to the Y_2O_3 percent. It is seen from the figure that the optimum ceramic composition is probably between 6 and 8 wt%Y_3O_2. Note that there is also more than one decade in difference between the lifetimes of specimens tested at approximately 990°C and 1100°C and that there may be a large distribution (up to 100% for the 12 wt%Y_2O_3 composition tested at about 1100°C) in the cyclic lifetime data. Some of the variance in the data may be attributable to the different substrates used (72), the relatively poor control in the production of identical coatings, and the different experimental methods for classifying failure.

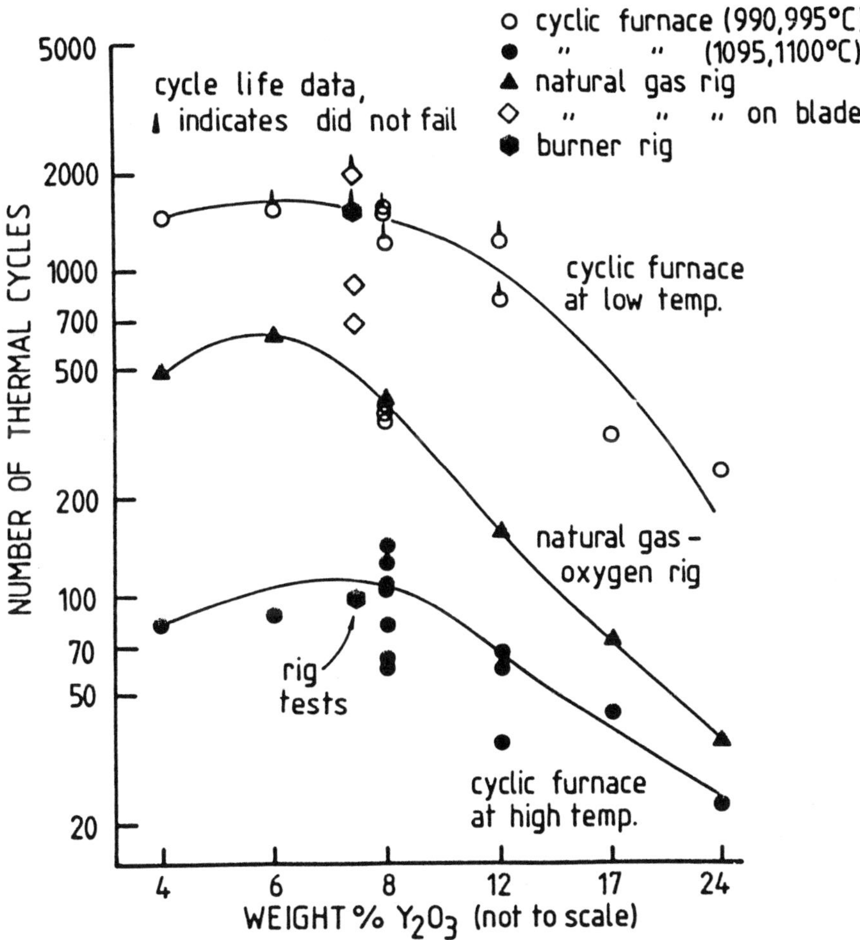

Figure 11. Collation of cyclic life data for TBC tests.

This small cross-section of results raises at least two important questions. First, it is important to carry out an accelerated test so that advanced material development can proceed rapidly and, if the test is simple, relatively inexpensively. If this were the only consideration, then cyclic furnace tests would seem to have a clear advantage over other methods since present day state-of-the-art coatings can be tested within 2000 hrs at 990°C, or, conservatively, within 750 hrs at 1100°C. However, cyclic furnace tests are qualitative and do not predict the ultimate service life of a coating. The burner rig tests are more suited to establishing the service life of the coated blade, but they are more time consuming and expensive to run. There is also a major problem of controlling the temperature and pressure conditions of each rig from run-to-run and rig-to-rig. The lifetime of coated blades which some manufacturers would like to achieve for a range of turbines is shown in Table 4 (24)(25).

Table 4. Design Specifications for Coated Turbine Blades

Application	Maximum Thermal Strain	Operating Temperature (°C)	Characteristics % of time	Overall Time (h)
Land based military aircraft	0.42	1180	10	1,000
Commercial transport aeroengine	0.20	1093	5	5,000
Base-load power generation gas turbine	0.08	820	90	25,000 to 30,000

A second important consideration is that the accelerated test must also reflect the operative mechanism of failure. Failure of the coating/substrate system is often ascribed to thermal fatigue, low temperature (650 - 750°C) or high temperature (750 - 950°C) hot corrosion, or to oxidation (>950°C) (24). The test methods, in turn, suggest possible failure mechanisms that the turbine materials may experience (73)-(76). For example, the oxidation behavior of the bond coat and substrate may lead to coating spallation (77)-(79), or hot corrosion (80) may adversely affect coating performance (1)(81)-(83).

At present it is difficult to compare the results between different workers who have used dissimilar materials. Thus, the potential of choosing the most suitable coating system/substrate combination for a particular application is not straightforward. The engine designers and materials engineers also wish to understand the reasons for coating spallation from the substrate. This has led to study of the factors which determine coating adhesion to the substrate, the operative failure mechanisms within coatings and fundamental investigations into the thermophysical properties of TBC's.

Thermal Expansion Tests. One way of understanding the adhesion property of a coating to a substrate under a changing thermal environment is to examine the thermal expansion difference between these materials. This is often referred to as the property of the "thermal expansion difference with respect to the temperature difference" or "ΔCTE/ΔT". Thus, strains (and, therefore, stresses) are developed within the coating system under temperature fluctuations due to the different coefficients of thermal expansion (CTE) of the substrate and TBC. The stresses are "developed upon heating to high temperatures or upon cooling to ambient temperatures" (15). Hence, a process of thermal fatigue can be envisioned since a series of alternating stresses are established. The ceramic TBC is subjected to a large tensile strain since the substrate, with a greater CTE, expands more than the coating. This strain changes signs (i.e., from tension to compression) at the substrate/ceramic coating interface. The magnitude of the TBC tensile strain increases with the testing temperature; thus, it may be expected with this oversimplified reasoning that coatings tested at high temperatures will have a reduced thermal cycling life. It is observed that the TBC/substrate combinations which exhibit the least strain from thermal expansion mismatches, also show the greatest cycle lives.

The material property of thermal expansion only allows strain mismatches between the coating and substrate to be evaluated. The assumption of a constant CTE at all temperatures during heating and cooling is also quite likely to be an over simplification. The thermal stresses within the TBC system can be determined if Young's modulus (E) and Poisson's ratio are known; and then some reference data is necessary to establish whether the yield point or ultimate tensile stress has been exceeded. None of the above materials properties have been unambiguously determined. It is important to remember that modelling or life predication studies will be in error if the material properties of the bulk, ceramic or bond coat components are assumed. The microstructural characteristics of thermally sprayed coatings are very different from the bulk materials and thus their thermomechanical properties are also expected to be structure dependent.

Acoustic Emission Tests. The previous section on strength measurements emphasized the difficulty in carrying out mechanical tests on coatings. Therefore, failure mechanisms have not been well characterized, especially the cracking behavior of the TBC during thermal fatigue studies. Acoustic emission (AE) tests (58)(59)(84)-(88) have been carried out to qualitatively assess these materials properties.

The tensile adhesion test has been used in conjunction with AE methods. These tests, carried out at room temperature (58)(59), discriminated between two different failure processes within coatings. These were distinguished by at least two count rate regimes which dominated at various times of the test. The count rate is directly proportional to the crack activity, and it was ascertained that many cracks were active during the initial loading of the tensile adhesion specimen. The crack interaction and growth during the test allowed some of these microcracks to evolve into several larger cracks, and the AE count rate was observed to decrease. These macrocracks lead to coating failure. A major problem with any AE count rate analysis is the uncertainity of quantitative measurements, such as the crack growth rate and the exact number of cracks and their size. When this information becomes available theoretical solutions to the TBC failure and thermal fatigue lifetime may also become possible.

Several AE tests have been performed under thermal cycling conditions (85)(86)(88). Different AE count rate distributions were observed on cooling TBC coated superalloy specimens from 1200°C. An apparently random count rate phenomenon was superimposed on a systematic response. These "large random count rates were presumed to evolve from macrocracking processes" (86), and this cracking process limits the TBC structural integrity. The AE studies are fraught with difficulties. For example, the experiments are difficult to carry out and it is common for them to have a high variability. Another factor is that so much information is generated during an experiment that data processing may become a means to an end in its own right rather than a study of phenomenological interpretation of TBC microstructural behavior. Regardless of these criticisms, it is felt that AE methodology is useful for comparing the thermal cyclic response of TBC's.

Overview of TBC's. The applications of TBC's are well-established; they have a long history that includes the thermal spraying of alumina onto burner cans of rocket nozzles and the manufacture of free-standing alumina radomes (89). The long term utility of this old technology and the more recent YSZ technology now rely on understanding the scientific basis of TBC's and TTBC's. The present work has detailed some thermal and mechanical testing methods which have been applied to TBC's to assess their adhesion to the substrate. These tests allow coatings to be ranked in an order approximating the coatings' performance under service conditions. It has been demonstrated that the testing environment under furnace cycling conditions is more severe than for rig testings. Cyclic furnace tests are used as an accelerated test method to develop suitable TBC compositions and to study the influence of substrate and coating deposition variables on coating life. These tests are the best at hand; however, it is important to ensure that exactly the same failure mechanism is operative during these tests as during service. Otherwise the results from these tests will be misleading.

The *mode* and *mechanism* of TBC failure should be distinguished. *Mode* is a general term which refers to the physical description of failure; for example, coating failure has been described as adhesive, cohesive, mixed mode or spallation. This description, by itself, does not give much insight into the failure process; that is, how microcracks initiate, then grow, coalesce, and interact to form macrocracks, together with the additional effects of oxidation, residual stresses and thermal fatigue, and mechanical stresses on these processes. This understanding of the fracture process is the *mechanism* of failure.

Some specific failure mechanisms may give rise to a characteristic failure mode such as biaxial compression and radial tension, leading to buckling and eventual spalling of the TBC. However, an understanding of the failure mechanism has the potential of permitting TBC improvement. Several studies have addressed the question of TBC failure mechanisms, and the logical extension of that work has been towards *life-predication* or *life modeling* studies. These are the eventual tools that the engine designer and TBC developer needs to use.

The mechanical testing of coatings is important because the derived materials property data is essential for modeling studies. At present there is no agreement concerning the value of the elastic modulus which has been reported as ranging from 0.032 to 462 GPa. In fact, all of these values may be correct for the particular technique by which the modulus was determined on the variety of differently prepared coatings. The modulus may be a difficult property to define for a material which exhibits pseudo-ductility. Thus, further studies are necessary to establish the precise manner of deformation of TBC's. The materials properties of these coatings are anisotropic.

The general view of the coating deformation process is that the individual lamellae slide over each other, causing a "pseudo-ductility" response in the coating. Monitoring of the acoustic emission response of coatings during thermal cycling experiments suggests that there are two distinct cracking processes. The macrocracking behavior, indicated by high acoustic emission count rates, is the predominant mechanism which leads to coating failure.

Some fundamental work on the mechanical behavior of coatings is necessary. This work should be focused on the failure processes which occur within TBC's. The application of these micromechanical behavior models to the thermal environment will promote a basic understanding of coating spallation from the substrate. Such knowledge will enhance the development of future TBC's. The acceptance tests used by industry, although useful in ranking coatings in terms of a particular property, present no fundamental knowledge concerning the materials properties of coatings. It is only when the phenomenological characteristics of the thermomechanical response of coatings is understood that coating development will progress substantially.

The next section addresses a non-ambient temperature application of plasma sprayed ceramic coatings; however, in this application the principal requirements are phase and electrical stability near liquid nitrogen temperature. A prime attribute of plasma spray technology, as for TBC and TTBC applications, is the ability to form thick, adherent ceramic coatings onto complex substrates.

5.0 PLASMA SPRAYED HIGH Tc SUPERCONDUCTORS

After the discovery of superconductivity above 90 K in the Y-Ba-Cu-O system (90), thick films of high-Tc superconductors were successfully formed by plasma spraying (91). Plasma spray technology offers great promise as a cost effective, scalable approach for depositing high "Tc", high "Jc" superconductive thick films. The term Tc, the critical temperature, refers to the ability of a material to remain superconducting at the specified temperature. Similarly, Jc is the critical maximum current flux of the material without reverting to "normal" electrical characteristics. An early result of this technique was the successful fabrication of a free-standing cylindrical, superconducting microwave cavity.

This section concentrates on air plasma sprayed superconductors in the Y-Ba-Cu-O system, though other thermal spray processes, including VPS and high velocity flame spray, have been applied to high Tc superconducting materials in the Y-Ba-Cu-O and Bi-Sr-Ca-Cu-O systems.

The feedstock powder (-140/+325 mesh or between 105 and 44 micrometers in diameter) for plasma spraying is typically prepared by mixing Y_2O_3, $BaCO_3$ and CuO in a ratio of 1:2:3 for the cations of Y, Ba, and Cu, respectively. Powder of this chemistry is often referred to as YBCO of 123 composition. Plasma spraying is generally carried out at low power levels, between 8 and 22 kW in air with argon as the primary gas. In a Stony Brook study, sprayed black coatings were formed to a thickness of 0.13 mm on a steel substrate which had previously been coated with NiCrAlY bond coat. However, the as-sprayed coating was not superconducting; rather it was a highly distorted, poorly crystallized material, which resulted from the rapid solidification associated with the spraying process (the cooling rate can be as high as 10^6 °C/s). The as-sprayed coatings were annealed in air to recover the superconducting properties. The annealing temperature is critical to the recovery process. The superconducting phase continuously decomposes to $BaCO_3$, Y_2O_3 and CuO when annealing is carried out at 600°C. On the other hand, the superconducting phase is replaced by several poorly identified phases when annealing is carried out above 1000°C. In the temperature range of 850 - 1000°C, the coatings recrystallized into the superconducting structure. Standard four-point AC resistance measurements indicate that the post-spray annealed coatings have a transition temperature of 88 K with a two degree wide transition.

5.1 Spray Parameter Optimization

In order to achieve dense, well-bonded coatings, a detailed study of plasma power level, spray distance, feedstock powder size, and stoichiometry of the powder were carried out in the Y-Ba-Cu-O system (92).

All of the coatings sprayed using the coarse powder (-140/+325 mesh) are very porous and laced with cracks. By comparison, the coatings prepared from fine powder (-325 mesh) have better particle flow on impact, resulting in little porosity and fewer cracks. Very fine particles (less than 10 micrometers) should be eliminated because they resolidify during spraying due to their small thermal mass. The degree of melting increases with spray distance and plasma power. However, resolidified particles appear in the coatings formed with fine powders at spray distances beyond 100 mm. Significant surface cracking is generally apparent in coatings sprayed at 35 kW. A study suggested that the optimal particle size distribution should be between 10 and 44 micrometers at a spray distance of between 75 mm and 100 mm and at a power of 25 to 30 kW (92).

Significant chemistry changes are observed in the sprayed coatings, especially those formed with fine powders. Copper loss is generally significant, as indicated by the Y_2O_3-BaO-CuO phase diagram, Fig. 12. The loss of copper becomes more dominant as plasma power level increases. Consequently, the coatings contain more Y_2BaCuO_5 (211) phase and $BaCuO_2$ (011) phase at the expense of $YBa_2Cu_3O_{7-x}$ (123) phase. This problem is usually overcome by the addition of excess CuO to the starting spray-dried powder. A powder cation composition of 1:2:4 instead of 1:2:3 is used for Y, Ba, and Cu, respectively. The excess copper effectively compensates for its loss during spraying.

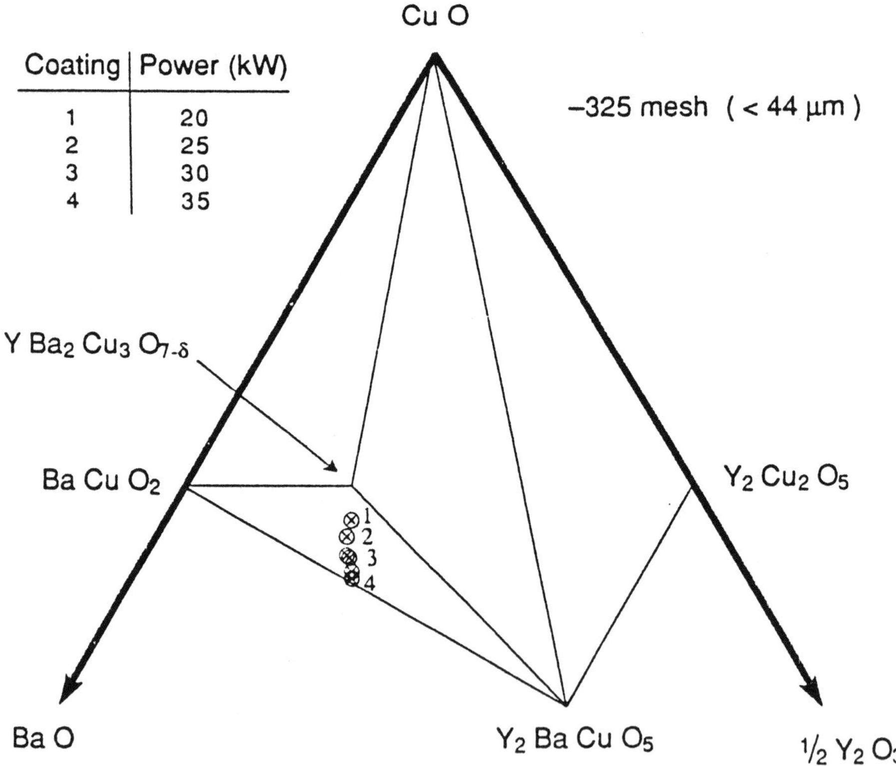

Figure 12. Chemistry changes during spraying of -325 mesh powder. The copper loss increases with increasing plasma power level (After Ref. 92).

Substrate temperature and thermal conductivity of the substrate influence both surface morphology and phase content of air plasma sprayed coatings in the Y-Ba-Cu-O system. Coatings sprayed onto cold substrates with a higher thermal conductivity value have almost no flow on impact and exhibit shrinkage cracks due to the high cooling rates. This results in poor interparticle bonding and non-equilibrium phases for the as-sprayed coatings. The phases present in an as-sprayed coating of this type are likely to be yttria, copper oxide, barium carbonate, single cubic phase with lattice parameter of 3 Å, and a substitutional phase ($YBa_{2-x}Y_xCu_3O_y$). On the other hand, coatings sprayed onto hot substrates exhibit more flow and less shrinkage cracks, because of the reduced cooling rates. The substitutional phase becomes more dominant, and the minor phases are (211), copper oxide and yttria. In order to reduce the shrinkage cracks and enhance interparticle bonding, substrate pre-heating is required for depositing the Y-Ba-Cu-oxide coatings onto metallic substrates.

The Y-Ba-Cu-Oxide coatings have also been deposited using VPS (93). The spraying was carried out within a chamber under 60 torr pressure with a partial oxygen pressure of 39 torr. Powders with particle sizes ranging between 44 and 105 micrometers were used and displayed complete melting at a power above 26.4 kW, leading to dense and strongly bonded coatings. The chemical composition of the coatings was very close to that of the original powder, indicating no copper loss during VPS. The substrate temperature reached 650°C, which was beneficial to the superconducting properties.

When using plasma spray techniques for the Y-Ba-Cu-O system, the as-sprayed coatings, without exception, are not superconductors. Therefore, it is essential to apply post-spray annealing to the coatings to restore the orthorhombic superconducting phase.

5.2 Post-Spray Annealing and Improving Superconducting Properties

Post-spray annealing is usually carried out in air or flowing oxygen environments in the temperature range between 900 and 970°C, followed by slow cooling. With the exception of VPS, the deposits produced by air plasma spraying (90)(94)(95) and flame spraying (96) show poor superconducting properties ($T_c < 90$ K and $J_c < 100$ A/cm^2). When poor superconducting properties are reported, the major x-ray diffraction peaks from the (013), (110) and (103) planes show incomplete splitting. This phenomenon may be related to the formation of the substitution phase, $RBa_{2-x}R_xCu_3O_y$, where R denotes rare earth elements such as La, Pr, Nd,

Sm, Eu and Y (97)(98). This phase has tetragonal structure when x is greater than 0.2 (99). The substitution phase of $YBa_{2-x}Y_xCu_3O_y$ can be formed either during air or oxygen annealing, or formed during air plasma spraying when sprayed onto a hot substrate. Once the phase is formed, it is not easily converted to the orthorhombic (123) phase.

Heat treatment in argon prior to oxygen annealing results in decomposition of the substitution phase to the desired orthorhombic (123) phase, which raises Tc and narrows the transition width (100)(101).

In Ref. 100, coating-substrate interactions were eliminated by removing the coating from the substrate prior to the heat treatment. The phase changes of the coatings after annealing were analyzed by X-ray diffraction and compared with patterns published by NIST for $YBa_2Cu_3O_7$ and $YBa_2Cu_3O_6$ (102). Susceptibility (AC) measurements were used to determine the superconductivity onset temperature, Tc, and the transition temperature width of the coatings. The onset temperature was defined as the temperature at which the signal decreased to 90% of the maximum value within the interval 77 - 100 K. Provided that the transition was completed by 77 K, the transition width was measured between the onset temperature and the temperature at which the signal was 10% above the minimum value.

Argon atmosphere annealing was attempted at three different temperatures (750°C, 850°C and 880°C). High temperature oxygen annealing was subsequently carried out at 930°C. The onset temperature of the sample pre-annealed in argon at 850°C/6hr (89.6 K) was higher than those which were pre-annealed at 750°C/6hr (87.3 K) or 880°C/24hr (89.1 K), as shown in Fig. 13. The transition width was found to be 3.8 K and 5.0 K for the samples pre-annealed at 850°C/6hr and 880°C/24hr, respectively. The transition was not completed by 77 K for the sample pre-annealed at 750°C/6hr. Pre-annealing in argon at 750°C/6hr produced a tetragonal phase whose lattice parameters matched those reported for the (123) tetragonal phase, with lattice parameters of a = 3.86 Å, and c = 11.84 Å. However, the intensities of several major diffraction peaks did not agree with the expected values, suggesting a defective crystal structure. After pre-annealing in argon at 850°C/6hr, the substitution phase separated into predominantly tetragonal (123), (211), and CuO phases. Both of the correct lattice parameters and relative intensities of the tetragonal (123) phase were observed. After oxygen annealing at 930°C/24hr, the samples pre-annealed in argon at 850°C showed better splitting of the major doublet peaks (013), (110) and (103) than the sample pre-annealed at 750°C. The sample pre-annealed at 880°C/24hr showed (211) as the major phase, which was not completely transformed into the orthorhombic (123) phase after subsequent oxygen annealing treatment, and its distribution was not homogenous. This probably resulted from the large size of the (211) particles, due to the fact that the rate of peritectic phase transformation was slow for the (211) and liquid phases transforming to the orthorhombic (123) phase. Consequently, there was more residual (211) phase distributed in the orthorhombic (123) matrix for the sample pre-annealed in argon at 880°C/24hr, resulting in a lower Tc value and wider transition width than the samples which were pre-annealed at 850°C/6hr. As determined by differential thermal analysis, the onset temperature of the phase transition in an argon environment from the substitutional phase plus CuO to (211) phase plus liquid phase was found to be 851°C. It seems that argon pre-annealing at 850°C is an optimum treatment in which the poorly superconducting substitution phase is destabilized without gross heterogeneous distributions of the remaining (211) and CuO phases.

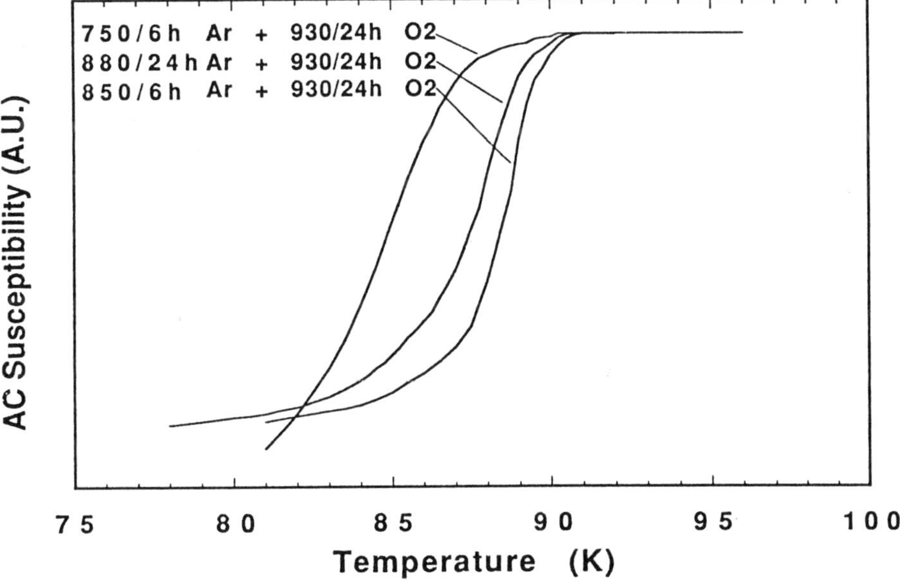

Figure 13. AC magnetic susceptibility vs. temperature for samples pre-annealed in argon at 750°C/6hr, 880°C/24hr and 850°C/6hr, followed by oxygen annealing at 930°C/24hr (After Ref. 101).

A final oxygen-annealing treatment was necessary to obtain the superconducting orthorhombic (123) phase. Therefore, the effects of oxygen annealing at 930°C, 950°C and 970°C for 24 hours after argon pre-annealing at 850°C/6hr were investigated (101). Figure 14 shows that the onset temperatures are 89.6 K, 90.4 K and 90.9 K for the samples oxygen annealed at 930°C, 950°C and 970°C, respectively. The most narrow transition width of 2.4 K was obtained from the samples annealed at 970°C. A 2.4 K transition width in AC susceptibility measurements is equivalent to a resistive transition width of less than 0.5 K. The observed differences in superconductivity can be correlated with the integrity of the crystal structure. Figure 15 reveals that a more distinct splitting of the doublet peaks (013), (110) and (103) can be achieved by increasing the annealing temperature. As the peak splitting increases, the unit cell becomes more orthorhombic (a and c lattices increase, while b decreases). Simultaneously with these changes in lattice parameters are an increased Tc and a sharpening of the transition.

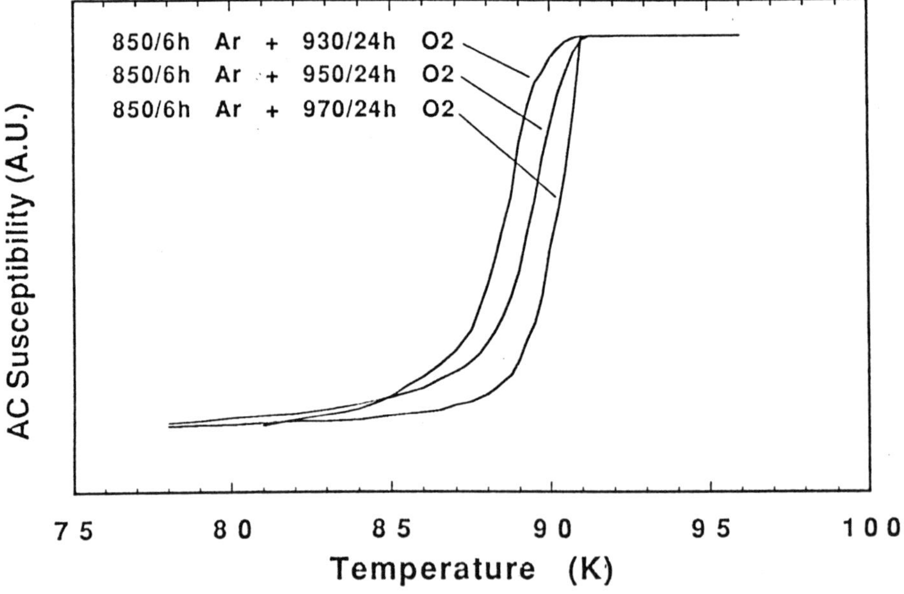

Figure 14. AC magnetic susceptibility vs. temperature for samples pre-annealed in argon at 850°C/6hr, followed by oxygen annealing at 930°C, 950°C or 970°C for 24 hours (After Ref. 101).

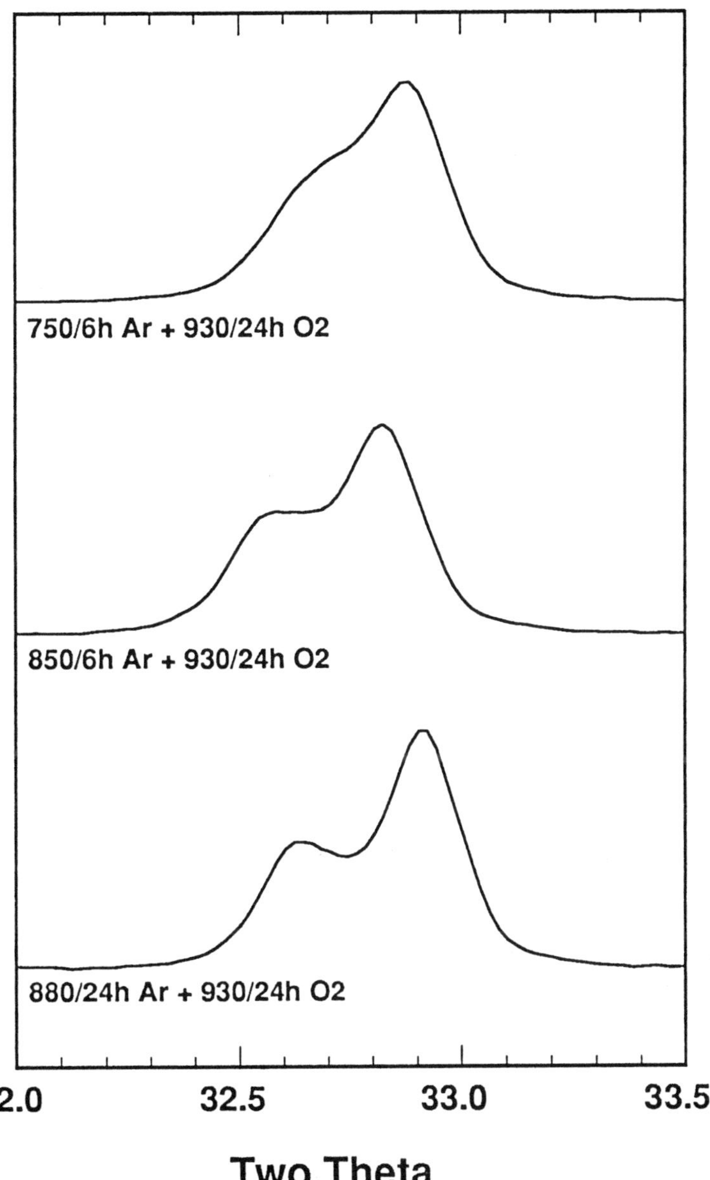

Figure 15. X-ray diffraction patterns for the samples pre-annealed in argon at 750°C/6hr, 850°C/6hr and 880°C/24hr, and followed by oxygen annealing at 930°C for 24 hrs, showing splitting of doublet peaks (013), (110) and (103) (After Ref. 101).

5.3 Texturing: Improving the Transport Critical Current Density

Although plasma sprayed deposits show high Tc values after the argon pre-annealing process, the transport Jc values generally remain below 100 A/cm^2, especially for sprayed coatings in zero magnetic field (92)(102)-(104). The critical values degrade severely when weak magnetic fields are applied (105).

However, the critical current density inside the individual grains is much higher. Based on the Bean model (106), the loop width of the magnetization curves and the average grain sizes were measured for the sprayed deposits pre-annealed in argon at 850°C and then annealed in oxygen at 930°C, 950°C or 970°C for 24 hours. The intragranular critical current density was thus estimated to be 2.4×10^6 A/cm^2 at 1 Tesla and 10 K (101). The optical micrographs of these samples are shown in Fig. 16. The difference in intragranular Jc and transport Jc values indicate the existence of weak links at the grain boundaries.

"Melt texture growth" methods have been applied to poly-crystalline sintered materials in the Y-Ba-Cu-O system to align the grains in the current flow direction, yielding improvements in transport critical current density by orders of magnitude (107). This technique was applied to the plasma sprayed deposits (108). After argon annealing of 850°C/6hr, the 1 mm thick deposit was removed from the substrate. Subsequently, the sample (25 mm long and 12 mm wide) was placed into a zone-heating apparatus, where it was heated by a pair of quartz halogen lamps. The sample temperature at the focal line of the lamps was maintained at 1100°C. The sample was suspended stationary while the lamps traveled upwards at a rate of 6.5 mm/hr. Oxygen, at atmospheric pressure, flowed over the sample during the texturing process. Post-oxygen annealing was carried out at 950°C/2hr and 500°C/24hr at a heating rate of 3°C/min and a cooling rate of 1°C/min.

Figure 17 shows a polarized light micrograph of the processed sample in three-dimensional view. The X-Z plane is parallel to the coating surface, which was facing the lamps during texturing. As revealed by inspection of the Y-Z plane, the plate-like (123) grains nucleated at both growth surfaces (X-Z planes), growing towards the center and making an angle of about 30° to the Z axis, which is parallel to the heating-zone direction-of-motion during texture growth. As observed in the X-Y plane, four to six plates, with about the same contrast, form bundles. These bundles are about 50 micrometers in diameter and 300 - 500 micrometers in length. The interface between the bundles is not always a low-angle grain boundary, but the interface between the plates within a bundle is a small-angle grain boundary.

Figure 16. Polarized optical micrographs of samples pre-annealed in argon for 850°C/6hr, and followed by oxygen annealing at *(A)* 930°C/24hr, *(B)* 950°C/24hr, and *(C)* 970°C/24hr (After Ref. 101).

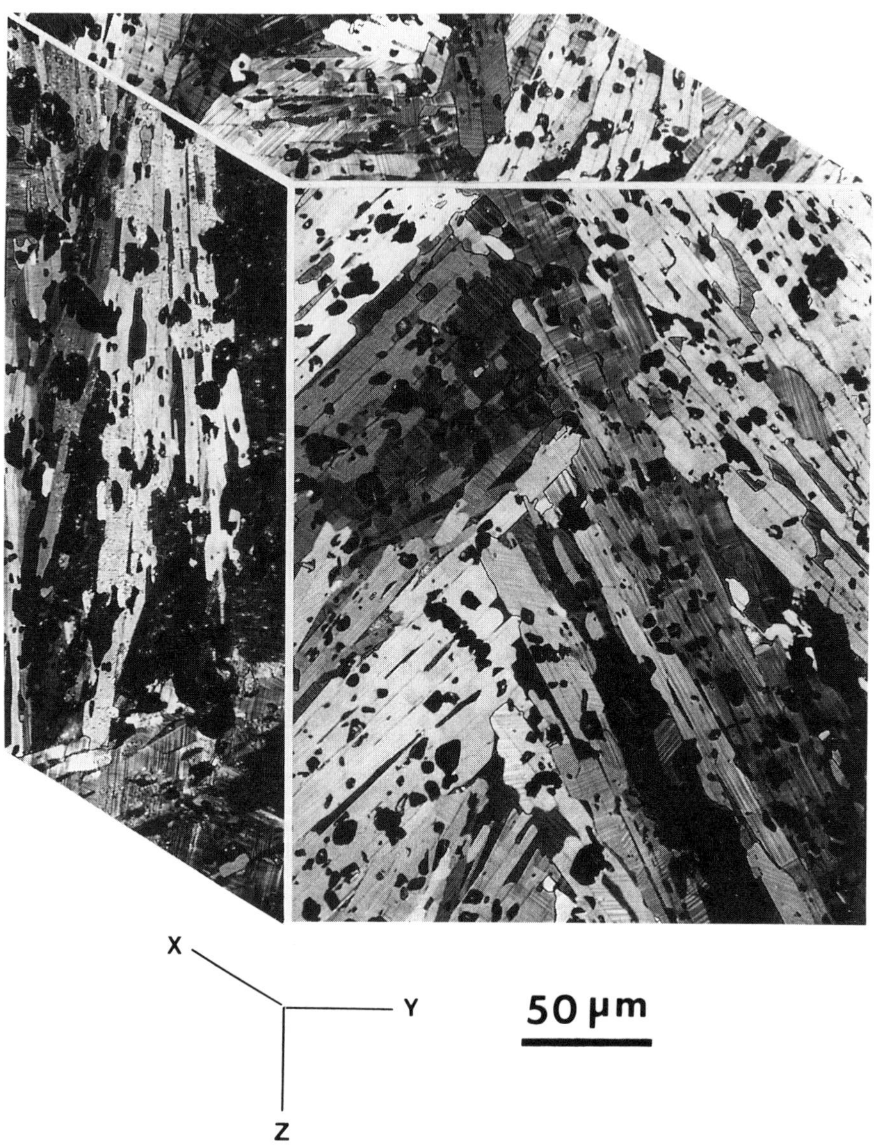

Figure 17. Polarized optical micrograph of a plasma sprayed and textured sample in three dimensions, showing aligned plated-like grains (After Ref. 108).

X-ray diffraction was performed on the deposit surface (X-Z plane). The diffraction pattern of the surface, Fig. 18, exhibits intense (001) peaks, indicating that the c-axis of the (123) grains is preferentially oriented perpendicular to the deposit surface.

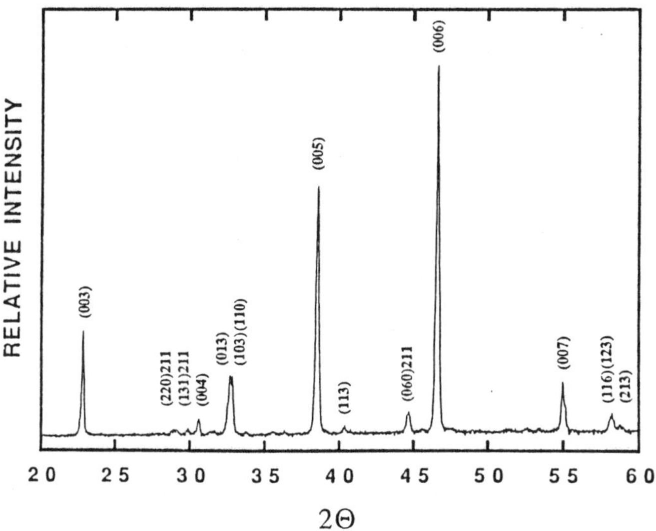

Figure 18. X-ray diffraction pattern of a plasma sprayed and textured sample of the X-Z plane, showing (001) preferred orientation (After Ref. 108).

Measurements of AC susceptibility indicate that the sample has a Tc of 90K and a transition width of 2.5 K. Using a continuous DC current source along the sample's Z-axis and the external magnetic field perpendicular to the sample's Y-Z plane, the transport critical current density of at least 5000 A/cm^2 at H = 0, 1 or 2 Tesla was measured at 77 K. Overheating of the current contacts prevented the application of higher current. The transport current is believed to be carried by the aligned grains within the bundles, where good coupling occurs between these grains.

Grain boundaries of 116 orientations were studied by transmission electron microscopy (109). More than 85% of the boundaries were small-angle grain boundaries, with misorientation angles smaller than 14°. Among these small-angle grain boundaries, about 60% of the boundaries had misorientation angles smaller than 5°. An example of a small-angle grain boundary with a misorientation angle of 3° is shown in Fig. 19.

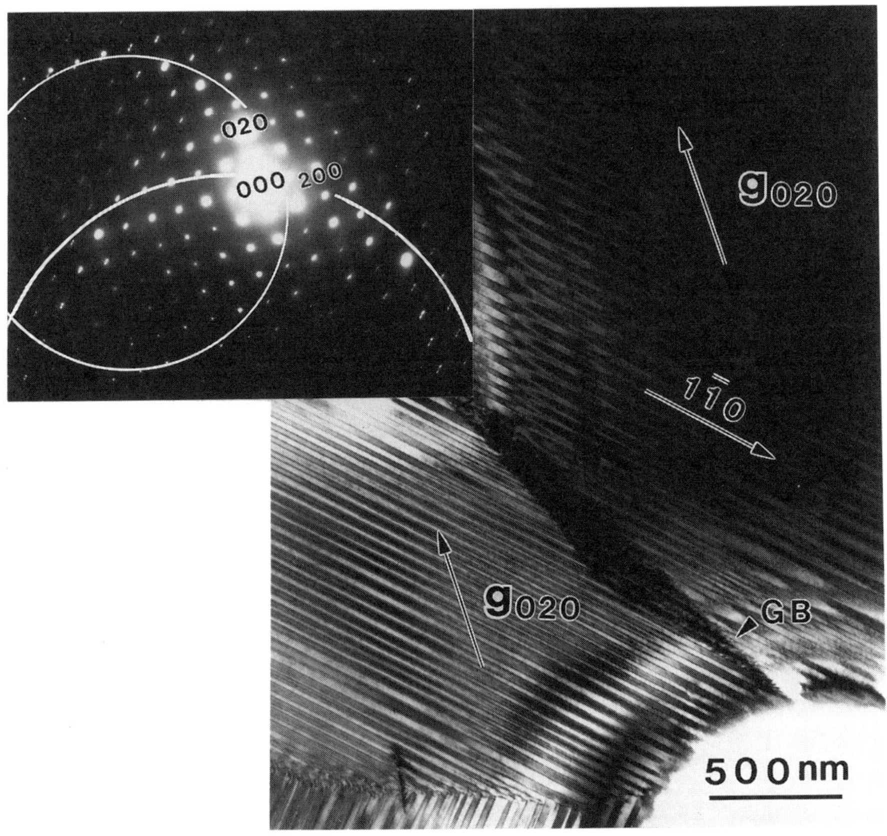

Figure 19. TEM micrographs of a plasma sprayed and textured sample, showing a small-angle grain boundary (as indicated by *GB)* with a misorientation angle of 3° (After Ref. 108).

5.4 Coating/Substrate Interdiffusion

When the Y-Ba-Cu-Oxide powders are directly sprayed onto an oxidized nickel substrate to form 0.1 mm thick coatings, the (123) grains are randomly oriented, even after melt texturing, as shown in Fig. 20 (110). The thermal gradient cannot be concentrated along the coating surface, due to the high thermal conductivity of the underlying substrate. Therefore, a thermal barrier coating of yttria partially-stabilized zirconia, was sprayed onto an oxidized nickel substrate before depositing the (123) coating. Heat conduction from the (123) coating to the substrate was substantially reduced during the texturing process. Consequently, the (123) grains are aligned along the coating surface, as shown in Fig. 21. The transport critical current density value is substantially improved, compared with the previous one; however, it is at least two orders magnitude lower than that of the substrate-free deposits. Energy dispersive X-ray analysis indicates that barium diffused through the zirconia layer and accumulated along the interface of the zirconia layer and the oxidized nickel substrate. A heterogeneous distribution of copper in the (123) coating was also observed. The net result of the diffusion of these elements through the zirconia layer is a loss in stoichiometry at certain regions of the (123) coating, resulting in non-superconducting phases at the temperature of liquid nitrogen. Microcracks might also develop during melt texture growth and annealing as a result of coating/substrate thermal mismatch stresses. This would, in turn, degrade the critical current density. Choosing a suitable material for a diffusion barrier, or increasing the (123) coating thickness might mitigate this problem.

The interdiffusion of the VPS-sprayed (123) coating and the substrates of yttria-stabilized zirconia, stainless steel, or nimonic alloy were studied at 950°C/1hr (93). For the yttria-stabilized zirconia substrate, a small amount of zirconium diffused into the (123) coating. For the stainless steel substrate, iron diffusion reached the midlayer of the (123) coating. For the nimonic substrate, nickel diffusion into the coating was restricted within the region closest to the interface between the coating and substrate. The highest critical current density of the coating was around 690 A/cm^2 at 77 K and 0 Tesla. Loss of superconductivity occurred when post-spray oxygen annealing was carried out at 970°C instead of 950°C, due to severe interdiffusion of these elements between the coating and substrate.

While high Tc superconductors are not being plasma sprayed as actively as they were during the early excitement of this field, nevertheless the (123) and related ceramics are continuing to be plasma sprayed and

Figure 20. Polarized optical micrograph of a plasma sprayed 123 coating, showing randomly oriented 123 grains on an oxidized nickel substrate (After Ref. 110).

Figure 21. Polarized optical micrograph of a plasma sprayed 123 coating, showing aligned 123 grains on the yttria partial stabilized zirconia layer and the oxidized nickel substrate (After Ref. 110).

studied. The apparent neglect of plasma spray by those industrialists interested in exploiting high Tc superconductors is to some extent due to a lack of appreciation of the versatility of plasma spraying. Nonetheless, work continues, and it is likely that plasma spraying will find a niche, for example, for depositing high magnetic field shielding, when large areas are required. Clearly, there is more to be done.

6.0 TEST METHODOLOGIES

Thermal sprayed coatings can be characterized relative to their metallographic microstructure, or their physical properties can be evaluated. In general, to obtain a true understanding of the coating-substrate system, both microstructure and properties should be examined. These are discussed below.

6.1 Characteristics of Coatings

Some of the characteristic features of plasma sprayed coatings were discussed above. Features common to both metal ceramic plasma sprayed coatings are porosity and surface roughness, which are closely related. These two characteristics are functionally dependent on feedstock particle size and spray parameters. In this regard, two generalizations may be made: *(i)* the larger the particle size, the greater will be both roughness and porosity; and *(ii)* the better the melting (the hotter the flame), the lower will be both roughness and porosity. Of course, particle size and meltability are related. Therefore, the best generalization is that better melting and high particle velocity yield the densest coating.

Surface roughness is easily measured by either an electro-mechanical profilometer or a laser scattering method. Recently, in-process optical devices have been introduced which will enable surface profiles to be fed back for process control (111). A useful paper on optical methods of surface profilometry is Ref. 112. Relative to roughness, it should be noted that the American National Standards Institute has developed a standard for the measurement of geometrical irregularities of surfaces of solid materials: ANSI-B46.1-1987; "Surface Texture - Surface Roughness, Waviness and Lay". An interesting review of this standard and related issues is found in Ref. 113.

Post-spray surface grinding, if carried out properly on a good coating, can significantly increase surface smoothness. It is not uncommon, in an effort to increase smoothness, to grind too aggressively leading to particle-pullouts and part damage. The AWS (1985) Thermal Spray Handbook, Ref. 114, is a good source of information on post-spray finishing and profilometry.

Porosity within plasma sprayed coatings can be process and/or feedstock material induced. Low velocity spraying will commonly yield resolidified particles *(unmelteds)* and attendant porosity. Other causes of porosity are from poor intersplat bonding, which generally occurs in flame and arc spraying. For metals, oxidation will be the cause of pores within the microstructure. Pores can be isolated or can be through-pores. Of course, the various available test methods will be sensitive to the type of porosity encountered in a given coating. There are essentially three methods of porosity measurements available: *(i)* mercury porosimetry; *(ii)* stereopycnometry; and *(iii)* cross-sectional metallographic image analysis. Mercury porosimetry is used principally with ceramic coatings (which are generally removed from the substrate for test), because there may be a problem with the mercury forming an amalgam with some metals. Using mercury porosimetry it is possible to determine both the volume and geometries of through-pores. Both mercury porosimetry and stereopycnometry measure only through-porosity, the latter being a rather simple but reasonably accurate (inert) gas displacement method.

Relative to closed pores, only image analysis is capable of characterizing pore structures in a reasonably accurate fashion. Unfortunately, for several reasons, metallography is not very accurate in determining porosity. First, in order to produce a metallographic cross-section, it is necessary to cut and mount the specimen and then subject it to metallographic polishing. This process frequently yields particle pull-outs, especially in coatings that have weak cohesive bonding. In addition, pores and inclusions are sometimes difficult to distinguish. Image analysis programs are commonly available, but none have been designed expressly for the analysis of cross-sections of sprayed coatings. The net result is that both isolated and through-porosity are commonly over estimated. But, worst of all, optical metallographic analysis, with or without attendant image analysis, tends to be subjective.

Other features of the plasma sprayed coatings involve phase content, chemical uniformity, the number and form of unmelteds, cracks, residual stresses, etc. Phase content is generally determined using optical metallography and x-ray diffraction techniques, the latter being far superior to optical methods. Phase content is complex in thermal sprayed materials due to the rapid solidification nature of the deposits; i.e., the splats contain metastable phases, as has been well-documented in the literature. Such metastable phases can influence physical properties (e.g., corrosion, mechanical) and, furthermore, are not stable to elevated temperature service. Phase changes within coatings can result in volume changes, giving rise to porosity and/or the formation of microcracks.

It is important to note that transmission electron microscopy (TEM) can be used to give a rather more detailed picture of phase content; unfortunately, however, of a very localized nature. TEM is used more commonly for fundamental research, whereas x-ray diffraction yields a good overall view of the phase content.

As discussed above, another very important feature of thermal sprayed coatings is the as-sprayed residual stress state. X-rays can be used to determine residual stress distribution within coatings in much the same manner as in bulk specimens. Formerly, the residual stresses associated with such coatings were evaluated by measuring the curvature of the coating-substrate system. Using curvature, and in some instances microstrain-gauges, it is possible to evaluate the macrostrain distribution within the deposit.

High intensity, well-collimated x-ray radiation is available through facilities such as the National Synchrotron Light Source at Brookhaven National Laboratory. Synchrotron radiation has been used to determine microstresses as a function of distance from the substrate for an oxide coating deposited onto nickel (115). The implications are significant for measuring phase content and stresses in situ in real time at high temperatures within a thermal barrier coating system.

Relative to coating characterization, the one further matter of some importance is the degree of chemical uniformity within the coating and chemical gradients associated with the coating-substrate interface, due either to the deposition process or, subsequently, in-service, such as for the thermal barrier coatings. A true chemically uniform feedstock particle will yield a chemically uniform coating. However, a composite particle or a mixed material feedstock, will yield an inhomogeneous coating. Such chemical characterizations are generally evaluated using EDAX attachments on SEM's. If higher resolution is desired, electron microprobe analysis is needed.

6.2 Properties of Coatings

As discussed in Sec. 4, much needs to be done in advancing test techniques of plasma sprayed ceramic coatings. Of relevance are mechanical, thermal and chemical properties and the development of laboratory tests for the evaluation of these properties. The following tests are used generally for evaluating a wide range of engineering properties of plasma sprayed ceramics.

Mechanical Tests

i. Tensile Adhesion Strength (Bond Test)
ii. Tensile Strength of Coating
iii. Shear Strength of Coating
iv. Compressive Strength of Coating
v. Cohesive Strength
vi. Microhardness and Macrohardness
vii. Superficial Hardness
viii. Abrasive Slurry Wear
ix. Lubricated Wear
x. Hard Particle Erosive Wear

Chemical Tests

i. Low Temperature Corrosion
ii. High Temperature Corrosion

Thermal Tests

i. Thermal Cycle/Shock Tests
ii. Thermal Conductivity Tests
iii. Thermal Expansion
iv. Thermal Gravimetric Analysis

These tests and the procedures for carrying out some of them are discussed in Ref. 114.

Further, good sources of information on test techniques are contained in the large numbers of published proceedings associated with the National Thermal Spray Conference and the International Thermal Spray Conference, e.g., Refs. 5, 6, 8, 9, 12, 33 - 38, 41 - 43, 57, 67.

ACKNOWLEDGEMENTS

The authors wish to thank a number of people and organizations for assistance in the course of their research on plasma spray ceramics. C.C.B. thanks Dr. R. A. Miller of NASA-Lewis Research Center for his support, interest and assistance in aspects of the work described in this review. H.W.

thanks Drs. D. Welch and M. Suenaga of Brookhaven National Laboratory for their support of the high Tc research discussed herein. H.H. extends his appreciation to a number of former graduate students who worked so ably to enable him to better understand these complex yet beautiful coatings.

REFERENCES

1. Herman, H., *Scientific American,* 256:113 (1988)

2. Herman, H., *Mat. Res. Soc. Bull,* 13:60 (1988)

3. Gross, B., Gryez, B. and Miklossy, K., *Plasma Technology,* Iliffe Books, London (1968)

4. Gerdeman, D. A. and Hecht, N. L., *Arc Plasma Technology in Materials Science,* Springer-Verlag, New York (1972)

5. Wilms, V. and Herman, H., *8th Intern. Thermal Spray Conf.,* pp. 236-243, Amer. Weld. Soc., Miami Beach, FL (1976)

6. Brown, L., Herman, H. and MacCrone, R. K., *Advances in Thermal Spraying,* pp. 507-512, Pergamon Press, Toronto (1986)
   Also: Brown, L., "The Dielectric Behavior of Plasma Sprayed Oxides." Ph.D. Thesis, State University of New York at Stony Brook (1987)

7. Zatorski, R. A. and Herman, H., *High Performance Ceramic Films and Coatings* (P. Vincenzinie, ed.), pp. 591-601, Elsevier, Amsterdam (1991)

8. Kingswell, R., Rickerby, D. S., Scott, K. T. and Bull, S. J., in: *Thermal Spray-Research & Appl.* pp. 179-185, ASM International, Cleveland (1991)

9. Chon, T., Bancke, G. A., Herman, H. and Gruner, H., in: *Thermal Spray-Advances in Coatings Technology.* pp. 329-334, ASM International, Cleveland (1988)

10. Wang, H. G., Fishman, G. S. and Herman, H., *J. Mat. Sci.,* 24:811 (1989)

11. Wang, H. G. and Herman, H., *Surface and Coatings Tech.,* 42:203 (1990)

12. Rhys-Jones, T. N., *Proc. Inter Thermal Spraying Conference,* Vol. 1, The Welding Institute, Cambridge (1989)

13. Stecura, S. *American Ceramic Soc. Bull.*, 56:1082 (1977)

14. Stecura, S. *Advanced Ceramic Materials,* 1:68 (1986)

15. Miller, R. A. and Lowell, C. E., *Thin Solid Films,* 95:265 (1982)

16. Liebert, C. H. and Miller, R. A., *Industrial and Engineering Chemistry Product Research and Development,* 23:344 (1984)

17. Miller, R. A., Levine, S. R. and Stecura. S., *The American Institute of Aeronautics and Astronautics 18th Aerospace Sciences Meeting,* January 14-16, 1980, Pasadena, CA., Paper AIAA-80-0302, Pub. AIAA, New York (1980)

18. Miller, R. A., *American Ceramic Journal,* 67:517 (1984)

19. Miller, R. A., Levine, S. R. and Hodge, P. E., *Superalloys-1980,* pp. 473-480, American Society for Metals, Philadelphia (1980)

20. Stepka, F. S., Liebert, C. H. and Stecura. S., *Transactions of the Society of Automotive Engineers,* 86:1487 (1977)

21. Liebert, C. H. and Stepka, F. S., NASA TP1425 (June, 1979)

22. Graham, J. A. S., *Surfacing Journal,* 16:63 (1985)

23. Burgel, R. and Kvernes, I. in: *Proc. High Temperature Alloys for Gas Turbines and Other Applications, Liege, Oct 86,* (W. Betz, R. Brunetanol, D. Coutsouradis et al., eds.), D. Riedel Publishing Company, Dordrecht (1986)

24. Burgel, R., *Materials Science and Technology,* 2:302 (1986)

25. Goward, G. W., *Materials Science and Technology,* 2:194 (1986)

26. Fisher, G., *Ceramic Bull.,* 65:283 (1986)

27. Miller, R. A., *Surface Coatings and Technology,* 30:111 (1987)

28. Liebert, C. H. and Stepka, F. S., *J. of Aircraft,* 4:487 (1977)

29. Siemers, P. A. and Hillig, W. B. NASA CR-165351 (August, 1981)

30. Merutka, J. P., *Ceram. Eng. Sci. Proc.,* 2:604 (1981)

31. Liebert, C. H. and Levine, S. R., NASA TP2057 (September, 1982)

32. Levine, S. R., Miller, R. A. and Hodge, P. E., *SAMPE Quarterly,* p. 22 (October, 1980)

33. Sickinger, A. and Sohngen, J., *Proc. 10th International Thermal Spraying Conf.,* pp. 140-145, (H. D. Steffens, ed.), Deutscher Verlag fur Schweisstechnik, Dusselforf (1983)

34. Clegg, M. A. and Mehta, M. H., *Thermal Spray: Advances in Coating Technology,* pp. 41-46, (D. L. Houck, ed.), ASM International, Metals Park, OH (1988)

35. Yonushonis, T. M., *Thermal Spray Technology: New Ideas and Processes,* pp. 239-243, (D. L. Houck, ed.), ASM International, Metals Park, OH (1989)

36. Novak, R. C., Matarese, A. P. and Huston, R. P., *Thermal Spray Technology: New Ideas and Processes,* pp. 273-281, (D. L. Houck, ed.), ASM International, Metals Park, OH, (1989)

37. Guillemot, J. M., Dehaudt, P. and Ducos, M., *Proc. 11th Inter Thermal Spraying Conference,* pp. 513-521, Pergamon Press, New York (1986)

38. Inwood, B. C., Meyer-Grunow, H. and Chandler, P. E., *Proc. 12th Inter. Thermal Spraying Conference,* Paper 91, Vol. 1, The Welding Institute, Cambridge (1989)

39. Sheppard, L. M., *Am. Ceram. Soc. Bull.,* 69:1012 (1990)

40. Miller, R. A., EPRI Report AP-5078 (1991)

41. Fukuda, Y. et al., *Proc. Inter. Symposium of Advanced Thermal Spraying Technology and Allied Coatings,* pp. 49-54, (Y. Arata, ed.), The High Temperature Society of Japan, Osaka (1988)

42. Chandler, P. E. and Quigley, M. B. C., *Proc. 11th Inter. Thermal Spraying Conference,* pp. 29-35, Pergamon Press, New York (1986)

43. Gustafsson, S., *Proc. 11th Inter. Thermal Spraying Conference,* pp.19-28, Pergamon Press, New York (1986)

44. ASTM, Designation C633-69, "Standard Method of Test of Adhesion or Cohesive Strength of Flame Sprayed Coatings," *19 Annual Book of ASTM Standards,* Part 17, pp. 636-642, American Society for Testing and Materials, Philadelphia, PA (1974)

45. German Standard DIN 50- 160, "Determination of Adhesive Strength in the Traction Adhesive Strength Test," German Welding Association (DVS), Beuth Verlag GmbH, Berlin 30 (August, 1981)

46. Japanese Industrial Standard H 8666, "Testing Methods for Thermal Sprayed Ceramic Coatings," Japan Standards Association, Tokyo 107 (1981)

47. AFNOR Standard; i.e, the French Tensile Adhesive Strength Standard.

48. Sickfield, J., *Adhesion Aspects of Polymeric Coatings,* (K. L. Mittal, ed.), pp 543-567, Plenum, New York (1983)

49. Berndt, C. C. and McPherson, R., *Trans. Inst. Eng. Aust.,* ME6:53 (1981)

50. Ostojic, P. and Berndt, C. C., *J. of Surface and Coatings Technology,* 34:43-50 (1988)

51. Hermanek, F. J., *Welding J.,* 57:31 (1978)

52. Hendricks, R. C. and McDonald, G., NASA TM-81743 (April, 1981)

53. Grisaffe, S. J., NASA TN D-3113 (July 16, 1965)

54. Siemers, P. A. and Mehan, R. L., *Ceram. Eng. Sci. Proc.,* 4:828 (1983)

55. Strangman, T. E., *Thermal Barrier Coatings for Turbine Airfoils,* Garrett Turbine Engine Company, Phoenix, Arizona; Repo 214892 (April 10, 1984)

56. Berndt, C. C. and Miller, R. A., *Ceram. Eng. Sci. Proc.*, 5:479 (1984)

57. Berndt, C. C., *Determination of Materials Properties of Ceramic Coatings Advances in Thermal Spraying,* pp. 149-158, Pergamon Press, New York (1986)

58. Shankar, N. R., Berndt, C. C. and Herman, H., *Ceram. Eng. Sci. Proc.,* 3:772 (1982)

59. Shankar, N. R., Berndt, C. C. and Herman, H., *Mat. Sci. Res.,* 15:473 (1983)

60. Hendricks, R. C., McDonald, G. and Mullen, R. L., *Ceram. Eng. Sci. Proc.,* 4:819 (1983)

61. Chang, G. C., Phucharoen, W. and Miller, R. A., *Surface and Coatings Tech.,* 30:13 (1987)

62. Padovan, J., Dougherty, D., Hendricks, R., Braun, M. J. and Chung, B. T. F., *J. Thermal Stresses,* 7:51 (1984)

63. Berndt, C. C., "The Adhesion of Flame and Plasma Spray Coatings." Ph.D. Thesis, Monash University (1980)

64. Berndt, C. C., *Fracture Toughness Tests on Plasma Sprayed Coatings, Advances in Fracture Research,* (S. R. Valluri, D. M. R. Taplin, P. Rama Rao, J. F. Knott and R. Dubey, eds.), 4:2545-2552 (1984)

65. Berndt, C. C. and McPherson, R., *Mat. Sci. Res.,* 14:619 (1981)

66. Ostojic, P., "The Adhesion of Thermally Sprayed Coatings." Ph.D. Thesis, Monash University (1986)

67. Berndt, C. C., Karthikeyan, J., Ratnaraj, R. and Yang Da Jun, in: *Proc. 4th National Thermal Spray Conference,* (T. Bernecki, ed.), to be published by ASM International, Metals Park, OH (1991)

68. Stecura, S. and Liebert, C. H., US Patent Application 686,449 granted 14 May 1976 and patent granted 25th October 1977

69. Stecura, S., NASA TM-79206, July, 1979

70. Miller, R. A., *J. Ceram. Soc.,* 67:517 (1984)

71. Stecura, S., NASA TM-78976 (August, 1978)

72. Stecura, S., NASA TM-81604 (October, 1980)

73. Miller, R. A., Argarwal, P. and Duderstadt, E. C., *Ceram. Eng. Sci Proc.,* 5:470 (1984)

74. McDonald, G. and Hendricks, R. C., *Thin Solid Films,* 73:491 (1980)

75. Miller, R. A. and Berndt, C. C., *Thin Solid Films,* 119:195 (1984)

76. Sevcik, W. R. and Stoner, B. L., NASA CR-135360 (January, 1978)

77. Miller, R. A., in: *High Temperature Protective Coatings,* (S. Singhal, ed.), The Metallurgical Society of the AIME (1982)

78. Gedwill, M. A., Glasgow, T. K. and Levine, S. R., NASA TM-82687 (September, 1981)

79. Gedwill, M. A., NASA TM-81567 (September, 1980)

80. Steinmetz, P., Duret, C. and Morbioli, R., *Materials Science and Technology,* 2:262 (1986)

81. Levine, S. R. and Miller, R. A. EPRI-AP-2618, The Electric Power Research Institute (September, 1982)

82. Taylor, T. A., Price, M. O. and Tucker, R. C. Jr, DOE/ET/13327T2, Union Carbide Corporation (January 22, 1982)

83. Taylor, T. A., Price, M. O. and Tucker, R. C., Jr, in: *Proc. of the Second Conference on Advanced Materials for Alternative Fuel Capable Heat Engines,* (J. W. Fairbanks and J. Stringer, eds.) (May, 1982)

84. Berndt, C. C., *Trans. ASME J. for Gas Turbines,* 107:142 (1985)

85. Shankar, N. R., Berndt, C. C., Herman, H. and Rangaswamy, S., *Ceram. Bull.,* 62:614 (1983)

86. Berndt, C. C. and Herman, H., *Thin Solid Films,* 108:428 (1983)

87. Almond, D., Moghisi, M. and Reiter, H. *Thin Solid Films,* 108:439 (1983)

88. Berndt, C. C. and Miller, R. A., *Thin Solid Films,* 119:173 (1984)

89. Ault, N. N. and Milligan, L. H., *Amer. Cer. Soc. Bull.,* 38:661 (1959)

90. Wu, M. K., Ashburn, J. R., Trong, C. J., Hor, P. H., Meng, R. L., Cao, L., Huang, Z. J., Huang, Y. Q. and Chu, C. W., *Phys. Rev. Lett.,* 58:908 (1987)

91. Neiser, R. A., Kerkland, J. P., Herman, H., Elam, W. T., and Skelton, E. F., *Mat. Sci. Eng.,* 91:L13 (1987)

92. Neiser, R. A., Ph.D. Thesis, State University of New York at Stony Brook (1989)

93. Tachikawa, K., *Appl. Phys. Lett.,* 52:1011 (1988)

94. Wen, L. S., Qian, S. W., Hu, Q. Y., Y, B. H., Zhao, H. W., Guan, K., Fu, L. S., and Yang, Q. Q., *Thin Solid Films,* 152:L143 (1987)

95. Karthikeyan, J., Sreekumav, K. P., Kurup, M. B., Patil, D. S., Anartapadmanabhan, P. V., Venkatramani, N. and Rohatgi, V. K., *J. Phys. D,* 21:1246 (1988)

96. Heintze, G. N., McPherson, R., Tolino, D. and Andrikidis, C., *J. Mat. Sci. Lett.,* 7:251 (1988)

97. Tsurumi, S., Iwata, T., Tajima, Y. and Hikita, M., *Jap. J. Appl. Phys.,* 26:L1865 (1987)

98. Iqbal, Z., Reidinger, F., Bose, A., Capollini, N., Eckhardt, H., Ramakrishna, B. L. and Ong, E. W., *Mat. Res. Soc. Symp. Proc.,* 99:907 (1988)

99. Takita, K., Katoh, H., Ainaga, H., Nishino, M., Ishigaki, T. and Asano, H., *Jap. J. Appl. Phys.,* 27:L57 (1988)

100. Gudmundsson, B., Wang, H., Neiser, R. A., Katz, B. and Herman, H., *J. Appl. Phys.,* 67:2653 (1990)

101. Wang, H., Gudmundsson, B., Neiser, R. A., Herman, H., Suenaga, M., Welch, D. O., *High Temperature Superconducting Compounds II,* (S. H. Whang, A. D. Gupta and R. Laibowits,eds.), p. 141, The Metal.Soc. (1990)

102. Wong-Ng, W., Roth, R. S., Swartzendruber, L. J., Bennet, L. H., Chiang, C. K., Beech, F. and Hubbard, C. R., *Adv. Ceram. Mat.,* 2:565 (1987)

103. Fraser, J. R., Finlayson, T. R., Smith, T. F., Hietze, G. N., McPherson, R. and Whitfield, H. J., *Mat. Res. Soc. Symp. Proc.* EA-14:167 (1988)

104. Konaka, T., Sankawa, I., Matsuura, T., Higashi, T. and Ishihara, K., *Jap. J. Appl. Phys.,* 27:L1092 (1988)

105. Ekin, J. W., *Adv. Ceram. Mat.,* 2:586 (1987)

106. Bean, C. P., *Rev. Mod. Phys.,* 36:31 (1964)

107. Jin, S., Tiefel, T. H., Sherwood, R. C., Davis, M. E., vanDover, R. B., Kammlott, G. W., Fastnatch, R. A. and Keith, H. D., *Appl. Phys. Lett.,* 52:2074 (1988)

108. Wang, H., Herman, H., Wiesmann, H. J., Zhu, Y., Xu, Y., Sabatini, R. L and Suenaga, M., *Appl. Phys. Lett.,* 57:2495 (1990)

109. Zhu, Y., Zhang, H., Wang, H. and Suenaga, M., "Grain Boundaries in Textured $YBa_2Cu_3O_{7-x}$ Superconductor", to be published by J. Mat. Res.

110. Wang, H., Herman, H., Orehotsky, J., Wiesmann, H. J., Zhu, Y., Moddenbaugh, A. R., Sabatini, R. L. and Suenaga, M., "Textured Growth Processing of HTSC in the Y-Ba-Su-O System," to be published by The Metal. Soc.

111. Mitsui, K., *Precision Eng.,* 8:212 (1986)

112. Brodman, R., *Precision Eng.,* 8:221 (1986)

113. Fultz, B. S., *J. Prot. Ctg. & Linings,* 2:5 (1985)

114. *Thermal Spray Handbook,* Amer. Weld. Soc., Miami, FL (1985)

115. Georgopoulos, P., Cohen, J. B. and Herman, H., *Mat. Sci. & Eng.,* 80:41 (1986)

6

Optical Thin Films

David G. Coult

1.0 INTRODUCTION

Optical thin films are widely used today in many diverse applications to control the way light is reflected, transmitted, or absorbed as a function of wavelength. They can be grouped into two major categories based on the application. In the first, the light travels parallel to the plane of the substrate with the films acting as waveguides in the emerging field of integrated optics. Here light signals could replace electrical signals in applications such as communications and computers (1). In the second application, the light travels perpendicular to the film plane for use as antireflection coatings, edge filters, high efficiency mirrors, beam splitters, etc. This, the more standard use of optical thin films, is the subject of this chapter.

The term "thin" is used to indicate a layer whose thickness (perpendicular to the substrate) is the same order of magnitude as the wavelength of interest, and the extent (parallel to the substrate) is a very large number of wavelengths. Typical layers might range in thickness from 8×10^{-8}m (800 Å) in the visible to twenty times that in the infrared. Filters are composed of a stack of such layers, alternating between high and low refractive indices, with typically 20 - 40 layers, although in some cases they may have one hundred layers or more. Thin film filters operate by interference of the light reflected from the various layers as the light passes through perpendicular to the substrate.

This chapter is a brief overview of the work done in this field. There is no way to be all-inclusive but every effort has been made to give proper recognition for the work described here.

2.0 OPTICAL THIN FILM DESIGN

When light strikes a film, it can either be reflected, transmitted or lost to absorption or scattering. If we consider a light beam incident on a homogeneous parallel-sided film, the amplitude and polarization state of the light transmitted and reflected can be calculated in terms of the angle of incidence and the optical constants of the three materials involved (2). Figure 1 depicts light from the incident medium of refractive index N_o passing through film material of index N_f, entering substrate material of index N_s. The incident medium is often air with an index of refraction assumed to be equal to 1.0, the index of vacuum. The film and substrate materials can be transparent or absorbing in which the optical constant (or complex index of refraction), N, is given by:

Eq. (1) $$N = n - ik$$

where n is the refractive index and k is the extinction coefficient. At optical frequencies, $\varepsilon = n^2$ where ε is the dielectric constant of the material. The extinction coefficient is related to the absorption coefficient, α, by the expression:

Eq. (2) $$\alpha = 4\pi k/\lambda$$

where α determines the intensity, I, transmitted through an absorbing medium by the exponential law of absorption (3):

Eq. (3) $$I = I_o e^{-\alpha x}$$

The reflectance and transmittance at the boundaries between these regions can be conveniently expressed in terms of the Fresnel coefficients. When the regions are absorbing, these terms are large and cumbersome but simplify into terms involving only n if the regions are transparent. Thus, if we assume that k = 0 and that plane waves strike a plane boundary at normal incidence, the reflectance between two regions is given by

Eq. (4) $$R = \left[\frac{n_o - n_s}{n_o + n_s}\right]^2$$

For example, the reflectance of uncoated window glass (with $n_s = 1.52$) in air (with $n_o = 1.0$) is 4.3%. The transmittance, T, through the surface would be

Eq. (5) $T = 1 - R = 95.7\%$

and through both surfaces would be 95.7% of 95.7% or 91.7%. The index range of optical glasses in the visible region is 1.4 - 1.9 and both high and low index glasses are used for complex lens systems to correct for various aberrations (4). By comparison, the transmittance through a single lens of n = 1.9 would be 81.7%. Complex systems may have many lenses cascaded together and the losses may total 50% or more. Infrared systems use very high index materials such as germanium which has a reflectance of 35% per surface. In addition to power lost, these reflections cause "ghost" images, thus it is necessary to put antireflection coatings on the lenses.

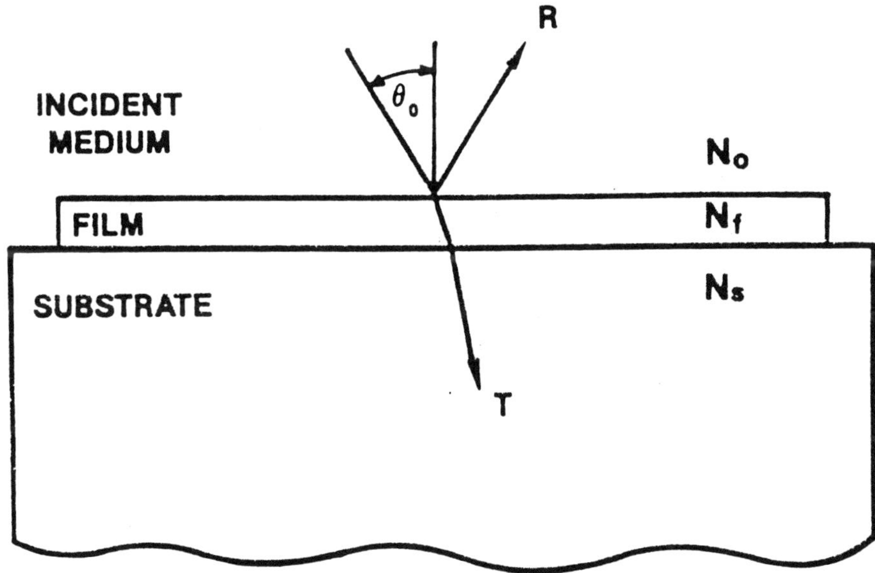

Figure 1. A single thin film.

2.1 Antireflection Coatings

Antireflection coatings operate on the principle of interference of the light reflected from the front and back surfaces of the films. The optical thickness of a thin film is defined as the index, n_f, times the physical thickness, d_f. In the case where $n_f d_f = \lambda/4, 3\lambda/4, 5\lambda/4$, etc., the equation for calculating the reflectance of a single layer as in Fig. 1 at normal incidence for the wavelength of interest, λ_o, simplifies to:

Eq. (6) $$R = \left[\frac{n_o - \dfrac{n_f^2}{n_s}}{n_o + \dfrac{n_f^2}{n_s}}\right]^2$$

This is a very useful formula for estimating how well a particular single layer coating will do. We can calculate the film index needed for zero reflectance at one wavelength as:

Eq. (7) $$n_f = (n_o n_s)^{1/2}$$

For a lens of n = 1.9, we get $n_f = 1.38$ which happens to be the index of refraction in the visible of MgF_2, the most widely used material for single layer antireflection coatings and one of the lowest index materials available. Thus, for glasses with an index lower than 1.9 (which is usually the case) we must either accept a little higher loss for the economy of a single layer MgF_2 coating or go to the additional cost of adding more layers of other materials. Adding more layers also gives the advantage of being able to achieve a lower reflectance over a much broader wavelength range. Figure 2 gives a comparison of commercially available single layer, double layer, and broadband antireflection coatings (5).

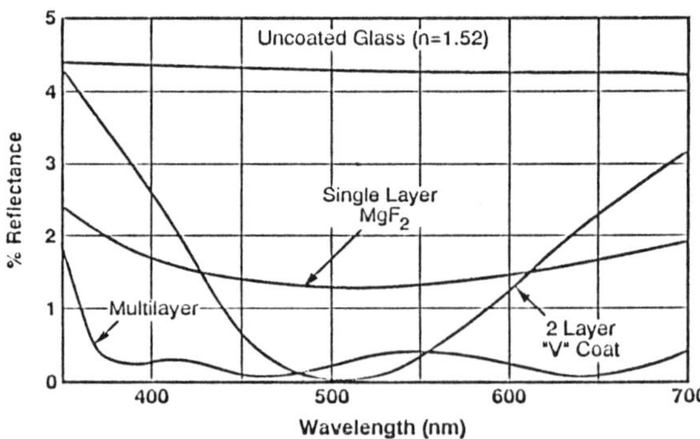

Figure 2. The spectral response of commercially available antireflection coatings (5).

2.2 Multilayer Stacks

The interference effect can be used by building up a stack of alternating high and low index materials to produce many interesting results. Figure 3 shows the construction of such a quarterwave stack in which the optical thickness n_jd_j of each layer is again equal to one quarter the wavelength of interest. The multilayer is completely specified if we know n_j, k_j and d_j for each layer, n_o for the incident medium plus n_s and k_s for the substrate. Given the angle of incidence, θ_o, we can calculate the reflectance, R, and the transmittance, T, as a function of wavelength.

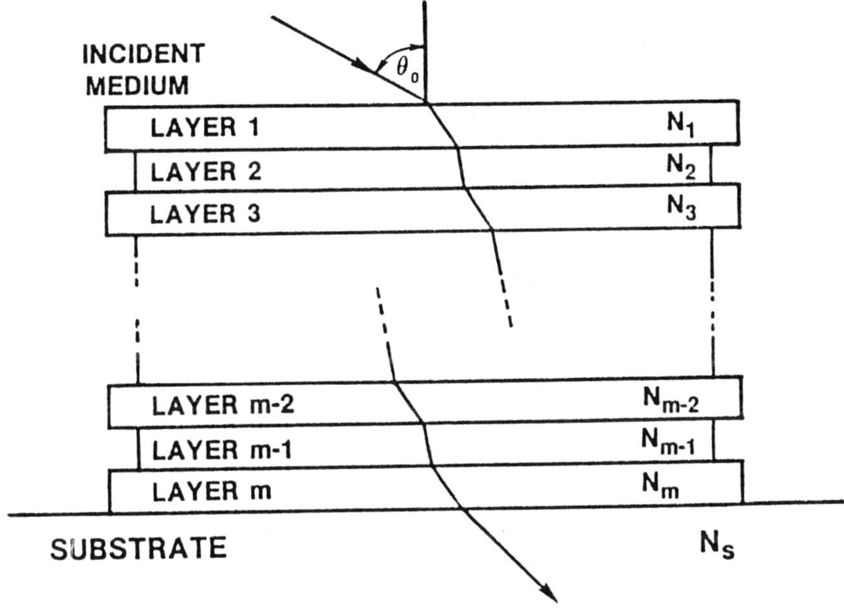

Figure 3. A multilayer stack.

Figure 4 shows the reflectance vs. wavelength of six and twelve layer stacks of Si/SiO_2. They are plotted as a function of λ_o/λ where the wavelength of interest, λ_o = 4nd. The multilayer has a characteristic stopband (or high-reflectance region) symmetric about the wavelength λ_o surrounded by long and short-wave pass regions characterized by many ripples in the passbands. The width of the high reflectance region is determined by the ratio of the high to low index, n_H/n_L. The higher the ratio, the wider the stopband. The maximum reflectance depends on the number of layers as well as the ratio n_H/n_L with R increasing with number of layers. The ripple in the passband is bounded by an envelope determined by n_H/n_L whereas the number of peaks depends on the number of layers.

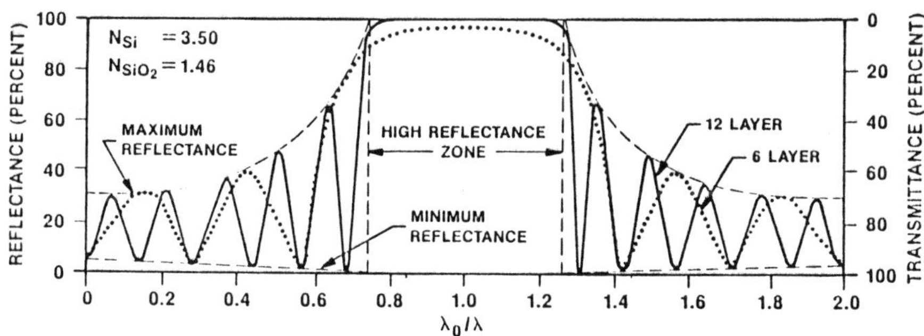

Figure 4. Reflectance vs. 1/λ for Si/SiO_2 multilayer stacks.

By suppressing the ripple on one side of the stopband we can create an edge filter. A long-wave pass (LWP) filter transmits the long wavelengths and efficiently reflects the short ones. On the other hand a short-wave pass (SWP) filter transmits the short wavelengths and reflects the long ones. A SWP filter is sometimes referred to as a high-pass filter meaning it passes high frequencies. Likewise, a LWP filter is sometimes called a low-pass filter and care must be used not to confuse these terms. The bandpass filter transmits a relatively narrow range of wavelengths and effectively rejects those on both sides.

Once a design has been made, the wavelength characteristics can be moved in wavelength by changing the layer thicknesses by the same ratio. For example, a LWP filter with an edge at 400 nm can be shifted to 800 nm by making all the layers twice as thick.

As the angle of incidence increases, two effects are seen. First, the effective thickness of the layers becomes smaller which causes the filter to shift to shorter wavelength. Since the effect is inversely proportional to the film index, at large angles the layers are no longer matched at the same wavelength and the spectral shape degrades.

The second effect at non-normal incidence is that the spectral characteristic becomes dependent on the polarization of the incident beam. This can be optimized and used to advantage for such components as polarizing beamsplitters. But in general, these angle effects are detrimental, especially for a non-collimated beam.

One of the most comprehensive volumes on the design and production of multilayer optical thin films is Thin-Film Optical Filters by Angus Macleod (6). Many other excellent references are available on thin film design (7)-(11). There are also a number of commercially available software programs for analysis and optimization of optical thin film designs (12).

3.0 THIN FILM MATERIALS

One natural limitation on the film designer is the finite number of materials, and therefore indices, available. For multilayer stacks, the usual goal is to have two materials as widely spaced in index as possible. As suggested in the last section, this gives desired properties using the fewest number of layers. For antireflection coatings, as many as four materials may be used and intermediate values of index are necessary. In all cases, low absorption is required.

Table 1 shows some materials commonly used in optical thin film deposition along with their indices. These are commercially available and are made specifically for this application. Some success has been achieved in producing intermediate values of index by co-evaporation. As previously mentioned, magnesium fluoride has been very widely used as a single layer antireflection coating or as the outer layer in multilayer antireflection coatings. Multilayer stacks have been made for years from zinc sulfide and either cryolite or magnesium fluoride although high stress in magnesium fluoride has limited the thickness of stacks made with it. These combinations continue to be used successfully, especially for benign environments or when protected with a cemented cover glass. More robust filters are often made from stacks of titanium dioxide and silicon dioxide. More detailed information on materials for optical thin films can be found in a review article by Ritter (13) or the catalogs of materials suppliers (14)-(16).

Within the field of optical thin films, the area that is in greatest need of work is materials research. Accurate measurement of the optical properties is of primary importance and has been pursued along with their relationship to deposition conditions. Good descriptions of the most common methods of measuring n, k and d have been described by Heavens (17), but determination of other basic materials properties is difficult and little is known about the thermal conductivity, stress, Young's modulus, etc. of these dielectrics. Some good articles on mechanical properties have been written

(18)-(23) with emphasis on the important topic of film stress evaluation. The study of film microstructure is also of utmost importance since it not only determines the optical properties, but also controls the mechanical properties and therefore environmental stability and long-term reliability of these films.

Table 1. Commonly Used Optical Thin Film Materials (6)

Material	Index	Transparent Region
MgF_2	1.38 at 550 nm	0.210 - 10 µm
SiO_2	1.46 at 500 nm	0.200 - 8 µm
ThF_4	1.52 at 400 nm	0.200 - 15 µm
Al_2O_3	1.59-1.63 at 600 nm	0.200 - 7 µm
CeF_3	1.63 at 550 nm	0.300 - 5 µm
PbF_2	1.70 at 1 µm	0.240 - 20 µm
MgO	1.70 at 550 nm	0.200 - 8 µm
Y_2O_3	1.82 at 550 nm	0.250 - 2 µm
SiO	2.00 at 550 nm	0.500 - 8 µm
HfO_2	2.00 at 500 nm	0.220 - 12 µm
ZrO_2	2.10 at 550 nm	0.340 - 12 µm
Ta_2O_5	2.16 at 550 nm	0.300 - 10 µm
CeO_2	2.18-2.42 at 550 nm	0.400 - 16 µm
TiO_2	2.20-2.70 at 550 nm	0.350 - 12 µm
CdS	2.27 at 7 µm	0.600 - 7 µm
ZnS	2.35 at 550 nm	0.380 - 25 µm
ZnSe	2.58 at 663 nm	0.600 - 15 µm
CdS	2.60 at 600 nm	0.600 - 7 µm
Si	3.50 at 10 µm	1.1 - 14 µm
Ge	4.25 at 10 µm	1.7 - 100 µm
Te	4.90 at 6 µm	3.4 - 20 µm
PbTe	5.50 at 10 µm	3.40 - 30 µm

From a reliability standpoint, the failure mechanisms are not well understood, therefore there are no theoretical models to predict degradation. Durability is often determined by visual observation and therefore tends to be subjective. The assumption is made that, since depositions are made by a batch process, all parts coated within one batch have similar durability and long-term aging characteristics. Environmental stability is tested by choosing a representative part (or witness piece) from a batch and subjecting it to some series of stresses such as elevated temperature and humidity, followed by an adhesion test, abrasion test, visual inspection and optical performance verification. Typical conditions are 24 hours at 50°C and 95% relative humidity. Other stresses might include thermal shock and resistance to water and salt spray. These tests as well as cosmetic qualities such as the number and size of scratches and digs allowed are often specified using applicable military standards (24)(25). Since the microstructure and materials properties are highly dependent on the deposition conditions, published results are often confusing and contradictory, and a much better understanding of the materials properties and their relationship to deposition conditions is sorely needed.

4.0 DEPOSITION PROCESS

There are many methods of thin film deposition as this book and others confirm (26)-(32). It is useful to note here that the optical thin film industry grew over the past two decades riding on the wave of the semiconductor industry due to larger capital available for research and development of equipment and processes for those devices. Thus, much of the deposition technology used in the production of optical thin films was adapted from semiconductor equipment. An interesting historical account dating to the 1600's was given by Strickland (33), and Jacobson (34) has collected a wealth of original papers on deposition of optical coatings dating to 1852.

The major technique in use today for the deposition of optical films can be termed physical vapor deposition (PVD) in which the material condenses from the vapor phase onto the desired substrate. There are many different ways the vapor can be produced, but thermal evaporation (or vacuum evaporation) is the most widely used technique today. It is carried out in a vacuum to prevent unwanted chemical reactions and to keep the film from becoming porous by incorporating gas during the condensation process. The substrate to be coated is typically placed in a vacuum chamber, pumped down in the range of 1×10^{-6} torr (10^{-9} atmosphere) and heated to 200 - 300°C. The coating material is then heated until it evaporates or sublimes, travels across the vacuum chamber, then condenses on the substrate.

The condition and cleanliness of the substrate is critical to the quality of the coating. Since the forces which hold the film onto the substrate are all short range, even a few molecular layers of contaminant will result in a film with reduced adhesion. Since the layers are only a few tenths of a micron thick, they will faithfully reproduce any irregularities in the surface and cannot be expected to smooth over imperfections left from grinding and polishing. As a matter of fact, imperfections and especially stains in the glass surface that might be unobservable are greatly enhanced by coating. Once the substrates are cleaned and loaded into the fixtures they should be put under vacuum and coated as soon as possible.

The fixture is generally placed at the top of the chamber with the substrates held in by gravity. To improve thickness uniformity, the holder is often a dome which is rotated about the vertical axis of the chamber. By using uniformity masking, thickness from one area of the fixture to another can be controlled to ± 2% or better. Tighter control may often be achieved with planetary tooling where smaller plates are rotated about their axes which are also moved about the chamber axis in a planetary motion. Some argue that the slight improvement in uniformity possible with the planetary is not worth the additional capital cost, complexity, particle generation and reduced throughput.

For vacuum evaporation, a high vacuum on the order of 10^{-6} torr is necessary so that evaporant molecules will travel across the chamber without encountering other molecules. This can be obtained with a variety of high vacuum pumps, the most common being the hot oil diffusion pump. It has been the workhorse of the industry for many years because it is economical, reliable and has a high throughput (35). If operated properly, it will not contaminate the chamber and substrates with oil. However, if oil contamination is seen, it is most likely from the mechanical pump used to "rough down" the chamber and back up the high vacuum pump (36). In recent years cryopumps and turbomolecular pumps have sometimes been used in place of diffusion pumps because of the perception that they will eliminate oil contamination.

Rough pumping takes on the order of 5 - 10 minutes with high vacuum pumping taking an hour or more. During this time the substrates are usually heated to improve adhesion and film properties. The heaters are either the rod type placed behind the substrates or, more often, quartz lamps placed below the substrates. In both cases, heating is by radiation in the vacuum. Since different materials absorb this radiated energy with different efficiencies, glass substrates, stainless steel tooling, etc., will heat at different rates and may be at different temperatures during coating. Thus we can't assume the temperature of a glass lens to be coated is the same as the stainless steel tip of a thermocouple probe near that substrate. It is difficult to attach a thermocouple to the substrate itself since it is rotating around the chamber, and if the probe was attached to a lens, the probe itself might still be heated at a different rate than the glass. Rather than worry about the exact temperature of the substrate, those doing coatings most often rely on finding a heating cycle that works and repeating it each time. Since temperature has a great effect on film properties and since it is so hard to measure, this might account for some of the discrepancies reported in the literature.

The two most widely used methods of evaporant heating are the resistance source and electron beam gun. In resistance heating, a large current of perhaps a few hundred amperes at a low voltage is passed through a boat or filament to evaporate or sublime the material. Many configurations of boats, etc. have been developed over the years for specific materials. A classic work that is still useful today is Vacuum Deposition of Thin Films by L. Holland (37).

The electron beam gun generates a stream of electrons of 6 - 10 kV up to a few amperes (38). Using magnetic fields, the beam is deflected typically 270°C to keep the filament out of the path of the evaporant which greatly extends the filament life. Electromagnets are used to move the beam around on the source material to control what area gets heated, as well as to raster the beam to avoid hot spots and keep the surface of the evaporant level. If this is not done, the surface of the evaporant changes shape and the distribution of deposited film changes.

The crucible is water-cooled copper which results in evaporation of the heated material from a molten pool contained in a "skull" of the same material. This keeps the film pure and allows for deposition of material which otherwise might react with a resistance source.

The advantages of the electron gun over resistance source are:

1. It is very versatile and can be used for almost any material such as very high melting point metal oxides that cannot be evaporated from a resistance source
2. Electron gun sources can be configured to hold a great deal of material and can be purchased with continuous feed for very large depositions
3. Films can be deposited that are often purer than those from a boat since reaction with the boat material is eliminated

Disadvantages of the electron beam gun are:

1. It is more complicated to install and maintain and is a factor of ten more expensive to purchase
2. Certain materials or particular film properties may be more easily obtained with the resistance source
3. Slightly more room is needed on the baseplate for the electron gun.

Filters made of high temperature refractory oxides such as TiO_2, ZrO_2, and Al_2O_3 have often been termed "hard coatings" as opposed to lower melting point "soft coatings" such as ZnS and MgF_2 which are deposited with resistance sources.

As implied earlier, the optical thickness (index times physical thickness) must be controlled accurately for each layer. There have been many schemes devised for controlling thickness, but the optical monitor and the quartz crystal microbalance are the two methods which are most widely used today.

The optical monitor measures the change in reflected or transmitted light from a substrate or glass witness slide located near the work, usually in the center of the chamber. Interference of light reflected from a thin film on a substrate causes the color to change with thickness similar to the colors seen in oil on puddles or wet pavement. Thus the simplest optical monitor is to visually watch the color of the reflected light change as the thickness increases. The eye is very sensitive to differences in color and very good control of thickness can be achieved for single layer antireflection coatings, especially for the visible. For more complicated multilayers, a monochromatic light source is used and the eye is replaced with a photodetector. As the thickness increases, the signal goes through maxima and minima corresponding to multiples of quarterwaves. Since the optical monitor measures optical thickness, it compensates for slight changes in index of refraction.

The quartz crystal monitor measures the change in mass of a vibrating quartz crystal as the coating builds up on it. The advantage is that the output signal varies linearly with thickness and thus it is easy to build this into a controller for both deposition rate and final thickness. There are a number of good controllers on the market that will ramp up source power to a preset level, soak for a given time, ramp to a second level and soak there, open the shutter allowing evaporant to reach the substrates, adjust the power to keep the deposition rate constant, then close the shutter and turn the source power down or off when the desired thickness is achieved.

The quartz crystal controller is excellent for metal films, but there is some debate over which method is best for dielectric optical thin films. Many have chosen to use the quartz crystal monitor to control the deposition cycle including rate, but use an optical monitor to control final thickness, especially for filters requiring high precision. Macleod (39) has done an excellent analysis of the various options and concerns.

5.0 FILM PROPERTIES

5.1 Effects of Deposition Conditions

The main parameters which affect the properties of the deposited film are substrate temperature, deposition rate and level of residual gases in the chamber (40). In the deposition of many refractory oxides such as TiO_2, ZrO_2, Y_2O_3, Ta_2O_5 and Nb_2O_5, the materials dissociate upon heating in vacuum and condense in an oxygen deficient state. To overcome this, oxygen is "bled" into the chamber at a level of 5×10^{-5} to 2×10^{-4} torr which reacts with the condensing atoms to form the desired compound (41). Some believe oxidation takes place at the evaporation source or en route to the substrate, but most authorities believe recombination takes place at the substrate. Thus to get one metal atom to combine with the appropriate number of oxygen atoms to form a stoichiometric oxide, there must be the proper evaporation rate, partial pressure of oxygen, substrate temperature, and possibly other conditions such as kinetic energies of the atoms, substrate surface conditions, level of residual gases, etc. Too little oxygen and the absorption goes up; too much oxygen and it becomes incorporated into the film making it porous. This process is widely used today and is called reactive evaporation.

As an example of how substrate temperature affects index of refraction, TiO_2 deposited on substrates at 200°C has an index of 2.22 at 750 nm whereas at 400°C the index is 2.40. Variation of index with oxidation state of Si is even more striking. SiO_2 has an index at 550 nm of 1.46. As the film becomes oxygen deficient, the index increases monotonically to 1.9 for SiO while pure Si has an index of 3.5 and is highly absorbing at visible wavelengths. Angle of incidence for the depositing material also has a marked effect on the index. Figure 5 shows the dispersion curves of TiO_2 deposited simultaneously on two substrates, one held normal to the evaporant and the other held at 30°C. The index at normal incidence is over 3% higher, increasing from 2.33 to 2.41 measured at 550 nm.

Even when we try to minimize variation, the index of conventionally deposited films can often vary by ± 0.1 (i.e., ± 5%), and thickness control of ± 1% is considered very good. By contrast, lens designers can specify refractive index and thickness to ± 0.01% realizing that this can be manufactured. In order to improve this situation, we must understand how the deposition conditions affect these properties.

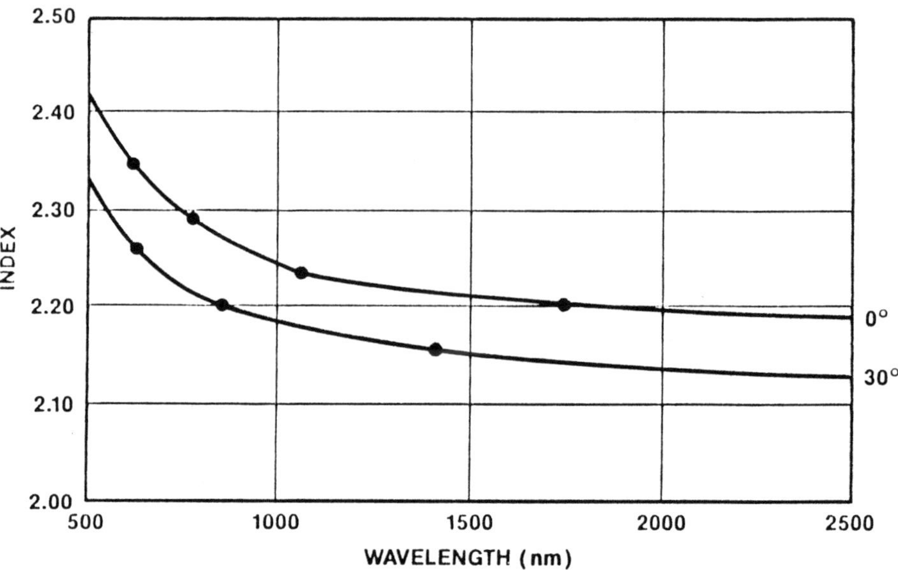

Figure 5. Index of refraction vs. wavelength for TiO_2 films.

5.2 Effects of Film Microstructure

Great strides have been made over the years in the area of thin film design. Filters of almost any shape can be theoretically designed (42). But the great challenge today is to be able to deposit films with properties that match the theoretical ones. In contrast to theory, actual films are inhomogeneous, less durable, and have an index of refraction that is lower with losses higher than that found in bulk samples. In addition, these properties are found to vary considerably with relative humidity and therefore temperature of the environment.

The film microstructure is crucial in determining many of these film properties. The single most important feature of optical films deposited by sputtering or evaporation for either dielectrics or metals is a columnar microstructure in which parallel columns of solid material several hundred angstroms in diameter can be found growing out of the substrate (43)(44). Figure 6 is a scanning electron micrograph of a multilayer film of TiO_2/SiO_2 showing this columnar growth.

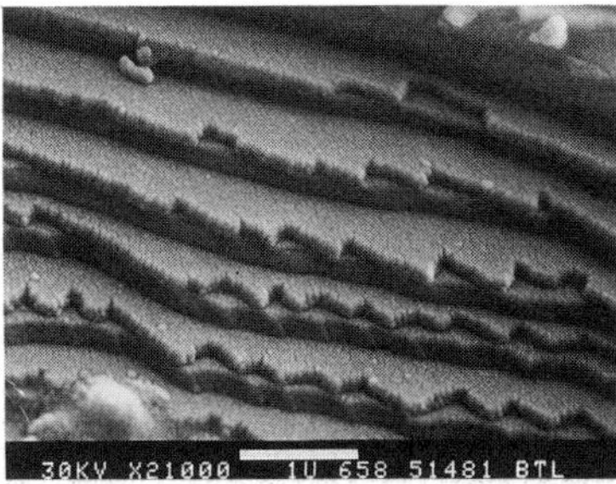

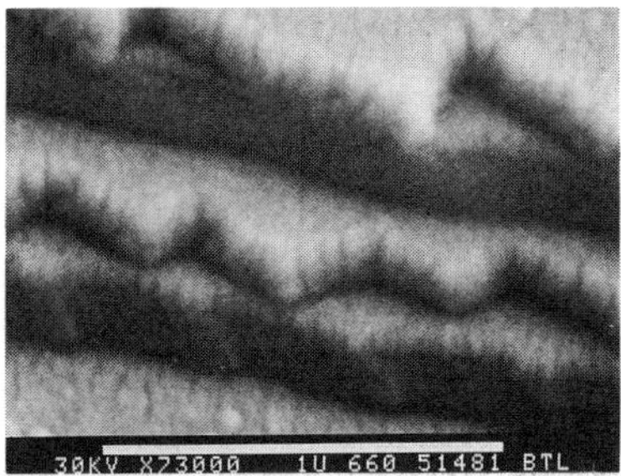

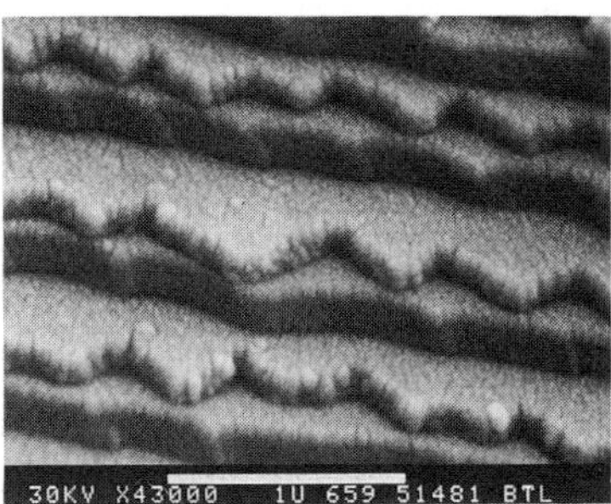

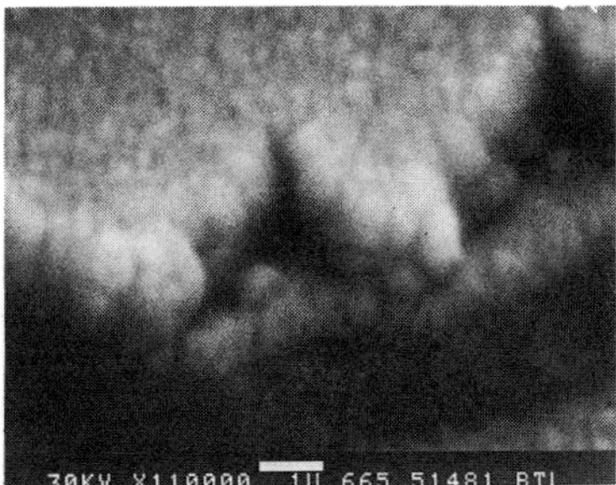

Figure 6. Cross-section of a TiO_2/SiO_2 quarter wave stack deposited on a glass substrate by conventional evaporation.

This columnar microstructure has a profound effect on the film properties. We can characterize how dense the films are by the packing density, p, where

$$\text{Eq. (8)} \qquad p = \frac{\text{Volume of solid part of film}}{\text{Total volume of film}}$$

For optical thin films we usually find $0.80 < p < 0.95$. The refractive index decreases as the packing density decreases. We can also see the reason for inhomogeneity in refractive index since the diameter of the columns and therefore the packing density may change as we move through the film. In addition, by capillary action the pores (with index equal to 1.0) readily take in or give up water (with index equal to 1.33) as a function of relative humidity. This causes a change in optical thickness and therefore optical performance that depends on the relative humidity of the environment.

The spectral performance of an optical coating changes with temperature on the order of 0.005%/°C which would, for example, shift the passband of a narrow bandpass filter centered at 1550 nm up in wavelength 8 nm on heating 100°C. This arises because: *(a)* the index of refraction decreases with temperature and *(b)* the coefficient of expansion causes the physical thickness to increase with temperature. It would be possible to design a filter with zero temperature dependence if these effects could be cancelled. Seeley (45) described such narrow bandpass filters using PbTe for the 4 - 20 micron wavelength range that have extremely small temperature dependence. Unfortunately, there are no such materials for the visible range.

In our studies of conventionally deposited TiO_2/SiO_2 multilayer edge filters, we baked them in a temperature controlled cell at 100°C in dry nitrogen with dew point < -70°C, then measured shift in cutoff wavelength as a function of temperature only. From 0 to 100°C in a dry environment, we measured a shift in an 850 nm edge filter of 2 nm for a temperature dependence of 0.003%/°C. However, upon raising the relative humidity of the nitrogen atmosphere to 90%, the films shifted to longer wavelength by 20 nm in a matter of minutes (Fig. 7). The spectral shape did not change significantly during this process indicating direct paths for the water through the entire stack. In contrast, Macleod (46) found that narrow bandpass filters made of ZnS and either MgF_2 or cryolite may take several weeks or months to shift with a striking change in performance during the transition. They attributed this to moisture travelling down pinholes then slowly percolating through the columnar pores of the layers.

We have found that in some cases edge filters made of Si/SiO_2 show no shift at the center of the substrate, but show significant shifts very near the substrate edge (Fig. 8). If the optical path was located near the edge or there were significant pinholes in the optical path, at normal operating humidity levels the filter would shift significantly in a few years.

Many have noticed spectral instabilities in films as a function of temperature, but since relative humidity is a function of temperature, care must be taken to separate temperature dependence from moisture sensitivity. Thus, if a filter is dry, heating from 0 - 100°C may increase the location of the edge 0.3%. But if the filter is wet, heating from 0 - 100°C may decrease the edge over 2%. These changes are not always reproducible and depend on the past history of the filter. Thus, from a practical point, it is best to expose filters to identical conditions and measure them after stablization at the same temperature and relative humidity as they will see in use.

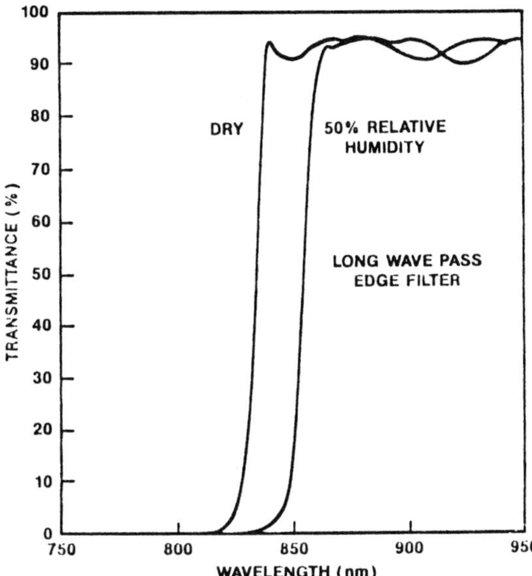

Figure 7. Shift with humidity in transmittance vs. wavelength for a TiO_2/SiO_2 edge filter.

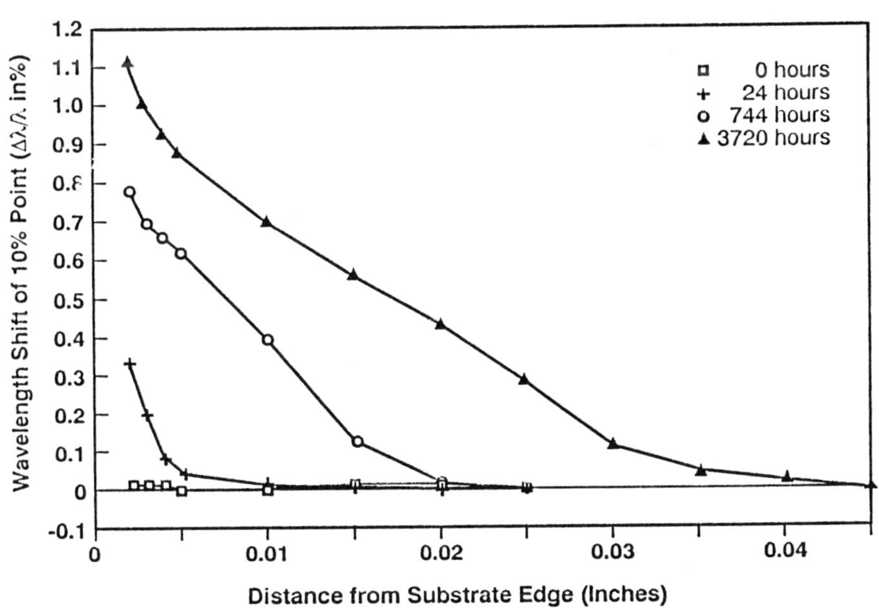

Figure 8. Wavelength shift of an Si/SiO_2 shortwave pass filter after storage in 85°C/85% R. H.

The columnar structure also has an effect on the film stress. The diameter of the columns generally gets smaller as they grow out from the substrate resulting in tensile stress. In a few materials such as ZnS, the column diameter grows with thickness resulting in increased packing density and compressive stress.

Movchan and Demchishin (47) proposed a structural zone model of growth in which the microstructure of films can be predicted from the ratio of the substrate temperature to the melting point of the evaporant (T_s/T_m). As a film condenses, the atoms move about on the surface depending on the energy available. If the major source of energy is thermal heating of the substrate, this model gives a way to gauge the adatom mobility. Zone 1 (low substrate temperature with $T_s/T_m < 0.3$) arises due to self-shadowing and consists of loosely packed, tapered crystallites. Zone 2 ($0.3 < T_s/T_m < 0.5$) is controlled by surface diffusion and consists of columnar grains. Zone 3 (substrate temperature near the melting point of the evaporant with $T_s/T_m > 0.5$) is controlled by bulk diffusion and is made up of equiaxed grains. Thornton (48) extended this model to include evaporation of complex alloys as well as including the effect of the partial pressure of inert or reactive gases due to sputtering (Fig. 9). Messier (49) has further improved this model to show how the microstructure evolves during growth (fractal model) as well as to separate the effects of bombardment-induced adatom mobility from thermal mobility.

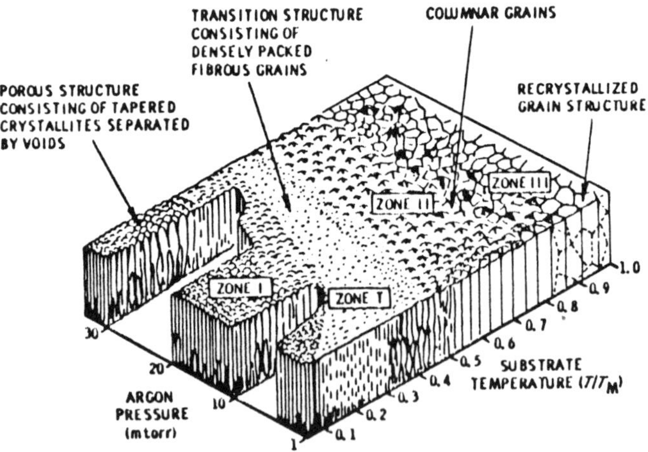

Figure 9. Zone-structure model (48).

Due to complexity of the deposition process, these models are largely qualitative at present. The boundaries between zones are gradual and not sharply defined, and not all zones are found in every material. For example, zone 3 is not present in high melting point materials. And pure metals do not have a transition zone. In general, when the ratio is less than 0.45, strongly columnar growth results. In conventional optical film evaporation, the ratio of substrate temperature to melting point is in the range of 0.10 to 0.30 and thus results in a marked columnar growth.

It has been empirically determined that the angle of incidence of the vapor, α, is related to the angle of the columns, β, by the "tangent rule" (Fig. 10):

$$2 \tan \beta = \tan \alpha$$

In addition to limited adatom mobility, an important factor in this formation of a columnar microstructure at low T_s / T_m is self-shadowing. Computer simulations (50)-(52) were done in which hard spheres were "evaporated" onto a substrate and allowed to move on the surface only until they found the nearest low energy pocket formed by three spheres in contact. These simulations agree very well with observed growth structures and follow the tangent rule. This is interesting since it only considers limited adatom mobility and resulting geometric shadowing and does not incorporate any other factors.

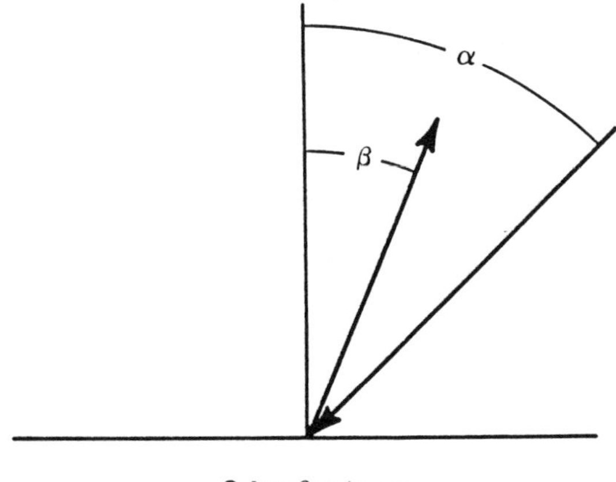

$2 \tan \beta = \tan \alpha$

Figure 10. Tangent rule which relates the angle of incidence of a vapor α, to the angle of the deposited columns, β.

A columnar structure is observed for many materials: high melting point elements like Cr, Be, Se, Ge; compounds like TiO_2, ZrO_2, Al_2O_3; non-noble metals evaporated in a residual atmosphere of O_2 like Fe and Al; and compounds such as CdTe, CeF_2, and PbS. It is due to a combination of limited adatom mobility plus self-shadowing. However, if the adatom surface mobility is high enough, the film will not have a columnar microstructure. Thus, Au and Ag films can be deposited at large angles on heated substrates and show no columnar anisotropy due to high surface mobility of the adatoms. Hodgkinson (53) gave a good review of these effects in optical films.

6.0 ATTEMPTS AT IMPROVED PROPERTIES

In order to eliminate this columnar growth and the properties associated with it, it has been necessary to understand the basic physics of the deposition process. As the film material begins to condense, the adsorbed atoms move about the substrate and group together to form nuclei. As more material arrives, the nuclei grow into columns as mentioned by a self-shadowing mechanism which is surprisingly independent of the material. In recent years researchers have reported on various methods for improving the properties of optical thin films by eliminating this columnar structure and improving the stoichiometry. All the methods involve increasing adatom mobility and can be roughly grouped into those that add energy at the substrate in some manner during otherwise conventional evaporation and those involving alternate deposition techniques, primarily sputtering.

6.1 Sputtering Techniques

Sputtering is a process whereby atoms and ions of argon or other gases from a plasma bombard a target and knock off (sputter) atoms of the target by momentum transfer. These atoms travel to the substrate where they are deposited. There are several different sputtering configurations depending on the substrate and material to be deposited. The simplest and most widely used configuration is the planar diode in which two planar electrodes 10 - 30 cm in diameter are spaced 5 - 10 cm apart (Fig. 11a). One holds the substrate and the other holds the target material to be deposited. The plasma may be sustained by either DC or RF power, and magnetic plasma confinement (magnetrons) may be used with either source. Magnetrons trap the free electrons in a magnetic field which reduces substrate heating and increases the efficiency with which electrons will ionize gases in the plasma zone resulting in the advantage of lower working gas pressure (Fig. 11b). Less common techniques involve using a grid to stabilize the discharge (triode) or diode sources with non-planar electrodes.

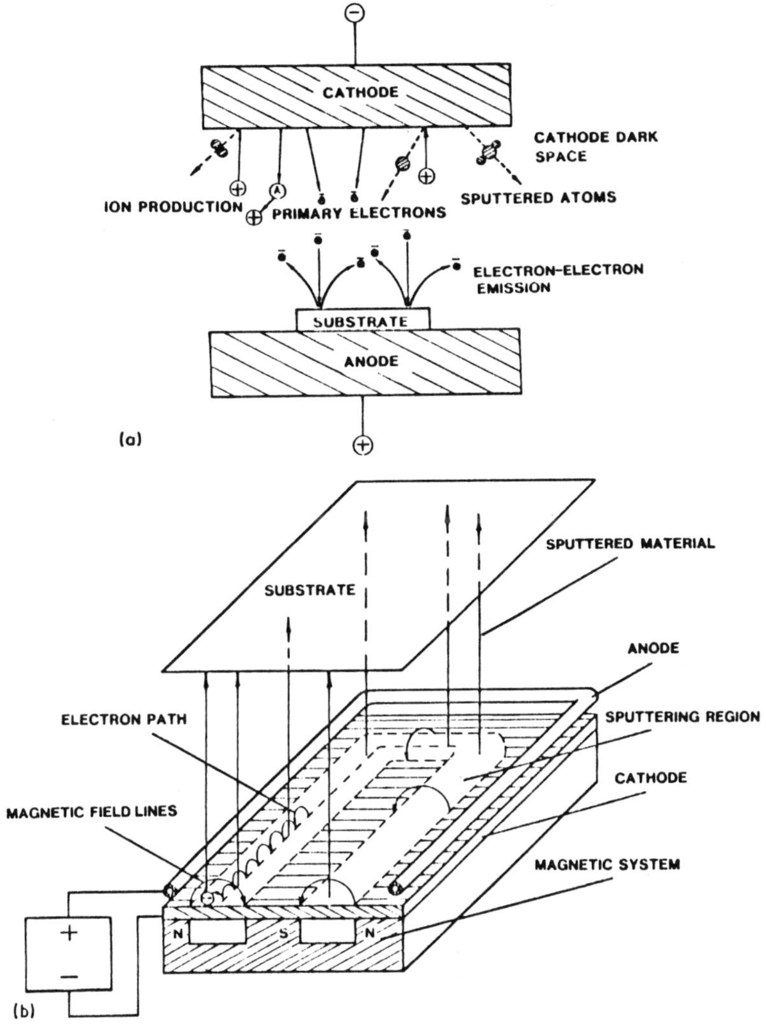

Figure 11. Schematic of (a) planar diode sputtering, and (b) magnetron sputtering (90).

DC Sputtering. In DC planar diode sputtering, a plasma is sustained as gas ions (usually Ar^+) strike the target creating secondary electrons which then form more ions. Magnetrons are often used and deposition rates for metals are high, but the charge which builds up on the surface of a dielectric repels any further ion bombardment, causing deposition rates of insulating materials to be very low.

Sputtering of metals in a reactive gas atmosphere is sometimes used to deposit oxide, nitrides and carbides. But the surface of the metal can quickly become "poisoned" with an insulating layer, causing the rate to drop which results in an even higher concentration of reactive gas which only further passivates the surface of the target. Because of this unstable situation, DC sputtering has seldom been used for dielectric thin films for optics. Motovilov (54) fabricated multilayers of Ta_2O_5/SiO_2 by DC sputtering which they claimed showed no spectral instability over a three year period. They attributed this to a reduction in the pores normally found in evaporated films, although measurements of packing density were not mentioned.

Schiller (55)(56) described a novel configuration for depositing oxides which keeps the simplicity and high deposition rate of DC magnetron sputtering of metals without poisoning the target. They physically separated the sputtering of the metal target and subsequent oxidation into two discrete regions of the vacuum system. Hartsough (57) claimed five angstroms of aluminum could be oxidized at a time using this method. Scobey (58) further developed this concept by adding an oxygen ion source to reduce the time needed to oxidize the metal layer as well as improve stoichiometry and microstructure.

In this configuration the substrates are mounted on the outside of a drum 3 - 4 ft in diameter, 2 - 4 ft high, positioned vertically in the vacuum chamber. Two DC magnetron sources with metal targets are placed outside the drum along with a linear ion source capable of producing 50 - 80 eV oxygen ions. These three sources are isolated by regions of slighly higher vacuum using baffles and differential pumping such that the sputtering sources do not become poisoned with oxygen.

To produce the desired optical film, the oxygen ion source and first metal source are turned on while the drum is rotated past at 50 - 90 RPM. A few angstroms of metal are deposited and rotated in front of the ion source where the layer is oxidized. This continues until the desired thickness is achieved. Deposition alternates between sources until the final design is completed. In situ substrate cleaning is usually done before deposition by rotating the substrates in front of the ion source for five minutes or so prior to coating.

This process has the tradename MetaMode™ (metal mode reactive sputtering) and has been described in detail by Seddon (59). He claims the film properties are similar to other reactively sputtered oxides with density and index lower than bulk values but higher than those of conventionally evaporated films. He listed typical values (with evaporated film values in parenthesis):

Material	Index of Refraction (550nm)	
SiO_2*	1.49	(1.45)
Al_2O_3	1.66	(1.60)
Ta_2O_5	2.18	(2.13)
TiO_2	2.47	(2.34)

* Contains 6% Al_2O_3

These films are claimed to adhere well, be robust and show no sensitivity to moisture.

RF Sputtering. RF magnetron sputtering has been used to produce optical films in a wide range of materials. Coleman (60) reported depositing films of TiO_2, ZrO_2, CeO_2, SiO and SiO_2. No data was given on the packing density although the values of refractive index suggest it was approaching unity except for TiO_2 where the packing density appeared to be lower. Humidity stability and film adhesion were excellent. Slusark (61) reported equivalent results for films of SiO_2 and Al_2O_3. Misiano (62) explored co-sputtering of SiO_2 with TiO_2 or CeO_2. Workers at Hitachi (63) reported a spectral shift of 2 nm (0.4% at 550 nm) after baking RF sputtered films of TiO_2/SiO_2 for one hour in air at 500°C. Holm (64) reported deposition of a TiO_2/SiO_2 bandpass filter by RF sputtering that had a humidity sensitivity "below detectable levels".

Pawlewicz and Martin have done a great deal of work in the area of RF sputtered optical films, primarily for solar cells and high damage threshold laser coatings. They have successfully sputtered oxides, nitrides, carbides, hydrides and II-VI compounds. In TiO_2 films they were able to control the phase composition from 40% rutile/60% anatase to 100% rutile with grain size from glassy (<10 nm) to 60 nm diameter (65). Phase composition had negligible effects on the index of refraction and laser damage threshold. However, grain size had a very dramatic effect on both these properties. For 100% rutile films a grain size of 60 nm gave an index of 2.4 at 1 µm

---

MetaMode is a trademark of Optical Coating Laboratories, Inc.

wavelength whereas for glassy films the index was only 2.0. They suggested that this was one explanation for the discrepancies often found in the literature. Humidity stability of their sputtered films was high.

Martin (66) reported that Si_3N_4/SiO_2 edge filters deposited by reactive RF sputtering showed spectral shifts less than 2 nm (0.3% at 600 nm) after fifteen months in 85°C/85% relative humidity. These results were promising since conventional films would shift in the range of 15 nm (1 - 3%).

Although sputtered films were produced more than ten years before evaporated ones, sputtering had historically been a slower, less flexible process and had not been used much over the years for production of optical thin films. Some of these limitations may be removed with the development of high rate magnetron sputtering (67) and the MetaMode process.

Ion Beam Sputtering. Another successful sputtering technique is ion beam sputtering (Fig. 12). Here the source material is sputtered at low chamber pressure by a beam of ions, usually Ar^+, produced by a Kaufman type ion source (68). Cole (69) reported <1 nm shifts and high index of refraction for ZrO_2, Y_2O_3, and Al_2O_3 sputtered with 2 kV, 100 mA argon ions. A variation of this technique called dual ion beam sputtering uses one ion source at high power to sputter the material while bombarding the growing film with another lower power beam. This seems to be successful for very high reflectance, very low loss mirrors for ring laser gyros. Wei (70) reported TiO_2/SiO_2 multilayer mirrors with scatter and absorption two orders of magnitude lower than conventional films. They also had superior adhesion and abrasion resistance.

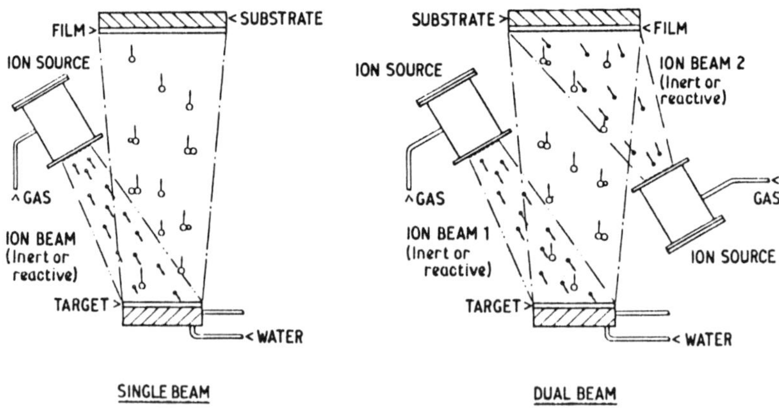

Figure 12. Schematic of ion-beam sputtering, single and dual beam techniques (90).

Allen (71) reported depositing films of calcium fluoride and magnesium fluoride by reactive dual ion beam sputtering that were stoichiometric and dense with packing densities up to 99%. Refractive index of magnesium fluoride was 1.453 ± 0.023 compared to the accepted values of 1.38 for thin film and 1.383 for bulk. Calcium fluoride index was 1.457 ± 0.011 compared to 1.25 for thin film and 1.433 for bulk. The higher film indices were attributed to the high packing density and perhaps a high density of bonding defects. Stresses were compressive compared to tensile for conventional films.

6.2 Evaporation Techniques

There is a large amount of equipment and experience in place based on conventional evaporation, whether thermal or electron beam gun. This has helped contribute to a number of techniques for improved films based on modifications of conventional evaporation.

Activated Reactive Evaporation (ARE). As mentioned previously, refractory oxides such as TiO_2 dissociate when heated and must be deposited in a background of oxygen to be fully oxidized on the substrate (Fig. 13a). Heitmann (72) partially ionized the oxygen (activated reactive evaporation) and produced films of TiO_2, SiO_2 and SiO_xN_y which were superior to conventional reactive evaporation (Fig. 13b). The index of the TiO_2 films was 2.2 to 2.3 at 550 nm which is lower than the bulk value indicating that the packing density was less than unity. However, he reported low absorption indicating that the film was stoichiometric TiO_2. Thus it appears his ion source made the oxygen more reactive but did not increase the adatom mobility enough to improve the packing density.

Kuster (73) and Ebert (74) improved on Heitmann's source by using the substrate tooling as one electrode. Most of the ions from Ebert's source have energy less than 100 eV. These ions improve the oxidation of the films, however there is a high energy tail which damages the film. Allen (75) has reported similar success with TiO_2 and SiO_2 using an Ebert source.

Ion Plating. These coatings are produced by evaporating the material through a glow discharge in which the substrate is biased negatively to attract the positive ions (Fig. 13c). An intense plasma with high ion current and relatively low ion energy surrounds the substrate. Improvements in adhesion and durability are due mainly to bombardment of the growing film by ions and energetic neutrals. This technique has been used very successfully to produce tough, durable coatings for tribological and other metallurgical applications (76) and has been recently adapted for optical thin films.

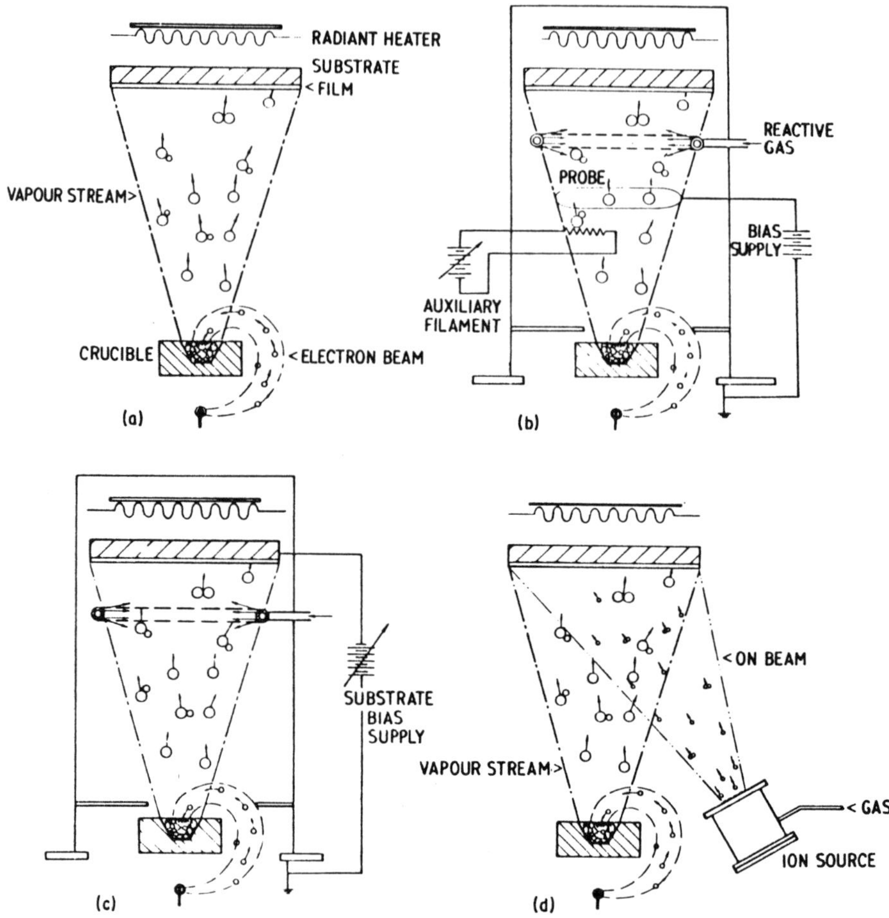

Figure 13. Ion-based methods of thin-film deposition: (a) vacuum evaporation, (b) activated reactive evaporation (ARE), (c) ion plating, (d) ion-assisted deposition (IAD) (90).

Seddon (77) reported depositing Al_2O_3, HfO_2, SiO_2, Ta_2O_5, TiO_2, and ZrO_2 by plasma plating. The films were dense and homogeneous with resulting higher index and lower scatter compared with conventional films. Several interesting applications were described including a Ta_2O_5/SiO_2 longwave pass filter which showed 0.06% shift in cutoff due to humidity compared with 1 - 3% for conventional films. Also mentioned was a four layer TiO_2/SiO_2 antireflection coating which withstood temperature cycling to 550°C compared with standard multilayer antireflection coatings which failed when cycled to 450°C.

Guenther (78)-(80) has presented many ion plating results using a commercially available system. Multilayers of TiO_2/SiO_2, ZrO_2/SiO_2, and Ta_2O_5/SiO_2 were smooth and dense with no voids up to 100,000x magnification. Surface roughness was low at approximately 0.2 nm RMS versus 6 nm RMS for conventional TiO_2 films 0.7 micrometers thick. In fact, surface roughness of a glass substrate was improved from 4.0 to 0.4 nm RMS after deposition of a twenty-four layer ZrO_2/SiO_2 multilayer stack. This is in contrast to conventional films which end up rougher than the starting substrate. This reduction in roughness helps account for a laser-induced damage threshold for single layers which is twice that for conventional reactive evaporation. However, multilayer films were not as good, perhaps due to reduction reactions between layers resulting in slight absorption at the interfaces. The films did not absorb water vapor. Guenther (81) also discussed extending the structure zone model for thin film growth to ion and plasma assisted processes which produce dense, vitreous films.

Ion-Assisted Deposition (IAD). This technique is very similar to activated reactive evaporation (ARE) where an electron beam gun is used as a standard deposition source while the growing film is bombarded with ions to improve film properties (Fig. 13d). The main difference is that for ARE the beam energy is generally much lower (in the range of 10 - 50 eV) compared with IAD where the beam energy ranges from the same level to ten times higher. A Kaufman source of the type used for ion-beam sputtering is often used for IAD.

Martin (82) deposited films of ZrO_2, TiO_2, and SiO_2 by IAD with the hope of increasing adatom mobility and therefore packing density with the goal of reducing sensitivity to moisture. They bombarded the films with 600 eV argon ions and also tried adding oxygen to the beam. They found the lowest absorption at a deposition rate of approximately 1 Å/s, ion current density of 16 $\mu A/cm^2$, and addition of some oxygen to the source gas. The result was the complete removal of moisture sensitivity in the films (Fig. 14). They also found the films had higher index closer to the bulk value and reduced inhomogeneity. Martin and Netterfield (83) have written an excellent review of the effects of IAD on optical thin films.

McNeil (84)(85) compared optical performance of TiO_2 and SiO_2 deposited with low versus high energy ion bombardment (30 to 500 eV). They found the highest transmittance was for the lowest ion energy and current density, especially for TiO_2. However, hydrogen (water) content was lowest for ion energy in the range of several hundred eV.

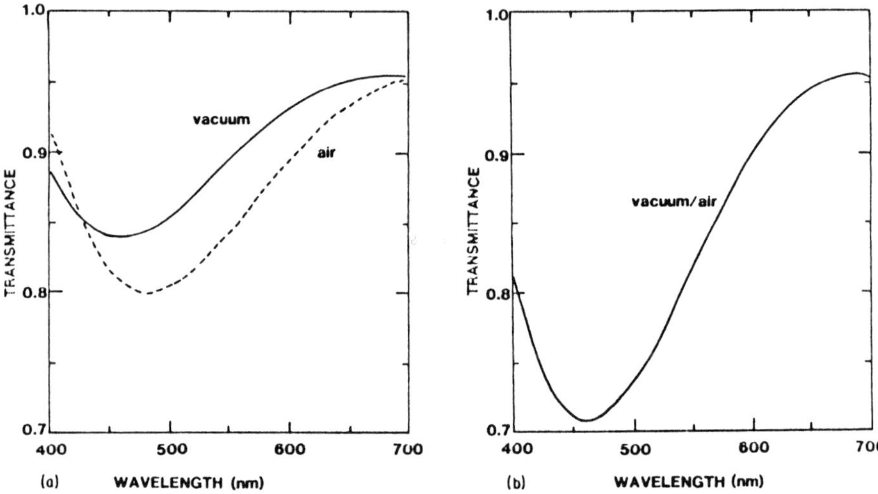

Figure 14. Transmittance over the visible spectrum of a ZrO_2 film deposited (a) without ion-assistance and (b) with oxygen-ion assistance. Solid line is vacuum data and broken line is in air (82).

Kennemore (86) used IAD to deposit hard, adherent films of MgF_2 at room temperature on plastic substrates. Flory (87) deposited dense, low absorption films of Ta_2O_5 and eleven-layer stacks of Ta_2O_5/SiO_2 which showed no spectral shift between vacuum and 40%RH air. The Ta_2O_5 was deposited with 250 eV Ar ions at 10 mA beam current although the conditions for SiO_2 were not listed.

Williams (88) reported a packing density of near 100% in Ta_2O_5 films deposited with 100 µA/cm², 30 eV oxygen ion bombardment. They also saw higher density, higher refractive index and lower absorption in films of Al_2O_3 and TiO_2. Substrate temperature had little effect on the properties of films prepared by IAD. Müller (89) has done computer modelling of IAD which concluded that densification is due mainly to momentum transfer, rather than localized heating as proposed by the "thermal spike" theory.

In all these ion assisted processes, the parameters that seem most important are the energy of the ions and the ratio of arriving ions to arriving evaporant atoms. The advantages are, of course, dense, high index films which do not change with relative humidity. Because packing density is high, the index tends to be more repeatable and run-to-run reproducibility is good. Higher film density also results in moderate to high compressive stress compared with conventional films which tend to have tensile stress. Deposition rate is comparable to the rate achieved by conventional electron beam-gun evaporation.

Since the substrates are bombarded with a high flux of ions, some substrates that are sensitive to radiation damage cannot be coated. Multilayer stacks may take hours to coat resulting in substrate temperatures up to 250°C eliminating some temperature sensitive substrates. Ion plating has the added advantage that, due to higher working gas pressure, the coating deposits uniformly on curved surfaces. However this results in the disadvantage that higher maintenance is required to keep unwanted coating material off high voltage insulators, etc.

A number of excellent references are available which describe these ion assisted methods and the materials properties of many dielectric, metal and semiconductor films (90)-(94). Many interesting results have been achieved by various methods, but a great deal still needs to be learned.

7.0 CONCLUSION

Optical thin films are critical to the performance of many optical systems in use today. The properties of present films are, in most cases, excellent although system requirements are demanding higher quality, primarily films that are fully densified. Several different techniques have been developed to deposit such films by increasing adatom mobility. The primary methods have been sputtering and variations of evaporation where energy has been added at the surface. These are beginning to become commercially available and will no doubt continue to do so more rapidly over the next few years.

REFERENCES

1. Macleod, H. A., in *Active and Passive Thin Film Devices* (T. J. Coutts, ed.), Academic Press, New York (1978)

2. Born, M., and Wolf, E., *Principles of Optics,* Pergamon Press, Oxford p. 51-70 (1970)

3. Jenkins, F. A. and White, H. E., *Fundamentals of Optics,* p. 231, McGraw-Hill, New York (1976)

4. Smith, Warren, *Modern Optical Engineering,* 2nd ed., pp. 57-86, McGraw-Hill Book Company, New York (1990)

5. Bausch & Lomb, Vacuum Coating Division, 1400 N. Goodman Street, Rochester, NY 14602

6. Macleod, H. A., *Thin-Film Optical Filters,* Macmillan Publishing Co., New York (1986)

7. Baumeister, P. W., in *Applied Optics and Optical Engineering,* Vol 1 (R. Kingslake, ed.) Academic Press, New York (1965)

8. Baumeister, P. W., in *Optical Design,* MIL-HDBK-141, Notice 1 (1968)

9. Cox, J. T. and Hass, B., in *Physics of Thin Films,* 2:239-304, Academic Press, (1964)

10. Liddell, H. M., *Computer-Aided Techniques for the Design of Multilayer Filters,* Adam Hilger Ltd, Bristol (1986)

11. Thelen, A., *Design of Optical Interference Coatings,* McGraw-Hill, New York (1989)

12. Baumeister, P. W., *Photonics Spectra,* p. 76 (July 1986)

13. Ritter, E., in *Physics of Thin Films,* 8:1049 Academic Press, New York (1975)

14. Balzers, 8 Sagamore Park Rd., Hudson, NH 03051

15. Cerac, Inc., Box 1178, Milwaukee, WI 53201

16. E. M. Industries, 5 Skyline Drive, Hawthorne, NY 10532

17. Heavens, O. S., in *Physics of Thin Films,* (G. Hass and R. Thun, eds.) p. 193-238, Academic Press, New York, (1964)

18. Ennos, A. E., *Applied Optics,* 5(1):51 (1966)

19. Hoffman R. W., in *Physics of Nonmetallic Thin Films,* (C. H. S. Dupuy and A. Chachard, eds.) Plenum Press, New York (1974)

20. Pulker, H. K. and Maser, J., *Thin Solid Films,* 59:65-76 (1979)

21. D'Heurle, R. M. and Harper, J. M. E., *Thin Solid Films,* 171:81-92 (1989)

22. Decker, D. L., *SPIE,* 1323:244-252 (1990)

23. Baker, S. P. and Nix, W. D., in *SPIE,* 1323:263-276 (1990)

24. Clover, J., *SPIE* 181:145-151 (1979)

25. Rancourt, J. D., *Optical Thin Films Users' Handbook,* Appendix B, McGraw-Hill, New York (1987)

26. Maissel, L. I. and Glang R., *Handbook of Thin Film Technology,* McGraw-Hill, New York (1970)

27. *Deposition Technologies for Films and Coatings,* (R. F. Bunshah, ed.), Noyes Publications, Park Ridge, NJ (1982)

28. Berry, R. W., Hall, P. M., and Harris, M. T., *Thin Film Technology,* Robert E. Krieger Publishing Company, Huntington, New York (1979)

29. Chopra, K. L., *Thin Film Phenomena,* Robert E. Krieger Publishing Company, Huntington, New York (1979)

30. Pulker, H. K., *Coatings on Glass,* Elsevier, Oxford (1984)

31. Vossen, J. L. and Kern, W., *Thin Film Processes,* Academic Press, New York (1978)

32. Vossen, J. L. and Kern, W., *Thin Film Processes II,* Academic Press, New York (1991)

33. Strickland, W. P., *SPIE,* 1323:2-7 (1990)

34. Jacobson, M. R., *SPIE Milestone Series 6* (1989)

35. O'Hanlon, J. F., *A User's Guide to Vacuum Technology,* p.189-205, John Wiley & Sons, New York, (1980)

36. Hoffman, D. M., *J. Vac. Sci. Technol.*, 16(1):71-75 (1979)

37. Holland, L., *Vacuum Deposition of Thin Films,* Chapman and Hall, London (1956)

38. *Physical Vapor Deposition,* (R. J. Hill, ed.), Airco Temescal, Berkeley, CA (1976)

39. Macleod, H. A., in *Applied Optics and Optical Engineering,* (R. R. Shannon and J. C. Wyant, eds.), 10:5-7, Academic Press, San Diego (1987)

40. Ritter, E., *Applied Optics* 20:21 (1981)

41. Auwarter, M., U.S. Patent #2,920,002, (Jan 5, 1960)

42. Dobrowalski, J. A., *SPIE,* 140:102 (1978)

43. Pearson, J. M., *Thin Solid Films,* 6:349-358 (1970)

44. Guenther, K. H., and Pulker, H. K., *Applied Optics,* 15:2992-7 (1976)

45. Seeley, J. S., Hunneman, R. and Whatley, A, in: *Symposium on Contemporary Infrared Sensors and Instruments*, 246:83, Proceedings of the SPIE (1980)

46. Macleod, H. A., and Richmond, D., *Thin Solid Films,* 37:163-167 (1976)

47. Movchan, B. A., and Demchishin, A. V., *Fiz. Met. Metalloved.,* 28:653-660 (1969)

48. Thornton, J. A., *J. Vac. Sci. Technol.,* 11:666-670 (1974)

49. Messier, A. P., Giri, R. A., and Roy, R. A., *J. Vac. Sci. Technol. A,* 2:500 (1984)

50. Henderson, D., Brodsky, M. H., and Chandhari, P., *Appl. Phys. Lett.,* 25:641 (1974)

51. Dirks, A. G. and Leamy, H. J., *Thin Solid Films,* 47:219-233 (1977)

52. Leamy, H. J., Gilmer, G. H., and Dirks, A. G., *Current Topics in Material Science,* (E. Kaldris, ed.), 6:317, North-Holland Publishing Co. (1980)

53. Hodgkinson, I., presented at *Optical Interference Coating Topical Meeting,* Tucson, AZ (April 12-15, 1988)

54. Motovilov, O. A., Lavrishchev, A. P., and Smirnov, A. N., *Sov. J. Opt. Tech.,* 41(5):2 78 (1974)

55. Schiller, S., Heisig, U. and Gopedicke, K., *Thin Solid Films,* 64:455-467 (1979)

56. Schiller, S., Heisig, U., Steinfelder, K. and Strumpfel, J., *Thin Solid Films* 63:369-375 (1979)

57. Hartsough, L., U.S. Patent 4,420,385 (December 13, 1983)

58. Scobey, M. A., Seddon, R. I., Seeser, J. W., Austin, R. R., LeFebvre, P. M. and Manley, B. W., U.S. Patent 4,851,095 (July 25, 1989)

59. Seddon, R. I. and LeFebvre P. M., *SPIE,* 1323:122-126 (1990).

60. Coleman, W. J., *Appl. Opt.,* 13:946 (1974)

61. Slusark, W., Lalevic, B., and Taylor, G., *Thin Solid Films,* 39:155 (1976)

62. Misiano, C., and Simonetti, E., *Vacuum,* 27:403 (1977)

63. Shimomoto, Y., Imamura, Y., Sasano, A., and Maruyama, E., *Surface Science,* 86:417-423 (1979)

64. Holm, N. E. and Christensen, O., *Thin Solid Films,* 85:71-75 (1981)

65. Pawlewicz, W. T., Martin, P. M., Hays, D. D. and Mann, I. B., *SPIE,* 325:105-116 (1982)

66. Martin, P. M., Pawlewicz, W. T., Coult, D. G., and Jones, J. S., *Appl. Opt.,* 23(9):1307 (1984)

67. Van Vouros, T., *Optical Spectra,* 11(11):30 (1977)

68. Kaufman, H. R., *J. Vac. Sci. Tech.,* 15(2):272 (1978)

69. Cole, B. E., Moravic, T. J., Ahonen, R. G., Smith, D. L., and Ehlert, L. B., *Technical Digest,* Opt. Soc. Am. (1984)

70. Wei, D. T., *1988 Technical Digest Series,* 6:294, Opt. Soc. Am., (1988)

71. Allen, T. H., Lehan, J. P., and McIntyre, L. C., *SPIE,* 1232:227-290 (1990)

72. Heitmann W., *Appl. Opt.,* 10:2414-2418 (1971)

73. Kuster, H., and Ebert, J., Nat. Bur. Standards, Special Publ. 568, p. 269 (1980)

74. Ebert, J., in *Optical Thin Films* (R. I. Seddon, ed.), SPIE, 325:29 (1982)

75. Allen, T. H., in *Optical Thin Films* (R. I. Seddon, ed.), SPIE, 325:93 (1982)

76. Mattox, D. M., *Electrochemical Technology,* 2:295-298 (1964)

77. Seddon, R. I., Temple, M. D., Klinger, R. E., Hart, T. T., and LeFebvre, P. M., *1988 Technical Digest Series,* 6:255, Opt. Soc. Am. (1988)

78. Guenther, K. H., *1988 Technical Digest Series,* 6:247, Opt. Soc. Am. (1988)

79. Guenther, K. H., *SPIE,* 1270 (1990)

80. Guenther, K. H., *SPIE,* 1323:29-38 (1990)

81. Guenther, K. H., *SPIE,* 1324:2-12 (1990)

82. Martin, P. J., Macleod, H. A., Netterfield, R. P., Pacey, C. G., and Sainty, W. G., *Applied Optics,* 22:178-184 (1983)

83. Martin, P. J. and Netterfield, R. P., in: *Handbook of Ion Beam Processing Technology,* (J. J. Cuomo, S. M. Rossnagel, and H. R. Kaufman, eds), Noyes Publications, Park Ridge, NJ (1989)

84. McNeil, J. R., Barron, A. C., Wilson, S. R., and Herrmann, W. C., *Applied Optics,* 23:552- 559 (1984)

85. McNeil, J. R., Al-Jumaily, G. A., Jungling, K. C., and Barron, A. C., *Applied Optics,* 24:486-489 (1985)

86. Kennemore, C. M. and Gibson, U. J., *Applied Optics,* 23:3608-3611 (1984)

87. Flory, F., Albrand, G., Montelymard, C., and Pelletier, E., *SPIE,* 652:248-253 (1986)

88. Williams, F. L., Boyer, L. L., McNeil, J. R., and McNally, J. J., *1988 Technical Digest Series,* 6:290, Opt. Soc. Am. (1988)

89. Müller, K. H., *SPIE,* 821:36-44 (1987)

90. Martin, P. J., *J. Mat. Sci.,* 21:1 (1986)

91. Macleod, H. A., *SPIE,* 652:222-234 (1986)

92. Gibson, U. J., in *Physics of Thin Films,* 13:109-150, Academic Press, NY (1987)

93. Rossnagel, S. M., and Cuomo, J. J., *Vacuum,* 38(2):73-81 (1988)

94. Cuomo, J. J., Rossnagel, S. M. and Kaufman, H. R., *Handbook of Ion Beam Processing Technology,* Noyes Publications, Park Ridge, NJ (1989)

7

Sol-Gel Derived Ceramic Coatings

Brian D. Fabes, Brian J. J. Zelinski and Donald R. Uhlmann

1.0 INTRODUCTION

According to a recent report by Kline & Co. (1), wet-chemically-derived coatings are the fastest growing segment of the global ceramics market. Within the world market for advanced ceramic coatings (estimated to grow from approximately $1.4 billion in 1989 to $4.3 billion in the year 2000) sol-gel coatings are expected to grow at an average annual rate of about 15%. Part of the driving force behind the rapid growth of sol-gel coatings is the ability to make coatings with unique properties. Semiconductor, metal, or organic doped oxide coatings, coatings with tailored porosity, and coatings which infiltrate and heal flaws are some examples. Many such coatings are difficult or impossible to make using more traditional coating techniques.

The most important driving force behind the growth of sol-gel coatings is cost. Sol-gel coatings can be prepared using simple, non-vacuum processes, which require only a small investment in capital, hence coatings which might be made by alternate techniques are made more cheaply using sol-gel processes. Antireflection, optical waveguiding, transparent conducting, electro-optical, piezoelectric, anti-corrosion, diffusion and oxidation barrier, abrasion resistant, electrochromic, and superconducting coatings are some examples.

This chapter reviews the sol-gel process and its application to making ceramic coatings. Unique features of sol-gel coatings are pointed out and some applications of sol-gel coatings are reviewed, including some which are commercial products and some which are still in the laboratory stage.

Finally, we summarize some of the outstanding issues which, in our opinions, have yet to be addressed and will be key to expanding the applications and understanding of sol-gel processing of ceramic coatings.

2.0 SOL-GEL PROCESSING

Sol-gel, also known as *wet-chemical*, processing of ceramics can refer to a multitude of reaction processes which employ a wide variety of chemical precursors to prepare many different products. Most processes can be categorized in one of three general approaches, which are drawn schematically in Fig. 1. In the first approach, a colloidal sol is prepared, and powders are precipitated from the sol (usually by changing the pH). The resulting powders are then dried and processed using traditional ceramic processing techniques. In the second approach, the particles in the colloidal sol are linked (instead of precipitated, as in the first approach) to form a gel, which is subsequently dried, to form a porous ceramic, and fired (if desired) to crystallize and/or densify the material. In the third approach, a gel is formed by polymerizing individual oligomeric units (instead of colloidal particles).

In both the second and third approaches, solutions can be cast, spun, dipped, or sprayed onto substrates prior to gelation, thus providing an extremely convenient method of forming coatings. Although colloidal particles can sometimes be used to form coatings, the vast majority of applications of sol-gel coatings comes from applying solutions to substrates prior to gelation, hence this chapter deals mainly with coatings which are formed using the second and third (i.e., gelation) methods.

The sol-gel process, as described above, is quite general, and can be applied to a wide range of precursors. The basic idea is to use the techniques of solution chemistry to build ceramic structures from the molecular level. This stands in contrast to traditional solid state ceramic processing techniques, which build structures from the micrometer, or tens of micrometers, scale.

2.1 Coating Chemistry

Oxides. Metal alkoxides (such as tetraethoxysilane, trimethylborate, and titanium ethoxide) and metal salts (such as barium carbonate, magnesium nitrate, and copper chloride) are the most commonly used starting materials for sol-gel processing of oxides. In large volume applications, considerations of cost usually make salts more attractive than alkoxides. Removal of the counter ions can sometimes be troublesome, however, so in applications sensitive to contamination, the alkoxides are often more attractive than the salts.

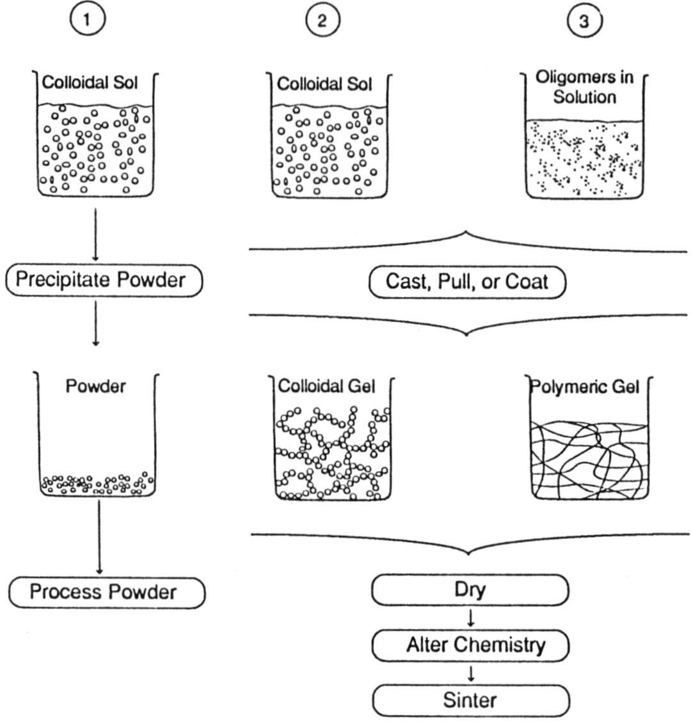

Figure 1. Schematic of three general sol-gel routes. After Ref. 2.

When alkoxides are used as the starting materials, the formation of ceramic coatings is based on hydrolysis and condensation, as shown in Eqs. 1 - 3. In the first reaction (Eq. 1), an alkoxide and water are placed in a mutual solvent and a suitable catalyst is added. Hydrolysis of the metal alkoxide bond (M-OR) results in the formation of a metal hydroxyl bond (M-OH), as shown in this reaction:

Eq. (1) $$M(OR)_x + H_2O \rightleftharpoons (RO)_{x-1}M\text{-}OH + ROH$$

In the next step, condensation between the hydroxyl and an alkoxide ligand,

Eq. (2) $(RO)_{x-1}M\text{-}OH + (RO)_xM \rightleftharpoons (RO)_{x-1}M\text{-}O\text{-}M(RO)_{x-1} + ROH$

or between two hydroxyl ligands,

Eq. (3) $(RO)_{x-1}M\text{-}OH + HO\text{-}M(OR)_{x-1} \rightleftharpoons (RO)_{x-1}M\text{-}O\text{-}M(RO)_{x-1} + HOH$

results is the formation of a metal-oxygen-metal bridge, which constitutes the backbone of any oxide ceramic structure. Continued condensation leads to an increase in the density of metal-oxygen-metal crosslinks until eventually gelation or precipitation occurs.

In solution, the rates, extents, and even the mechanisms of the reactions shown in Eqs. 1 - 3 are profoundly affected by the electronegativity of the metal, size of the alkyl ligand on the metal, solution pH, type and concentration of solvent, concentration of water, temperature, and pressure. Each, in turn, can affect the course of structure (hence property) development in the gels and ceramic materials made therefrom. In one sense, this abundance of significant variables makes sol-gel processing a nightmare to control. On the other hand, the multitude of variables provides the great flexibility in material structure and properties which can be achieved with sol-gel processing.

The formation of multicomponent solutions adds further complexity. The goal often is to achieve homogeneous mixing of the components on the atomic scale; but different alkoxides do not, in general, have the same reactivities, hence mixing of different alkoxides with water usually results in preferred hydrolysis and condensation of one species over the others; the result is chemically inhomogeneous gels. Several techniques have been used successfully to increase the homogeneity of multicomponent gels (3). These include: *(i)* matching the reaction rates of the alkoxides by increasing the size of the alkyl group on the faster reacting species and decreasing the size of the alkyl group on the slower reacting species; *(ii)* partially hydrolyzing the slower reacting species prior to mixing with the faster reacting alkoxide; *(iii)* adding water very slowly, so that the rate-limiting step is water addition and not alkoxide reactivity; and *(iv)* formation of so-called double alkoxides (4). These alkoxides, often actually coordination compounds, contain two cations in the alkoxide molecule, so mixing is assured on the atomic scale. It is important to note that the successful implementation of any of these techniques requires knowledge or at least some enlightened intuition of the mechanisms and rates of the underlying chemical reactions.

When metals salts are used as precursors, the details of the chemistry are different from those described above. Much of this chemistry has been developed and reviewed by Schroeder (5)(6), and will not be described here. The goal of metal salt use is still to form metal-oxygen-metal bridges in solution, so much of what follows applies to both alkoxide and salt-derived coatings.

Non-Oxides. Sol-gel approaches can also be used to synthesize non-oxide materials. Sulfides, for example, can be made by reacting organometallic precursors with H_2S instead of H_2O to generate a sulfide based network. Such reactions have been carried out by Johnson et al. (7), who reacted diethylzinc, triethyaluminum, and diethymagnesium with H_2S to form ZnS, $ZnAl_2S_4$, and MgS powders. Using a similar approach, Guiton et al. (8) reacted $[EtZn(SBu^t)]_5$ with H_2S to form ZnS whiskers.

Recent work on sulfides has focused on the generation of semi-conductor doped oxide glasses. Bagnall and Zarzycki (9) introduced aqueous salts of Cd and Se into wet oxide gels and exposed the gels to a variety of conditions which promoted reaction of the metal with thioacetamide. These treatments resulted in the formation of chalcogenide precipitates in an oxide matrix. Nogami et al. (10) exposed a dried porous gel containing ZnO, CdO, and PbO to H_2S at 500°C to produce the metal sulfide precipitates. Finally, Tohge et al. (11) introduced thiourea directly into the sol-gel solution containing cadmium nitrate. The thiourea and Cd salt reacted at 350°C to form the chalcogenide semiconductor cluster.

In most of these systems, little work has been done to extend the approaches used for synthesizing powders, whiskers and precipitate-doped monoliths to the generation of ceramic films. In principle this extension would be straightforward because smaller diffusion distances are required for the synthesis of thin films.

Nitrides, carbides and borides can be made by wet-chemical routes which involve the controlled decomposition of preceramic precursors. Since these techniques are based on decomposition reactions rather than direct polymerization of the ceramic network, they are not discussed here. The interested reader is referred to Seyferth et al. (12), who have discused these techniques in detail.

Finally, Fabes et al. (13), Pantano et al. (14) and Kamiya (15) have made oxynitride coatings by reacting undensified sol-gel films with NH_3 at elevated temperatures. The high porosity of the sol-gel films allows NH_3 to penetrate much deeper than it does into dense coatings, such as thermal oxide films on silicon. Since nitridation occurs at high temperature, a competition is established between the rate of ammonolysis and the rate of densification, and the extent of nitridation turns out to be sensitive to the details of the prefiring treatment and firing schedules of the coatings. As shown in Fig. 2 (13), with proper pretreatment and firing schedules, it is possible to nitride silica coatings fully to form oxynitride coatings thicker than 1000 Å.

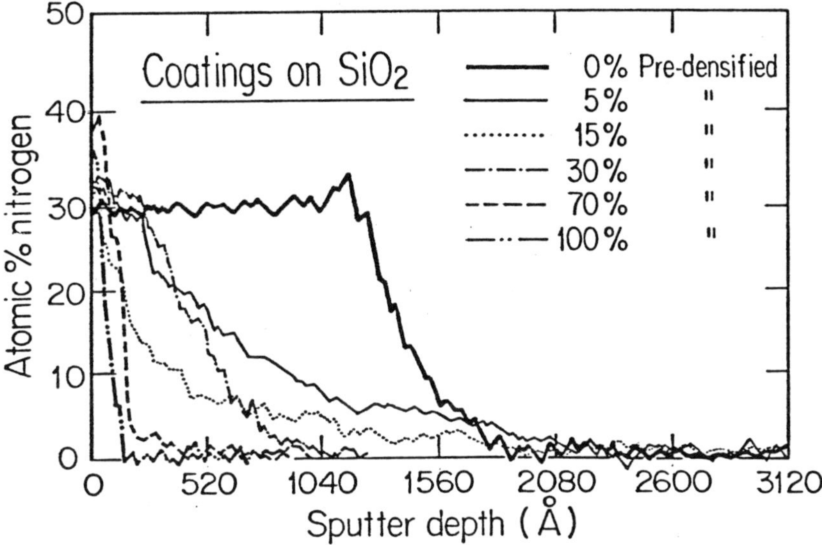

Figure 2. Effect of coating pretreatment on incorporation of nitrogen in thermally nitrided TEOS coatings. Predensification is a result of various heat treatments which the coatings underwent before high temperature ammonolysis. From Ref. 13.

Infiltration of porous gels with reactive fluids (gases or liquids) seems to present a wealth of possibilities for forming coatings with unique chemistries and properties. This potential, however, remains largely unexplored.

2.2 Drying and Firing

Once the solution has gelled, a sol-gel system is by no means stagnant. The oxide and solvent phases are intimately mixed, and the high solid/liquid surface area (and associated surface energy) drives the network to shrink, a process known as *syneresis.* Condensation reactions between unreacted hydroxy groups also continue to take place, resulting in further shrinkage.

If the system is open, the solvent will evaporate. For coatings, the ratio of surface area to volume is so large that evaporation takes place extremely rapidly—sometimes instantaneously. At some point during evaporation, the outside surface of the gel dries out, and a meniscus forms within. The meniscus produces a capillary tension, P, given by

Eq. (4) $$P = 2\Delta\gamma/r$$

where $\Delta\gamma$ is the difference between the solid/vapor and solid/liquid surface energies and r is the radius of the pore. Since the pores in the gels are typically very small (typically a few nanometers), the resulting tensile stresses can be very large. For a typical gel, with $r = 3$ nm and $\Delta\gamma = 150$ ergs/cm^2, the capillary stress turns out to be 100 MPa, which is often high enough to fracture the gel.

A detailed model of the shrinkage caused by the capillary stresses during drying has been developed by Scherer (16)-(20), who assumed that contraction is driven by the large differences in surface energies in the porous material, while the rate of contraction is limited by the viscosity of the solid phase and by the rate of transport of the liquid through the pores. Using this model, the capillary stress is more complicated than represented by Eq. 1. However Eq. 1 describes the maximum stress which develops in a drying material, and as such, is a guide to the stresses which must be tolerated by the drying gel.

For bulk gels, the shrinkage which occurs during drying often leads to cracking. For coatings less than about 0.5 µm thick, on the other hand, cracking is seldom observed, even though the capillary stresses of $2\Delta\gamma/r$ are still present. Since strong adhesion between the coating and substrate prevents relaxation in the plane of the coating, the ability to make thin crack-free coatings is related to the difficulty in relieving strain energy when a crack propagates through a well-adhered coating. Several researchers are currently exploring these issues (see Refs. 21 and 22).

Once dried, coatings are seldom used in the porous state. Often they are fired *(i)* to burn out the residual carbon; *(ii)* to alter the chemistry of the coating (for example, to form Si_2N_2O coatings by reacting SiO_2 with NH_3); *(iii)* to densify the coating; *(iv)* to crystallize the coating; or *(v)* to enact any combination of these. The conversion from the dried gel to the final, desired state is often tricky, since competition between the various processes may require sequential completion of one process before another. For example, if all of the residual carbonaceous species are not removed before the pores close, the coating may bloat with further firing. Similarly, it is important to infiltrate the coating fully with ammonia before pores close when forming oxynitride coatings. On the other end of the process, if a dense, crystalline coating is desired, it is important to densify the coating before crystallization takes place, since the kinetics of densification of crystalline materials are typically orders of magnitude slower than those of the corresponding amorphous materials.

3.0 COATINGS VIA SOL-GEL PROCESSING

3.1 Special Solution Requirements

To be used for coating, solutions of partially polymerized alkoxides are usually applied to substrates after hydrolysis. In some cases, unhydrolyzed alkoxides can be applied to the substrates, and water from the atmosphere is used for hydrolysis. When prehydrolyzed solutions are used, it is important to recognize that the solutions are live, in that hydrolysis and condensation reactions continue to take place while the solutions age; solution viscosity, hydroxyl content, molecular weight, oligomer morphology, and other properties are evolving continuously. Thus there often (but not always) exists a particular window of aging times for which coatability is optimal. On the long time end, this window is usually bounded by the gelation time, at which the viscosity increases dramatically. On the short time end, the boundary is less well defined. In general, lightly crosslinked, linear polymer species are preferred for producing quality coatings (23)(24), so some degree of hydrolysis and condensation are prerequisites to producing quality coatings. Unfortunately, quantitative studies of the effects of oligomer morphology on coatability are scant. Moreover, as discussed above, the development of polymer morphology is extremely sensitive to the metal cation type and concentration, solvent, pH, and water content.

Depending on the application, coating solutions need to be stable (i.e. need to have a coatability window) between a few minutes, for exploratory use, up to a few weeks or months, for most commercial uses. Often it is possible to increase the solution stability by storing in an inert atmosphere, so that further hydrolysis is minimized. Replacing some of the alkoxy groups with non-hydrolizable ligands can also increase solution stability. This has been demonstrated by Melpolder and Coltrain (25) for TEOS solutions with various amounts of phenyltriethoxysilane additions (Fig. 3); increasing the concentration of the unhydrolyzable phenyl ligands to only 6 mol% of the total ligand content increases the stability time from about 100 to nearly 1000 hours.

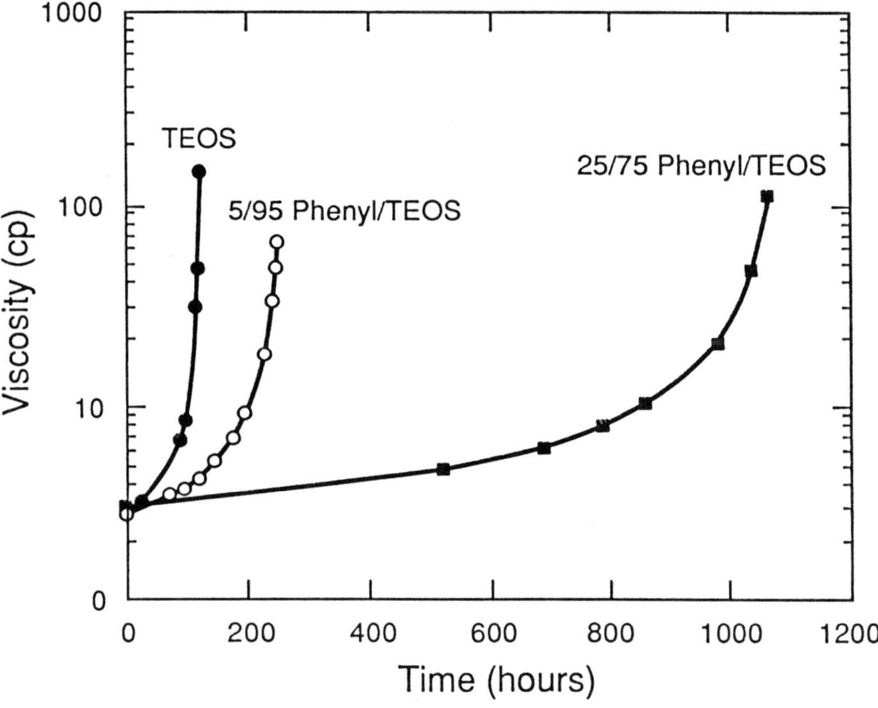

Figure 3. Effect of phenyl-modified TEOS on the stability of sol-gel solutions at 50°C. From Ref. 25.

3.2 Coating Techniques

Once a suitably stable solution has been synthesized, sol-gel coatings can be applied by several techniques such as dip, spin, spray, or roller coating. In dip coating, the substrate is immersed into the coating solution, and a film is made either by removing the substrate from the solution or by draining the solution. In spin coating, an excess of coating material is placed on a stationary or slowly spinning substrate. The substrate is then rapidly accelerated to a few thousand RPM, producing a large centrifugal force which throws off all but a thin layer of material. In spray coating, the solution is forced through a nozzle by high pressure, forming an aerosol, which is accelerated toward the substrate with an inert carrier gas. The drops of solution then coalesce on the substrate to form the coating. Finally, in roller coating, the coating solution is forced onto the substrate with a roller, which may or may not contact the substrate surface and which sometimes contains grooves to carry a reservoir of coating material. Many other coating techniques—such as premetered coating (26), meniscus coating (27), flow coating (27), jet stripping (28), and physical vapor deposition—can be used with sol-gel solutions. These, however, are all essentially variations on the four basic techniques described above. It is this variety of simple, non-vacuum techniques, applicable to a wide variety of substrate geometries, which constitutes one of the important advantages of sol-gel coatings.

Dip and spin coating are by far the most widely used techniques. Because of its simplicity, dip coating has been used commercially in a variety of applications. The physics of film formation by dip coating has been studied by a number of researchers (26)(29)-(34), and equations relating the thickness of the final coating to solution and process parameters have been derived.

Deryagin (29)(30) examined the problem of predicting the thickness of material which remains on a substrate after the substrate has been dipped into a coating solution. By balancing the downward force of gravity with the viscous drag pulling upward, Deryagin derived an expression for the coating thickness, h:

Eq. (5) $$h = [\eta U/\rho g \sin(\alpha)]^{1/2}$$

where h and ρ are the viscosity and density of the solution, U is the speed of withdrawal of the substrate from the solution, g is the acceleration of gravity, and α is the acute angle between the horizontal fluid plane and the substrate. More complicated models (32)-(34) take into account the effects of solvent concentration, surface tension, solution curvature, and other effects, but their application in predicting, a priori, the thickness of sol-gel coatings is currently of limited value. First, the constants in these equations are generally known to no better than about ± 20%. Second, the solution parameters are constantly changing in the case of sol-gel coatings. Third, the viscosity of a given sol-gel solution tends to change from day to day as the solution ages. Moreover, the solution viscosity depends exponentially on temperature, so small swings in temperature can have a large effect on the final coating thickness.

Thus, it is almost always necessary to calibrate the thickness produced using a given coating solution for a given set of coating conditions on a given day. Once a calibration is made, the desired coating thickness can be obtained by changing the coating parameters (e.g., withdrawal rate) as predicted by Eq. 4. This process can be used to produce films with highly reproducible thicknesses, easily within ± 5%.

Spin coating is a much newer technique than dip coating. Nonetheless, because it has become the technique of choice for deposition of photoresist in the electronics industry, the physics of spin coating are well established.

The spin coating process has some notable characteristics. First, it is inherently a batch process, but since it can be very rapid, this is not necessarily a limitation (26). Second, only flat objects, with an axis of symmetry perpendicular to the plane, can be spin coated conveniently. In addition, only one side of a substrate can be coated at a time. (Depending on the application, this can be an advantage or a disadvantage.) Third, spin coating produces coatings which are exceptionally even; for example, for an 860 nm coating on a 35 cm diameter substrate, Meyerhofer found less than 0.6% variation in thickness over the entire sample (35). This is often the overriding advantage which dictates the use of spin coating, especially in applications where evenness is important.

Emslie et al. (36) were the first to derive an expression for the change in thickness with time for a film on a rotating substrate. By balancing the outward (centrifugal) and inward (viscous drag) forces and applying a conservation of mass condition, they derived an equation for the coating thickness, h, as a function of time, t (assuming no evaporation),

Eq. (6) $$h = \frac{h_o}{[1 + (4\rho\omega^2 h_o^2 t/3\eta)]^{1/2}}$$

where h_o is the initial coating thickness before spinning, ω is the spinning speed, and η and ρ are the viscosity and density, respectively, of the coating solution. Equation 6 represents the key to the high uniformity of coatings produced by spinning; the inward and outward forces balance each other at each point across the substrate in a way that there is no radial dependence in the thickness. Even for initial conditions where h_o is not constant, the solutions to the differential equation relating the opposing forces tend towards Eq. 6, i.e., towards an even coating (26).

In spin coating sol-gel (and most polymer) solutions, the thickness does not approach zero as the spin time is increased past a few seconds. Instead, the viscosity tends to rise as evaporation concentrates the solution, until the centrifugal force is overwhelmed, and thinning is accomplished only by evaporation. It is this transition from spin-off to evaporative thinning that controls the final thickness obtained.

Meyerhofer (35) assumed a power law dependence of viscosity with concentration and derived an expression for the final thickness,

Eq. (7) $$h_f = c_o \left[\frac{3\nu_o e}{2(1 - c_o)\,\omega^2} \right]^{1/3}$$

where c_o is the initial polymer concentration, ν_o is the initial kinematic viscosity ($\nu = \eta/\rho$), and e is the evaporation rate.

Eq. 7 has been tested extensively for polymer solutions by measuring the thickness as a function of spin speed. This has also been done for sol-gel solutions by Melpolder and Coltrain (25) and Pantano et al. (37). In most cases, it is found that the coating thickness varies with the one-half power of the spin speed, rather than with the two-thirds power, as predicted by Eq. 7. Since the rate of air flow onto a spinning disk is proportional to $\omega^{1/2}$, Meyerhofer proposed that the evaporation rate (e) is proportional to $\omega^{1/2}$ (35), which would result in an apparent square root dependence of h on ω.

If modified to take into account the spin-rate dependence of e, Eq. 7 can be used to predict the thickness of a sol-gel coating applied by spinning. For precise work, however, small changes in convection around the sample or variations in temperature may produce unacceptable changes in the coating thickness. Thus, as with dip coating, it is often preferable to calibrate the thickness of a given coating solution and then use Eq. 7 to predict the effect of changes in the coating parameters on the final thickness.

Equation 7 is valid only for a coating solution in which the viscosity is independent of shear rate. This is a reasonable assumption for dilute, low molecular weight solutions and sols. For solutions approaching the gel point, however, it is likely that the viscosity will be affected by shear rate. For such solutions, the viscosity of the outer regions will decrease during spinning, producing a gradient in coating thickness; the coating should therefore be thicker near the middle than at the outer region of the sample. This effect has been mentioned by others (38), but, to our knowledge, has not been observed experimentally for sol-gel coatings.

The importance of evaporation rate and coating atmosphere must be emphasized. To produce even coatings, the evaporation rate must be the same over the entire area of the coating. This means that the spinning sample must be isolated from external air currents, so that the natural, even convection is not disturbed. Seemingly minor variations, such as a change in the thermal diffusivity between the substrate and spinning chuck, must be held constant across the entire sample. Thickness variations greater than 50 %, caused by loss of thermal contact between the sample and spinning chuck in the regions above small (1mm) vacuum grooves, have been observed in 1000 Å SiO_2-TiO_2 coatings (39). To prepare reproducible coatings, the evaporation rate must be constant from one run to the next.

While controlling the coating atmosphere to the degree necessary for producing even coatings can be bothersome, such control should allow one to program the drying schedule and control the course of evaporation during coating. This provides possibilities for altering the microstructural development in sol-gel coatings. To our knowledge, however, such programmed drying of sol-gel coatings has not been explored.

3.3 Unique Advantages of Sol-Gel Coatings

The most important advantage of sol-gel processing, with respect to most other processes for forming ceramic coatings, is cost. As with more costly techniques such as CVD and plasma coating, a wide variety of films can be made on a plethora of substrates. Non-equilibrium compositions, such as high titania content silica-titania coatings can also be made. Sol-gel processing allows this to be done using simple, non-vacuum methods, which are generally less expensive than methods which require a vacuum.

In addition to being generally a less expensive alternative to vacuum processes, sol-gel processes provide the possibility of producing unique structures, hence properties, in materials. Most notable is the ability to dope the solutions, prior to gelation, with almost any other material. Metal, semiconductor, and non-linear active dopants have been added to sol-gel solutions to make two-phase materials with unique properties. It is also possible to include non-hydrolyzable organic moieties, resulting in polymer/ceramic composites, known as *ormocers, polycerams, or ormosils.*

Unlike most ceramic coating techniques, sol-gel coatings are applied as low viscosity solutions, hence they can be used to smooth over rough surfaces, as in the planarization coatings used in IC chip fabrication. Being applied as low viscosity solutions also allows the solutions to penetrate and partially heal cracks in ceramics. This has been shown to increase the tensile strength of glass when rods or plates are coated with sol-gel solutions (40)-(43).

Since sol-gel coatings are formed first as fragile, porous networks, which strengthen as they densify, forming complicated patterns by embossing (stamping) them with a master die is possible. It is also possible to densify only selected regions in a large coating using confined radiation (such as from a laser). Both of these applications are discussed in greater detail in Sec. 4.2.

Finally, the porosity itself can be useful. Porous sol-gel coatings can be infiltrated with fluids much more extensively than the more dense coatings which result from traditional coating processes. Moreover, unlike vapor deposited coatings, the characteristics (geometry, interconnectivity, etc.) of the porosity of sol-gel coatings can be tailored by altering the processing conditions. Yoldas and Partlow (44), for example, produced sol-gel coatings with graded porosity to make broad band antireflection coatings. This is, however, is only a simple example of the great potential of tailored porosity in fabricating coatings with new and unique properties.

4.0 APPLICATIONS

Cracking and the high cost of precursor solutions have limited the use of sol-gel processing in many applications. Cracking is seldom a problem for thin films, however, and the volume of solution needed for a given coating is small. Thus, coatings represent the most natural application of the sol-gel process, and wet-chemically-derived coatings are used, or have been proposed for use, in a wide range of applications. These include coatings in electronics and optics, passivation coatings, catalytic coatings, coatings with controlled porosity and pore chemistry, and coatings to increase the strength of glasses and ceramics.

In this chapter we do not survey the progress which has been made in each of these areas. Rather, we direct attention to electrical and optical applications, as these are areas which provide significant opportunities for new coating technologies. Since considerable effort has been directed to tailoring wet-chemical methods to prepare electrical and optical coatings, examples from these areas also illustrate the potential of sol-gel processing for tailoring the properties of coatings.

The reader who is interested in other applications is directed to the "Proceedings of the International Workshops on Glasses and Ceramics from Gels", published in the *Journal of Non-Crystalline Solids,* volumes 82 (1986), 100 (1988), 121 (1990), and 1992 (to be published) as well as the "Better Ceramics through Chemistry" series, published in the *Materials Research Society Symposium Proceedings,* volumes 32, 73, 121, and 180, and the *Proceedings of the Ultrastructure Processing Conferences,* published by Wiley. *Sol-Gel Technology for Thin Films, Fibers, Preforms, Electronics and Specialty Shapes,* published by Noyes, and *Sol-Gel Optics,* published by SPIE, volume 1328 (1990) also contain many papers and discussions of applications of sol-gel coatings.

4.1 Electrical Applications

The use of coatings in electrical applications is an area of considerable importance. Such coatings include ferroelectric thin films, spin-on glasses, dielectric layers (e.g., Ta_2O_5, silicon oxynitride), high temperature superconductors, fast ion conductors and transparent conductive coatings.

Ferroelectric films are used as an example in this section. The characteristics of such films prepared by wet-chemical methods are compared with those of films prepared using vapor deposition techniques.

The applications of ferroelectrics (FE) cover a broad range, and are based on their piezoelectric, pyroelectric, electro-optic and FE properties. These applications include transducers, high dielectric constant capacitors, IR sensors, optical shutters, microwave materials and non-volatile memory devices. FE films are typically produced by evaporation or sputtering, and sometimes by the use of powder slurries. With such methods, control of stoichiometry is difficult, and the high temperatures required to establish satisfactory FE properties limit the choice of substrates and can impair the underlying device profiles.

The lead titanate family of FE's includes lead titanate ($PbTiO_3$), lead zirconate titanate (PZT) and lead lanthanum zirconate titanate (PLZT). In combination with various dopants, they constitute the most versatile class of perovskite FEs, and have achieved widespread commercial use. Wet-chemical synthesis of lead titanate films generally uses a mixture (45) of Pb salts such as acetates with Zr/Ti alkoxides (46). In some cases, they are reacted in methoxyethanol to form complex alkoxides. All-alkoxide synthetic routes have also been reported (47).

Representative values of the properties which have been obtained for PZT and PLZT films using sputtering and wet-chemical methods are given in Table 1. From the data presented there, it is clear that bulk ceramics and thin films are characterized by different dielectric properties, and that the properties of thin films themselves cover a broad range.

A lead-deficient pyrochlore phase is often detected in as-deposited sputtered $PbTiO_3$, PZT and PLZT films (49)(50)(52)(64)-(67). Post-deposition annealing at elevated temperatures is used to transform this phase into the perovskite structure and thereby improve the FE properties. Lack of reproducibility and non-stoichiometry of the films pose serious problems, especially during multitarget sputtering/multisource evaporation. At elevated temperatures the volatility of lead is a matter of concern. Incorporating excess lead in the targets is often employed, but this approach is both system- and set-up- specific.

Epitaxial effects are often observed in sol-gel derived films. Examples include PLZT on sapphire (67)(68), PZT on platinum (69), and $PbTiO_3$ on MgO (70)-(72).

The pH of the precursor solution plays a poorly understood role in affecting the final properties of chemically derived films. There exists an

Table 1. Ferroelectric Properties of Various Lead Titanate Thin Films

Method	Dielectric Constant E_r	Dissipation Factor tan δ	Permanent Polarization P_r ($\mu C/cm^2$)	Coercive Field E_c (kV/cm)	Composition (Zr/Ti) (La/Zr/Ti)	Ref
Rf mag sputtering	400		40	150	45/55	48
Rf mag sputtering	300-500	0.02-0.2	12.5	90	65/35	49
Rf mag sputtering	350			35	90/10	50
DC mag sputtering	194	0.1-0.3	12	40	52/48	51
DC mag sputtering	820	0.1-0.3	70		52/48	
Rf diode sputtering	751		20.4	23-3	52/48	52
Sol-Gel Methods on Various Substrates						
Pt	2000	0.1	18.3	37.5	53/47	U of A/AML
Pt			36	40	53/47	53
ITO	260	0.1-0.7	6.6	26.7		54
Pt	1800	0.02	5	8	8/65/35	55
Si(100)	130	0.17			7/65/35	56
Au	225	0.004			7/65/35	56
Si			2.2	7.5	50/50	57
Pt	1200	0.01-0.02			50/50	58
Pt	1000	0.01-0.02			44/56	58
Pt (Amorphous PZT)	35-50	0.005-0.001				58
Pt	500	0.024	10	37	53/47+2 at % Nb	59
Pt	300	0.03	21.8	81.6		60
Stainless Steel	400	0.4			50/50	61
Pt	500	0.06	12	60	50/50	62
Bulk Ceramics	625		33	6.5	50/50	57
Bulk Ceramics	750	0.004	45	17	53/47	63

optimal degree of hydrolysis of the solutions, which depends on the composition, concentration, precursors and post-deposition processing. The presence of adequate amounts of water lowers and shrinks the temperature window where pyrochlore is found. Extensive hydrolysis produces amorphous, porous films which do not densify even at high temperatures (73).

From Table 1, the dielectric constants of chemically derived films range from 35 to 1800, and the dielectric losses (tan δ) from 0.001 to 0.7. The FE properties are functions of film thickness, grain size, applied DC bias, amplitude and frequency of the probe signal, as well as the measurement geometry and specifics. In addition, the electrode itself can effect the microstructures and properties of FE films.

In light of these considerations, comparisons of the properties of FE films such as that presented in Table 1 should be used with care. Subject to this reservation, the loss of crystalline films is seen to be higher than that of amorphous films. This is likely associated with the motion of domain boundaries in the crystalline films. The highest dielectric constants are obtained for films deposited on Pt. More generally, the choice of substrate is important because of substrate-film reaction and interdiffusion, and various approaches—such as rapid thermal annealing or the use of intermediate passivation layers—are employed to reduce or eliminate such effects.

Among FE ceramics, $BaTiO_3$ has been the most extensively investigated. Representative properties of $BaTiO_3$ films prepared using various methods are tabulated in Table 2 (74). It is seen that considerable variability exists in the FE properties which have been obtained—undoubtedly reflecting differences in processing and measurement methods. The dielectric constants of films are seen to approach or even exceed those of bulk samples, but the polarization remains substantially below bulk values. Films with dielectric constants below 1400 (the bulk value) exhibit polarizations closest to those of bulk samples. Internal stresses within the grains, presumably causing the high dielectric constants, also produce a clamping effect on polarization.

A number of wet-chemical synthesis routes for making $BaTiO_3$ films have been reported. Perhaps the simplest involves mixing alkoxides of Ba and Ti (75)-(78). With this approach, the Ba alkoxide tends to hydrolyze and condense much more rapidly than the Ti alkoxide, and this difference in rate leads to the formation of island-like heterogeneities during processing. To reduce or eliminate this problem, a number of approaches have been explored. These include using different alkoxide species with balanced reactivities; using double alkoxides; lowering the reaction temperature;

Table 2. Properties of $BaTiO_3$ Films from Ref. 74

Method/Substrate	Dielectric Constant E_r	Dissipation Factor Tan δ	Remanent Polarization P_r ($\mu C/cm^2$)	Coercive Field E_c (kV/cm)
RF Sputtering/Si	300			
RF Sputtering/Si	10-110		1	
RF Sputtering/Si	120	0.15		
RF Sputtering/Pt			6	27
RF Sputtering/Pt	1680	0.03	0.8	3
RF Sputtering/Pt	16-1900	0.005-0.065	0.32	3
RF Sputtering/Si	10-40		1.28	
DC Sputtering/Pt	1700	0.018	0.2	
Evaporated/Pt	1050-1350		3-4	
Polycrystalline Bulk Ceramics	**1400**		8-10	
Monodomain Single Crystal			26	
Sol-gel/Si Crystalline, 600C, 4 hrs.	53		0.339	0.080
Sol-gel/Pt Crystalline, 1200C max. temp.	200-1000	0.03-0.1	3.1	

substituting the Ba alkoxide with metal salts (79)-(81); preparing a polymeric resin via glycols and/or oxalates; and using complexing agents or chelates such as 2,4-pentanedione (82). Such chelates act to stabilize the alkoxide species in the precursor solution by partial substitution of the reactive alkoxy groups.

The as-deposited films of $BaTiO_3$—like those of the $PbTiO_3$-based compositions—are typically amorphous. On heating, they crystallize at temperatures in the range of 500-600°C (81)-(83). Unlike $PbTiO_3$-based FE's, $BaTiO_3$ does not exhibit the formation of a pyrochlore phase, but invariably crystallizes to the perovskite phase.

The as-deposited amorphous films typically have higher dielectric strengths than those of crystallized films (81), and sol-gel derived amorphous films exhibit higher dielectric strengths than sputtered films. This is likely due to their high chemical homogeneity and low incidence of defects, and perhaps to the lack of a columnar structure in the sol-gel derived films (unlike typical sputtered films).

Doped $BaTiO_3$ films have been prepared where the dopants were introduced as alkoxides in solution (76). This ease of introducing dopants provides the ability to obtain desired combinations of properties.

In summary, wet-chemical methods have been applied to the synthesis of a wide range of FE films. In many cases, encouraging results have been obtained. Issues such as fatigue and aging (changes in the FE properties with cycling and time) must be addressed before chemical processing of FE films becomes a wide-scale commercial reality.

4.2 Optical Applications

The optical applications of chemically derived films also cover a broad range. Representative approaches to several of these areas, as well as recent progress and outstanding problems, are discussed below.

Antireflection Coatings. Two main approaches have been used to reduce reflection losses at the surface of an optical medium. In the first, an antireflection (AR) film—whose optical path length is $\frac{1}{4}\lambda$ of the incident radiation—is deposited on the surface. The refractive index, n, of the AR coating is intermediate between the indices of the body and the external medium. The second approach is based on the fact that reflection is caused by the sharp index discontinuity at the interface. In this approach, the sharp interface is essentially eliminated by providing a film with a controlled gradient in n. This is accomplished by controlling the pore structure and composition of the film.

Sol-gel methods have been used to exploit both approaches for preparing AR coatings (84). Single-layer, double-layer, multi-layer and gradient index coatings have been deposited on a variety of materials. These coatings utilize the inherent flexibility of wet-chemical methods to manipulate the chemical composition and film microstructure. As examples, porosity can be controlled by altering the size and structure of the condensed species prior to deposition; the density of the films can be varied by varying the method of deposition; and the index of the films can be modified by processes such as etching and thermal treatment.

A variety of chemically derived single-layer AR coatings have been explored. These include Ta_2O_5 (85)(86), SiO_2-TiO_2 (85)(87) and SiO_2-B_2O_3-Al_2O_3-BaO (84,88). Such coatings have been deposited on plastics as well as glasses. Two-layer and multi-layer coatings have also been prepared by wet-chemical methods (85)(89)-(91). Three-layer coatings prepared in this way have yielded ~ 99.6% transmission over the range of 1.1 - 1.3 μm (92).

It should be recognized, however, that multilayer coatings deposited from solution often craze and peel because of shrinkage-associated stresses. To avoid such problems, stacked porous layers have been deposited from colloidal suspensions (93). Because of an interest in transmission/reflection over only a range of wavelengths, the stacks consisted of alternating layers of high index and low index material of tailored thicknesses.

Stacks of various combinations of oxides have long been used, typically prepared by vapor deposition methods, to produce optical interference filters (dielectric mirrors) (94). For a TiO_2-SiO_2 stack of 12 layers tailored for reflection about 550 nm, the results shown in Fig. 4 were obtained using wet-chemical methods to deposit the layers (95). Using similar methods, stacks with 16 layers have also been produced, as has a 37 layer stack which approximates a rugate filter (96).

In the wet-chemical approach to gradient-index AR coatings, a porous layer is first deposited and exposed to a chemical etching treatment. The etchant increases the scale of the pore structure and preferentially etches one or more of the glass cations to produce a gradient in composition (hence n) through the film. Since the pore structure dominates the etchant interaction, many glass compositions are available for use in this way.

An alternative approach to gradient-index AR coatings involves coating a series of SiO_2 sols of differing particle sizes (TMOS-based sols aged for differing time periods). This method can potentially overcome problems associated with the use of leaching to produce a coating with graded porosity (97)(98).

Finally, taking advantage of the use of sol-gel solutions as vehicles for

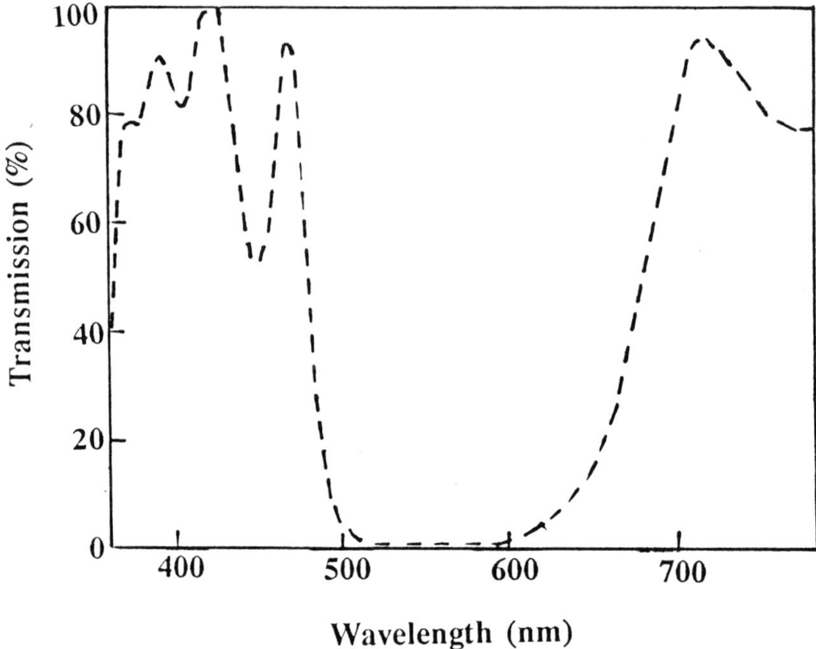

Figure 4. Optical transmission spectrum for a sol-gel TiO_2-SiO_2 12-layer stack. From Ref. 95.

colloidal particles, AR light-scattering coatings have been prepared by spray-coating an alcoholic solution of colloidal SiO_2 particles suspended in a partially condensed Si alkoxide. The optical quality of the coatings is comparable with that of acid-etched glass surfaces (91).

Planar Waveguides. Planar waveguides are used in integrated optical circuits to route signals between different input, output and processing components of a chip. Since light is guided in the plane of these coatings, these films must be of reasonable optical quality. Index uniformity must be sufficient to achieve losses on the order of 1 dB cm^{-1}, with a reproducibility in index of at least 1%.

A number of investigator teams have prepared planar waveguides using wet-chemical methods (99)-(104). Despite the wide range of potential compositions, only a limited number have been explored for this application. Most work has focused on the SiO_2-TiO_2 system. Not surprisingly, the optical properties of waveguides made from these solutions depend strongly on environmental factors, such as coating atmosphere and temperature.

Recent works by Weisenbach et al. (105) and Martin et al. (106) have shown that the coating characteristics and optical properties of sol-gel derived SiO_2-TiO_2 films are influenced significantly by the choice of titanium precursor, extent of hydrolysis, addition of leveling agents, and post deposition heating schedule.

Optical loss in most chemically derived planar waveguides is typically in the range of 0.5 - 1 dB cm^{-1} which is well suited for planar waveguide configurations. As discussed by Roncone et al. (107), the mechanisms of loss in these materials are not understood satisfactorily at the present time, but are dominated by scattering from volume index inhomogeneities.

Surface Patterning. Many optical applications are based on the use of surface relief features in or on thin films. Specific geometries are used to create a range of optical components including couplers, distributed feedback gratings, filters, lenses, beam splitters and mirrors. Two creative approaches have been used to generate such surface relief structures in wet-chemical films. In the first, embossing is used to create a negative of the master in the film. The technique is based on the malleability of sol-gel films at suitable stages of processing. The process has been used for some time to form surface relief structures in polymer films. However, the sol-gel process now provides the opportunity to manufacture these gratings in dense, hard dielectric materials. Roncone et al. (108) used a master grating with a 0.52 μm period and a peak-to-trough depth of around 200 nm to form an embossed grating in a SiO_2-TiO_2 waveguide. As shown in Fig. 5, the quality of the replication was excellent. Grooved disks (109)(110) for optical data storage have also been made by embossing undensified sol-gel films.

The shape and quality of the embossed pattern depends on a number of process parameters, including pressing pressure, solution chemistry, master composition and film heat treatments. For example, to avoid sticking between the film and master, SiO_2-TiO_2 waveguides need to be prebaked before being embossed (108). As illustrated in Fig. 6, very different patterns can be formed in the same film when the extent of prebake is changed. For moderate amounts of prebaking, the film appears to deform by flow under the master. For longer prebakes, gratings are formed by a mechanism which appears to be more brittle in nature. Finally, for even longer prebake times, no deformation is seen after embossing. The details of the deformation processes in each of these regions have not been studied.

The second approach to patterning sol-gel films is based on the use of laser light to densify local regions of the film. Here, selected areas of a film are densified by exposing an undensified coating to a laser which is scanned over the coating in the desired pattern. One of two methods—direct or indirect writing, as illustrated in Fig. 7—is usually used in this approach (104). In the first method, a wavelength that is absorbed directly by the coating (e.g., the 10.6 μm line of a CO_2 laser for an oxide coating) is used to heat the coating directly. The densified trough formed by heating a SiO_2-TiO_2 coating with a CO_2 laser is shown in Fig. 8.

In the second, indirect writing method, the sample is exposed to a wavelength that is not absorbed by the sol-gel coating but is absorbed by the substrate (e.g., the 1.06 μm line from an Nd:YAG laser for oxide coatings on a silicon substrate). The laser light heats the substrate, which then heats the sol-gel coating, indirectly, from the bottom (111). Alternatively, a wavelength that is transparent to both the film and the substrate, but is absorbed by a metal film on top of the sol-gel coating (e.g., the 1.06 μm line from an Nd:YAG laser for a Au/Pd-coated oxide coating), can be used to heat the sol-gel film indirectly from the top (112)

In all laser densification methods, the densified regions are much less reactive (chemically) than the undensified regions, so that the unexposed areas can be etched away, leaving densified material in the pattern of the laser scanning.

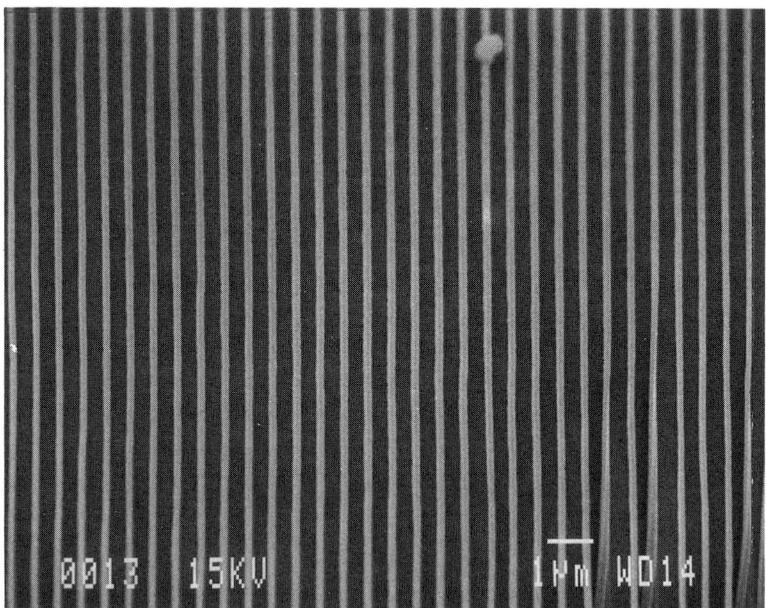

Figure 5. Embossed grating formed in a SiO_2-TiO_2 sol-gel waveguide using a master with 0.52 μm period and peak-to-trough depth of about 200nm. From Ref. 108.

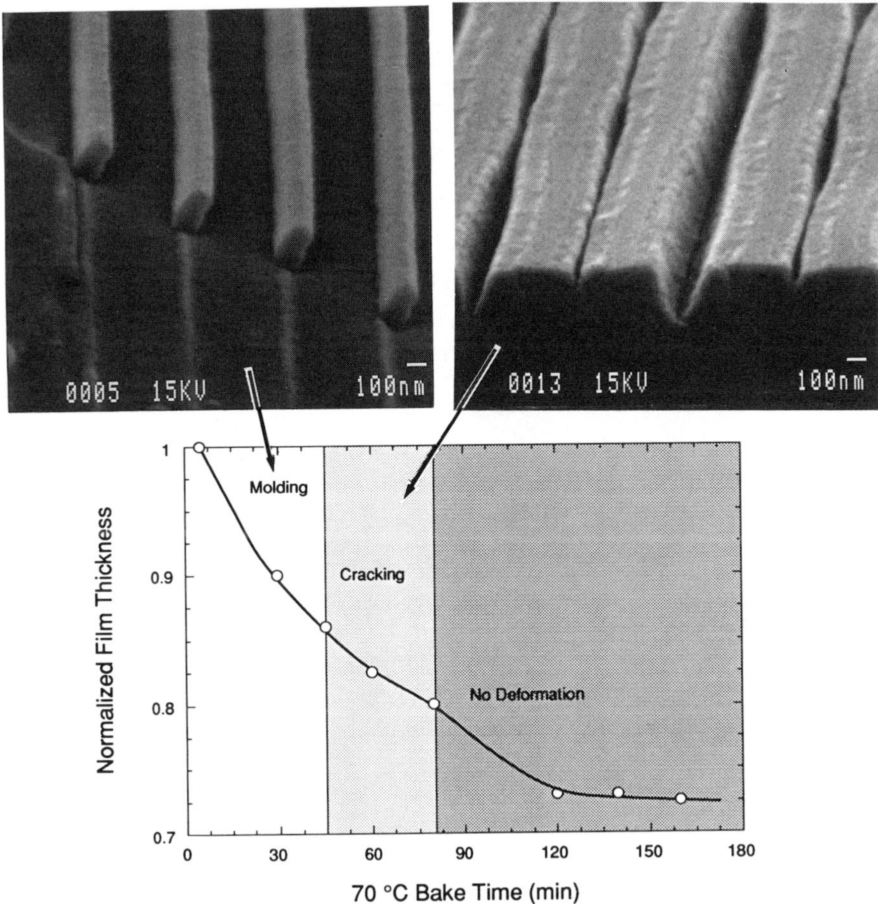

Figure 6. Effect of prebaking times on embossing of SiO_2-TiO_2 waveguides. From Ref. 108.

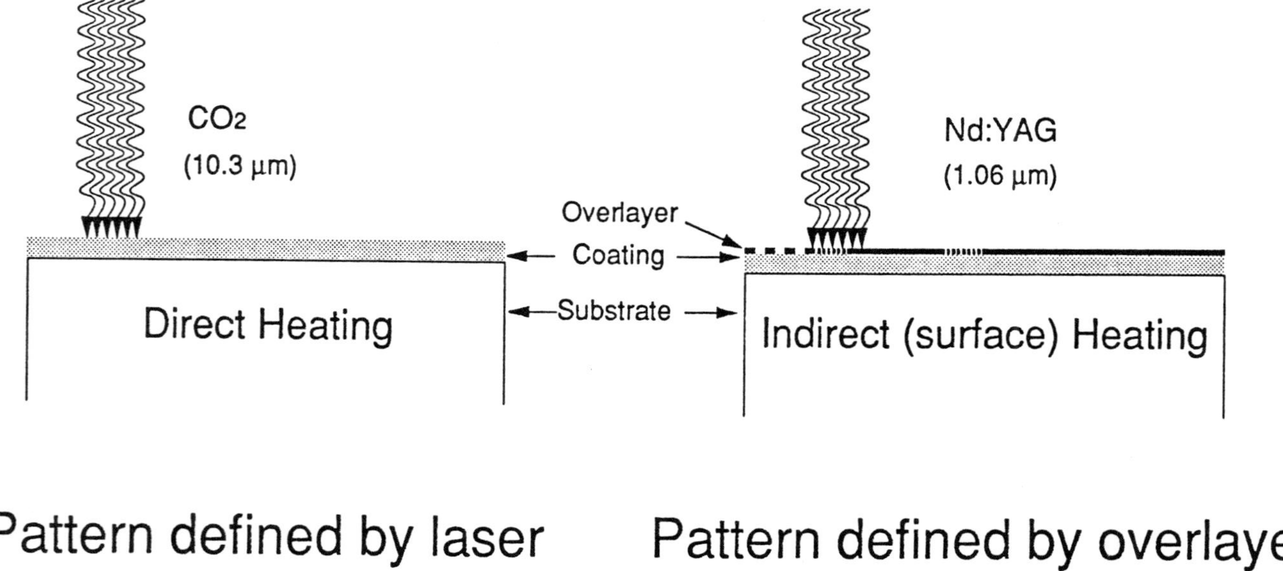

Figure 7. Schematic illustration of direct (left) and indirect (right) methods of laser densification of sol-gel coatings. From Ref. 104.

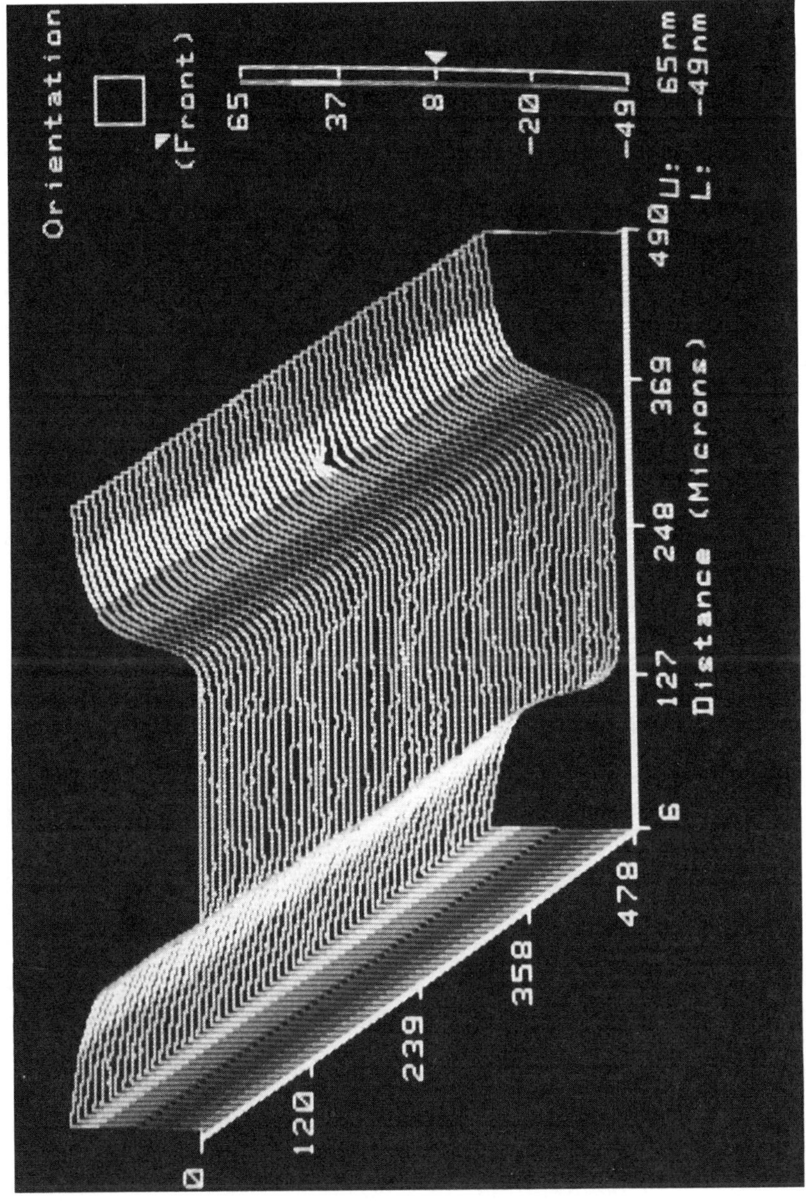

Figure 8. Trough formed by CO_2 laser in SiO_2-TiO_2 coating.

The physical properties (e.g., density, refractive index and absorption edge) of laser-densified coatings tend to differ from the properties of the same materials when they are processed using conventional (furnace) heating (113)(114). It is likely that such differences are due to the extremely short time scale (tens of milliseconds) (114) in which heating takes place during laser processing. The hydroxyl content of laser densified SiO_2 films, for example, is higher than that of SiO_2 films fired to the same densify using a furnace. In one case (115), it was shown that carbon is trapped in laser densified SiO_2-TiO_2 coatings, while in another case, dense SiO_2 coatings were made essentially carbon-free (116).

The detailed effects of the differences in time scale on the mechanisms of burnout, densification, and related processes have not been investigated. If these differences can be understood and controlled, laser densification has the potential to lead not only to a simple, rapid method of patterning ceramic coatings, but also to providing coatings with new and unique properties.

Colored Coatings and Reflective Coatings. A variety of colored coatings have been prepared by wet-chemical methods (117)-(125), as have a range of reflective coatings. Of particular interest here are the commercial products of Schott Glaswerke and Central Glass/Nissan Motors. The IROX product of Schott consists of coatings of Ti alkoxide + colloidal Pd particles which are applied from solution to glass plates (90). The plates are exposed to the H_2O-containing atmosphere to set the coatings, and are fired at 400 - 500°C to densify them. The Central Glass/Nissan Motor product consists of alkoxide-derived SiO_2-TiO_2 coating patches on glass windshields (126). After densification, the patches provide abrasion-resistant areas of sufficiently high reflectivity for heads-up displays.

Electro-optic Materials. Electro-optic (EO) thin film materials are increasingly important in the development of optical device technology. Examples include optical phase retarders, optical and electro-optic switches, spatial light modulators, and devices based on optical phase conjugation. EO materials also have potential for optical data storage and computing.

Many of the important inorganic EO materials are ferroelectric. For in-plane applications such as planar waveguides, single crystals or perhaps oriented epitaxial films are desired to avoid boundary scattering. In cases where phase matching is required, selectively oriented single-crystal films are necessary. For through-plane applications, in contrast, the optical path length is small, and polycrystalline films can be used and may be preferred.

The ability to tailor EO coefficients through compositional (hence structural) variations can be quite important. Taking PLZT as an example, optimization of the quadratic EO coefficient—which is related to n_2 and χ^3—leads to PLZT's with near-cubic structure, while optimization of the linear EO coefficient—which is related to n, and χ^2—leads to PLZT compositions with a pronounced tetragonal character.

EO films covering a range of composition have been prepared by a variety of techniques, including but not restricted to sputtering, spray pyrolysis and wet-chemical methods. As a general observation, the EO coefficients of films, however deposited, are notably lower than those of bulk samples. The lower values for the films may reflect chemical interactions between film and substrate, greater porosity in the films, differences in phase distributions between film and bulk, and substrate clamping effects on the quadratic coefficients of films.

Waveguide losses in sputtered films are in the range of 10 - 20 dB/cm^{-1} (168)(127)-(129). The losses for epitaxial films are typically larger than those for single-crystal films, likely reflecting scattering from grain boundaries in the epitaxial films. The higher values for the single-crystal films compared with amorphous films are likely due, at least in part, to scattering from domain boundaries—although it remains to be established whether this represents the only important source of loss in single-crystal films. In the case of relaxer FE's, relatively high losses may be expected because of the presence of large numbers of microdomains. For both normal and relaxer FE's, poling would be expected to have a considerable effect on loss, but this effect has yet to be demonstrated experimentally.

Various EO thin films have been prepared using wet-chemical processing. These include PZT (46), PLZT (130), $LiNbO_3$ (131)-(135) and SBN ($Sr_xBa_{1-x}Nb_2O_6$) (136). Wet-chemical synthesis of PLZT films is usually based on the use of Pb and La salts, such as acetates, in combination with Zr and Ti alkoxides.

For sol-gel derived 8/65/35 PLZT films* on sapphire, preferred orientation was observed with the (001) plane parallel to the substrate when annealed at 750°C. The degree of preferred orientation decreased with further heating to 850°C (130). Schwartz et al. reported (111) orientation for chemically derived PLZT films on platinized silicon (137).

The linear and quadratic EO coefficients of both sol-gel and sputtered PLZT films are tabulated in Table 3 (138). The data on sol-gel derived PLZT films are limited. The results in Table 3 indicate, however, that the properties of sol-gel derived films compare very favorably with those of sputtered films. The results shown in Fig. 9 illustrate second harmonic generation (SHG)

* The x/y/z notation refers to the formula $Pb_{1-x}La_y(Zr,Ti_z)_{1-x/4}O_3$

Table 3. Electro-Optic Coefficient of PLZT Thin Films from Ref. 139

COMPOSITION (x/y/z)[a]	LINEAR COEFFICIENT, r_c 10^{-11} m/V	QUADRATIC COEFFICIENT, R $x10^{-16}$ $(m/V)^2$	DEPOSITION TECHNIQUE
8/65/35	3.0	0.5	Sol-gel
0/60/40 (PZT)	2.4	0.01	Sol-gel
28/0/100		0.8	Sputtering
9/65/35		1.0	Sputtering
28/0/100		0.6	
14/0/100	2.8		Sputtering
21/0/100	8.1		Sputtering
9/65/35		0.6	Sputtering

[a] Here x/y/z refers to the formula $Pb_{1-x}La_y(Zr,Ti_z)_{1-x/4}O_3$.

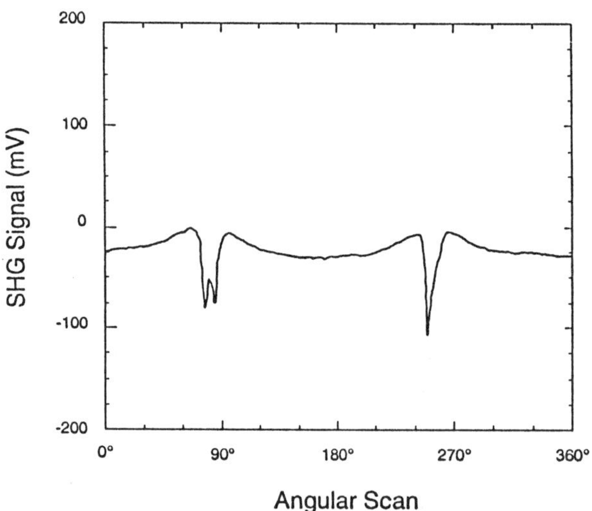

Figure 9. SHG from chemically derived PT film.

from a chemically derived PT film. The magnitudes of the maximum observed SHG signals for PLZT films of various compositions heat treated at 650°C and 750°C are shown in Table 4 (139). Also shown there are data on $LiNbO_3$ and $BaTiO_3$ films prepared by wet-chemical methods. As seen from the data in Table 4, the maximum SHG intensity increases with increasing heat treatment temperature, and is higher for $PbTiO_3$ than for any of the La- and Zr-doped materials. It is also seen that the incorporation of Zr has a considerable effect in decreasing SHG.

Table 4. Second Harmonic Generation (SHG) from Various Films (0.4 µm thick) on Silica from Ref. 139

Composition (mV)	Maximum SHG Signal (mV) Treated at 650C	Maximum SHG Signal (mV) Heated at 750C
PLZT[(a)] 0/0/100	100	265
PLZT[(a)] 7/0/100	25	122
PLZT[(a)] 15/0/100	15*	120
PLZT[(a)] 0/65/35	None Detected	20
PLZT[(a)] 9/65/35	15*	20
$BaTiO_3$		180[(b)]
$LiNbO_3$ on Sapphire		520[(c)]

(a) x/y/z PLZT: $Pb_{1-x/100}La_{x/100}(Zr_{y/100}Ti_{z/100})_{1-x/400}O_3$

(b) Heated at 900C

(c) Heated at 800C

* Within detection limits

The effect of heat treatment temperature on SHG from 7/0/100 PLZT films is shown in Table 5 (139). A rather dramatic increase in SHG between 700°C and 750°C, due to densification of the films, is observed.

For materials such as $BaTiO_3$ whose Curie temperature lies not far above ambient, heating induced by high intensity laser beams can lead to a decrease in the SHG signal—and even to its disappearance as beam heating raises the temperature above the Curie point (where the material is centro-symmetric) (139).

Table 5. Effect of Heat Treatment Temperature on Second Harmonic Generation (SHG) from 7/0/100 PLZT Films (4000 Å) from Ref. 139

Annealing Temp. (C)	Maximum SHG Signal (mV)	Phases
500	13*	Pyrochlore + Perovskite
650	21	Perovskite
700	38	Perovskite
750	122	Perovskite

† x/y/z PLZT: $Pb_{1-x/100}La_{x/100}(Zr_{y/100}Ti_{z/100})_{1-x/400}O_3$

* Within detection limit

Sol-gel derived $LiNbO_3$ films have been prepared, some exhibiting c-axis preferred orientation on (0001) sapphire (131)-(135). No such behavior is observed, however, for films deposited on Si (134)(135). $LiNbO_3$ films can crystallize at temperature as low as 250°C if an underlying fully crystallized film is first deposited. When single-crystal $LiNbO_3$ is used as a substrate, epitaxial films are obtained (140). The refractive index of sol-gel $LiNbO_3$ films is 2.10 - 2.25 (141), comparable to that of sputtered film, but less than the single-crystal value, presumably due to residual porosity. Hirano has reported making a successful waveguide using a sol-gel derived Ti-doped $LiNbO_3$ film (135). No data on attenuation were reported. The losses are expected to be high due to grain boundary scattering. There are no reports on the EO coefficients of these films. The dielectric constant has been measured as about 22, and the value of tan δ is highly dependent on the choice of metal contacts used.

Dyes in Gels. The low temperatures at which wet-chemical processing can be carried out have opened a number of attractive opportunities for incorporating thermally sensitive organic molecules in the forming chemical networks. Using this approach, it has been possible to obtain improved thermal and photochemical stability compared with solution or polymeric matrices. Using dyes as molecular probes, the approach can also be used to obtain new insight into wet-chemical processing and gel structure.

To date, the incorporation of selected organic dyes in chemically derived hosts has been used in a range of applications. Tunable solid state lasers (142)-(147), and sensors (148) are two noteworthy examples.

Progress in the area of tunable solid state lasers has been significant and is summarized most effectively in the paper by Dunn (149). A variety of dyes, including rhodamine 6G, rhodamine B and oxazine-4-perchlorate, have been incorporated in chemically derived oxides based on Al_2O_3 (142)-(144)(150), SiO_2 and TiO_2 (151). The results indicate a reduction or elimination of dye aggregation as seen in solution, as well as enhanced thermal stability and photostability.

A range of dyes, including rhodamines, coumarin and BASF-241 (a perylene dye), have been incorporated in polymer-modified oxides (ormosils, ceramers, polycerams) (152). Notably enhanced photostability compared with the same dyes in solution or in organic polymers is again observed. While the origin of this improved photostability remains to be conclusively established, it seems likely to be associated with interaction of the dye with the inorganic framework, as well as with the effective isolation of the dye molecules in the structure. Recall, in this regard, the improved photostability observed for dyes in strictly inorganic matrices. If this suggestion is correct, it would imply preferential adsorption/association of the dye molecules at the interfaces between the organic and inorganic components of the material, and would suggest interesting new routes to the control of lasing/photochemical properties. In any event, further improvements in photochemical stability will be required if wet-chemically-derived tunable dye lasers are to become an effective reality.

Chernyak et al. (148) have explored the use of oxazine-170 in wet-chemically-derived glasses as reversible sensors for ammonia and acids. These sensors are based on spectral shifts with changes in the environment from acidic to ammoniac. While the present authors have some concern about the viability of the system employed here, they remain positive about the prospects for wet-chemically-derived materials in sensing applications. Considerable attention must, however, be directed to the overall cost of the system employed, to the uniqueness of the information obtained, and to the tailoring of chemistry and microstructure of the synthesized materials to obtain exceptional combinations of properties.

Electrochromic Films. WO_3 and MoO_3 represent the classic electrochromic materials, and films of these materials have been synthesized by a variety of wet-chemical techniques (153)-(158). The coloring efficiency of chemically derived WO_3 gels has been shown to be comparable to that of coatings prepared by other methods (155). The microstructures of chemically derived WO_3-MoO_3 films have not, however, been characterized in satisfactory detail—much less the dependence of structural features and their stability with time and temperature on chemistry and process history.

Organic/Inorganic Composites as Non-Linear Optical Materials. A number of organic/inorganic composite materials with interesting χ^3 behavior have been synthesized. These include the introduction of fluorescein and its derivatives into wet-chemically-derived SiO_2 and into low-melting lead-tin fluorophosphate glasses (159), as well as the synthesis of composites based on chemically derived SiO_2 + poly-p-phenylenevinylene (160). The χ^3 values of the latter composites were found to vary with composition, and materials could be prepared with $\chi^3 \geq 5 \times 10^{-11}$ esu. Such materials could also be prepared as promising waveguiding films.

Turning from χ^3 to χ^2, it is recognized that isotropic materials or crystals having a center of symmetry should have $\chi^2 = 0$. In this light, results obtained in our laboratory on 2-methyl-4-nitroaniline (MNA) and p-nitroaniline (pNA) incorporated in organically modified oxide films seem worthy of note (139)(161). The matrix was synthesized from tetraethoxysilane (TEOS) and tetramethoxysilylpropyl-substituted polyethyleneimine (MPEI), following the procedure outlined in Fig. 10. Second harmonic generation (which is related to χ^2) was observed on excitation by a Nd:YAG laser. The SHG signal was found to increase with increasing film thickness for a given composition of the film (Fig. 11), and with increasing dye concentration for a given film thickness (Fig. 12). For comparison with the data shown in this figure, crystalline MNA powder, about 100 µm thick, gave an SHG signal of approximately 1100 mV.

The SHG signal was also observed to increase with increasing TEOS content for a given ratio of MNA to MPEI. This is significant since it represents an increase in SHG under conditions where the concentration of the dye is being diluted by the addition of increasing amounts of TEOS. It very likely reflects the importance of silica-polymer interfaces in promoting alignment of the active dye molecules.

All of the data in Figs. 11 and 12 were obtained on optically clear, x-ray amorphous films. When the MNA/MPEI wt. ratio was increased to 0.284 from the range shown in the figure, significant crystallization of MNA on the free surface of the film was observed. Heating a sample of composition 2 TEOS/MPEI/0.2 MNA at a temperature some 15 - 20°C above the melting point of MNA and subsequent quenching of the sample led to notable volatilization of MNA, and to a concomitant decrease in the SHG signal. The sample did, however, display significant SHG (~290 mV), and it was suggested that the observed SHG signal was not associated with small crystals of MNA in the film.

To explore this suggestion, a film with the composition 2 TEOS/MPEI/ 0.24 pNA was prepared and found to yield an SHG signal of about 550 mV.

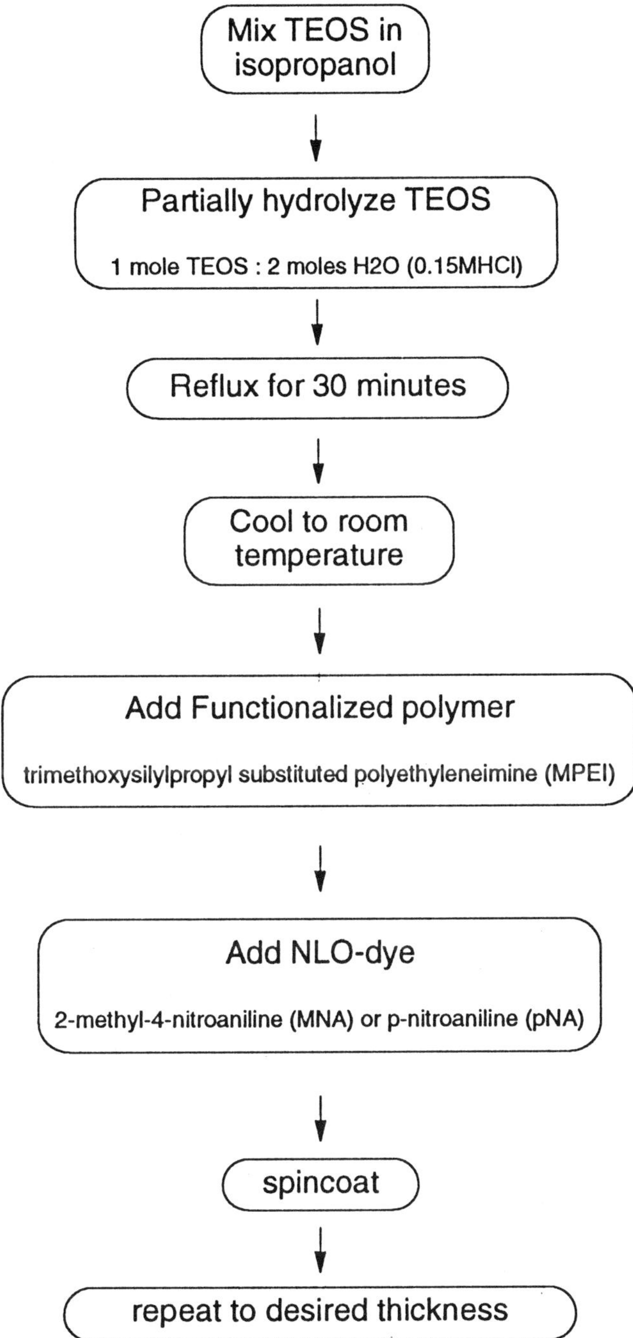

Figure 10. Synthesis procedure for NLO doped polyceram films. From Ref. 139.

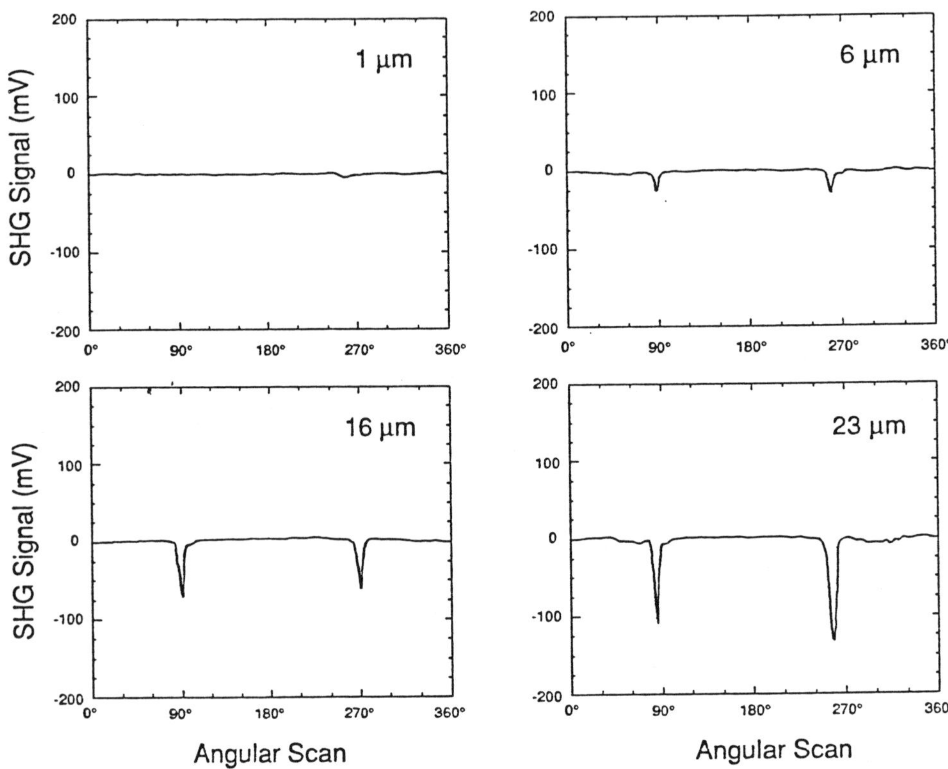

Figure 11. Effect of film thickness on SHG for pNA doped polyceram films (1 TEOS: 0.5 MPEI: 0.120 pNA). From Ref. 139.

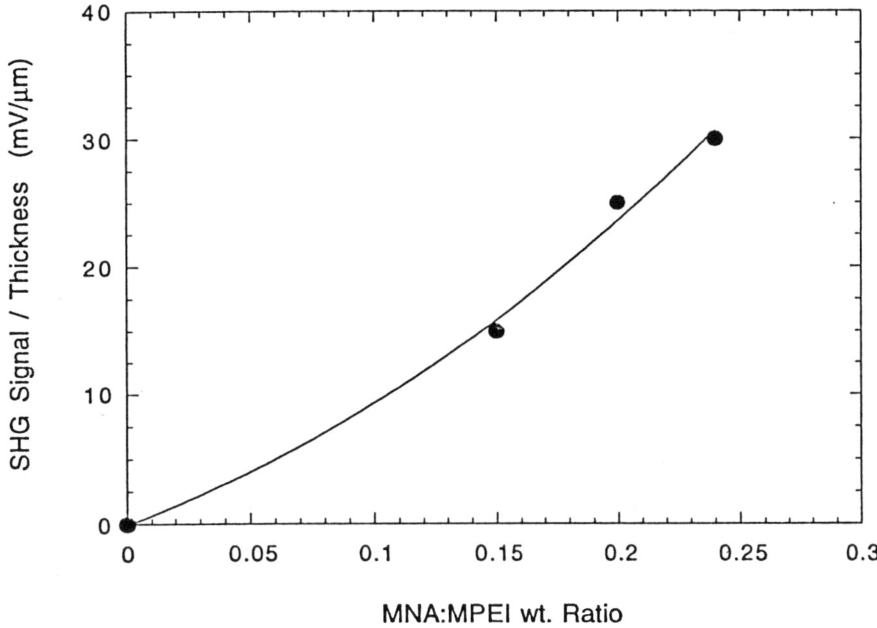

Figure 12. Effect of increasing dye concentration on SHG for MNA doped polyceram films (1 TEOS: 0.2 MPEI). From Ref. 139.

Since pNA (unlike MNA) crystallizes to a centro-symmetric structure, the observed SHG in the pNA-doped films (and inferentially in the MNA-doped films as well) is unlikely to be due to crystallization (precipitation) of the active dye. Rather, it seems likely that the SHG reflects molecular alignment of the optically active species in the sample.

It is suggested that this alignment takes place at the interfaces between the organic and inorganic components of the structure; and this in turn suggests that such interfaces themselves exhibit orientation (presumably reflecting the effects of the substrate and drying behavior). It is well known (162) that molecules such as MNA and pNA absorb strongly on SiO_2 surfaces, and such behavior in chromatography columns is well established. The present results therefore suggest that study of SHG from active molecules in organic-modified oxides, besides reflecting highly interesting optical behavior, may also provide important insight into the structure of the films.

More recent work in our laboratory, carried out on both inorganic (PLZT) and organic-modified oxide (MNA-pNA/MPEI/TEOS) films which have been corona poled, indicate that poling can have a considerable effect in increasing SHG. In the case of the inorganic films, the principal effect of the poling is likely that on orientation of the ferroelectric domains; while with the polyceram films, the principal effect likely involves orientation of the active molecules in the films. In the latter case, there is an issue of the stability of the poling-induced orientation against relaxation over time under ambient or near ambient conditions.

5.0 OUTSTANDING PROBLEMS

At present, the wet-chemical approach is used to synthesize a number of ceramic thin films for commercial applications. However, the broad expansion of this technique into a wide range of electronic and optical applications is currently limited by the inability to prepare routinely thick, carbon-free coatings. Productive research in other important areas, such as precursor characterization, the impact of deposition conditions on film properties, and low temperature densification will lay the groundwork for the continuing development of wet chemistry as a viable means to generate ceramic coatings.

5.1 Film Cracking

Most sol-gel derived films crack upon firing if the film thickness exceeds about 0.5 - 1.0 µm. This cracking occurs because processing-induced stresses in local regions of the film exceed the inherent strength of the gel or film skeleton. These stresses are produced during drying, solvent removal and carbon burnout, as well as during sintering or consolidation of the film. During these processes, the inherent strength of the film skeleton is also changing. Thus the competition between skeletal strength and film stress is dynamic, being a strong function of coating solution formulation and process history. This is demonstrated by the results of Garino (22) shown in Fig. 13, where the critical cracking thickness for SiO_2 films is plotted as a function of processing temperature. In this figure, the critical cracking thickness is the delineating thickness above which films were observed to form extensive cracking and R is the molar ratio of water to TEOS used in the formulation.

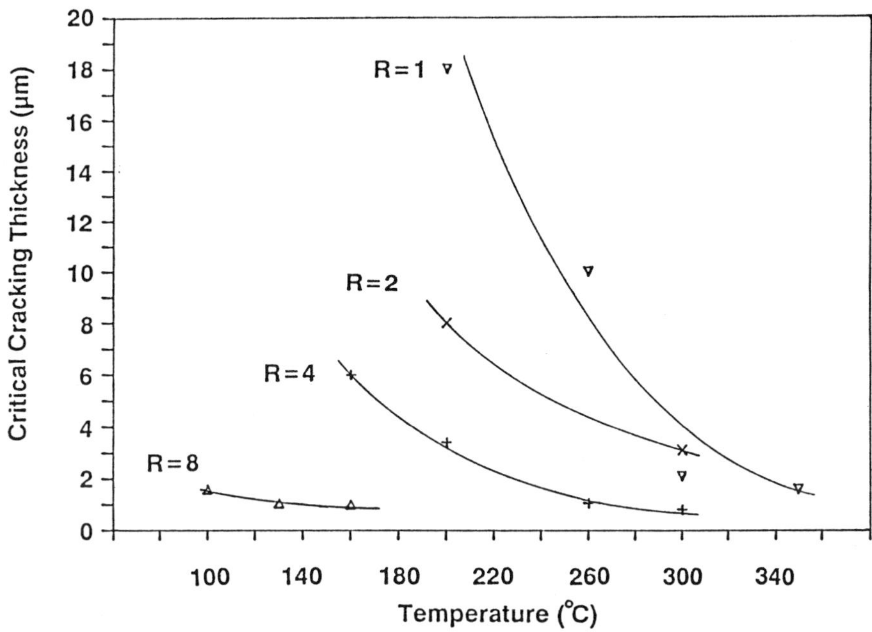

Figure 13. Effect of hydrolysis ratio (R = moles H_2O per mole TEOS) and temperature on the critical cracking thickness of silica sol-gel films. From Ref. 22.

Films with R values of 1 and 2 are tacky at room temperature and do not gel until heated to 140°C. Very thick films with these water contents can be made if they are not fired above 250 - 300°C. When R is large, films of thickness greater than 1.5 µm crack at low temperatures during the drying process. Garino offers several explanations for the decrease in critical cracking thickness with water content. As the water content in the solvent decreases, the surface tension of the liquid in the pores decreases. This reduces the capillary stresses which develop on drying. Also, for the films with low values of R, gelation does not occur until the film is at temperatures above the boiling points of the solvent phase. Consequently, no solvent is present to produce capillary stresses after gelation. Note that firing to high temperatures causes all films with thicknesses greater than 0.5 - 2 μm to crack. Apparently the skeletal structure is not sufficiently strong in these films to tolerate the stresses generated when the residual chemical species such as higher boiling point organics are removed.

Interestingly, Hench et al. (163)(164) have successfully generated large monoliths (bulk pieces) of SiO_2 by optimizing the use of drying control chemical additives (DCCA's). Typical examples of these additives include formamide, glycerol and oxalic acid. These and other additives help to minimize monolith cracking by reducing the solvent evaporation rate and by forming a more uniform pore size distribution in the gel network. This minimizes the differential drying stresses and also results in more uniform growth of the network during aging. In addition to SiO_2, DCCA's have been used to prepare monoliths of SiO_2-TiO_2 and Li_2O-Al_2O_3-SiO_2-TiO_2 (164), but much work remains to be done to establish their effectiveness in other multicomponent systems.

Despite advances in the area of monolith preparation, almost no exploration of the use of DCCA's in thin film synthesis has taken place. It is unclear whether the formulations which work for monoliths will be effective in generating films thicker than about 1.0 μm. This is because the constraints imposed by the rigid substrate during drying and consolidation lead to the generation of additional tensile stresses in the film which are not present during densification of monoliths. Since thin films dry so quickly, little or no aging of the gel network can occur to assist in strengthening the skeletal network. The inherent gel film weakness, combined with the enhanced stresses due to constrained shrinkage may limit the applicability of strategies useful in avoiding cracking in gel monoliths. Clearly more work needs to be done to characterize the mechanical properties and consolidation behavior of films during both drying and sintering, with the aim of increasing the network strength and reducing stress development during processing.

One approach to stress reduction during drying involves incorporating organic constituents into the film structure. The organic component contributes added flexibility to the gel network, allowing it to flow and relax during subsequent processing, and thereby alleviate stresses. Much research is currently directed towards the exploration of organic additions to bulk wet-chemically-derived materials (165)-(172); but little research is directed towards optimizing the behavior of coatings, particularly for organic constituents used as transient additives which are burned out at higher temperatures.

An alternate approach to thick, ceramic film synthesis is suggested by the work of Yoldas (173), who made monolithic pieces of Al_2O_3 by careful consolidation of a colloidal aluminum hydroxide sol. When the sol is dried, the colloidal particles come in close contact and develop sufficient strength at contact points to resist drying stresses. In these materials, the pores which control the drying and consolidation processes are much larger than for alkoxide derived coatings. As a result, the magnitude of the induced processing stresses is much smaller for colloidal gels. Similar results have been obtained by Shoup during the preparation of SiO_2 gel monoliths (174).

The same concept can be applied to solution-derived thick films. For example, Garino and Bowen (175) made 1 - 50 µm thick, crack-free, ceramic films of SiO_2, Al_2O_3 and ZnO-Bi_2O_3 using colloidal dispersion and deposition techniques. These films consolidated by a variety of mechanisms including viscous sintering, liquid phase sintering and solid state sintering. In each case, stresses which developed during these processes were resisted by the strong necks which formed between colloidal particles to produce the crack-free coatings. These promising results suggest that considerable potential exists for the synthesis of crack-free films using colloidal coating solutions.

5.2 Removal of Residual Species

Many of the current and proposed applications of sol-gel materials require that they be free of all carbon and hydroxyl groups, particularly for electronic and electro-optic applications. These species are unavoidable because they are inherent in the sol-gel processing. If high temperatures can be used, the species can easily be removed via oxidations and chlorination treatments, but one key advantage of the sol-gel process is that much processing can be done at low temperatures—making it compatible with other low temperature processes and materials. This advantage is lost if high temperatures are required to remove residual carbon or water. One factor which aids in the removal of these species from thin films is that the diffusion distance is small. As a result, the reactants and products involved in carbon removal are more easily transported through the film than in bulk pieces.

To remove carbon and water, the system must be designed so that the bonds which retain these species are broken at temperatures which are below those where sintering to closed pores occurs. As a result, further understanding of the nature of the bonding which retains these species at the skeletal surfaces and interior must be developed. In other words, more work on understanding decomposition reactions and the influence of the skeletal network chemistry on these reactions is required. Also needed is further insight into the relationships between film consolidation, which leads to pore closure and carbon entrapment, and the thermolysis reactions. Attention also must be focused on the role of the substrate in determining thermolysis behavior. The chemical nature of the substrate likely has a significant influence on the quantity of residual species retained in regions of the film located near the substrate-film interface.

5.3 Precursor Characterization and Aging

At present only limited information concerning the hydrolysis and condensation behavior of alkoxides (other than Si) is available. Of particular interest are studies involving chemical modifications of the alkoxides to improve stability in solution. For example, acetylacetonate is often added to coating solutions to reduce the functionality of one of the chemical species in solution. Ideally this should lead to more linear polymeric species in solution which are better suited for coating applications. Recent years have seen a resurgence of studies on the characterization and behavior of alkoxides and modified alkoxides. For examples, see the review by L. G. Hubert-Pfalzgraf (176) and recent papers in *Better Ceramics Through Chemistry IV* (22). Such studies need to be continued and expanded.

Of particular interest in this area is the impact of solution aging on coating behavior. As mentioned earlier, for typical alkoxide systems, the hydrolysis and condensation reactions used to generate homogeneous solutions proceed even as the solution is stored on the shelf. This leads to a variability in molecular weight and degree of crosslinking of the polymeric species in the solution. As a result, the solution viscosity increases with time, leading to variations in coating thickness and properties.

Efforts should be focused on developing techniques for minimizing aging in coating solutions. The work of Melpolder et al. (25) illustrated earlier in Fig. 3 shows that significant increases in solution stability are achieved by reducing hydrolysis and condensation through the capping of functional groups on the alkoxides. This approach, and others, deserve further development and attention.

5.4 Impact of Deposition Conditions on Film Properties

Advances in this area will significantly improve the reproducibility of the coating process. The work of Brinker et al. (177), shown in Fig. 14, illustrates the sensitivity of refractive index to spin speed for spin-coated multicomponent silicate films. The index increases with spin speed, suggesting that the shearing action of the spinning process influences the manner in which the polymeric species align and consolidate during the deposition and drying stages of the process. Note that these are dried films which have not been consolidated at high temperatures. Figure 15 shows the effect of spin speed on refractive index of SiO_2-TiO_2 films fired to 500°C in our lab following the procedures in Ref. 105. These films do not demonstrate a dependence of refractive index upon spin speed. Unlike the films of Fig. 14, these films are fully consolidated; and any spin speed induced variations in film microstructure apparently have been eliminated by heating to 500°C. More work covering a broad range of compositions and deposition conditions must be conducted to determine relationships between processing and properties for both as-deposited films and films which have undergone post-deposition heat treatments.

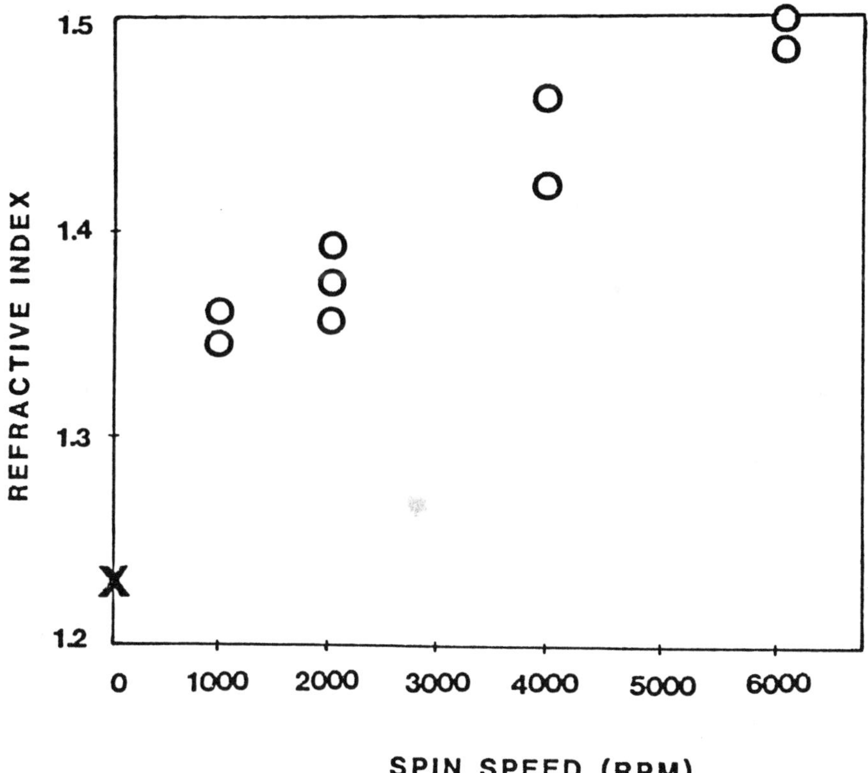

Figure 14. Refractive index versus spinning frequency for films prepared from multicomponent silicate precursors aged at 50°C and pH 3 for two weeks prior to deposition. From Ref. 177.

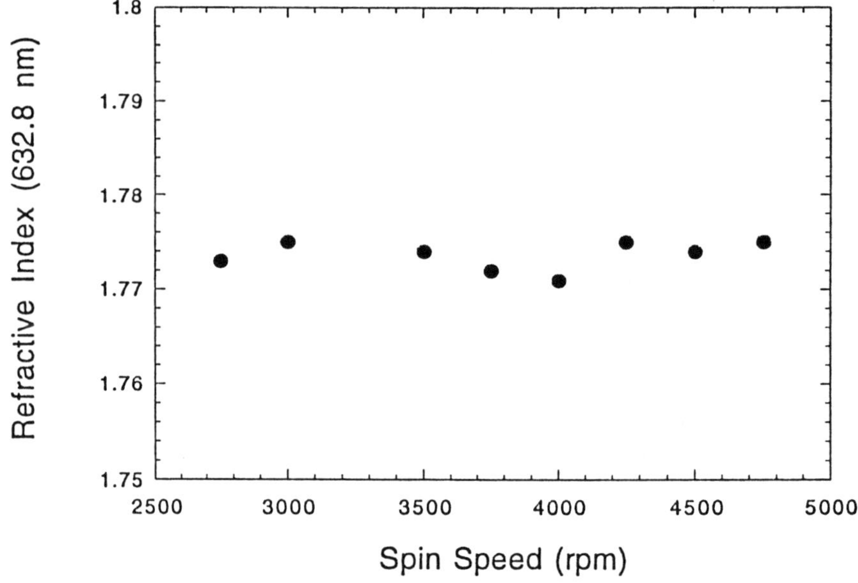

Figure 15. Effect of spin speed on refractive index for SiO_2-TiO_2 films fired to 500°C.

The results in this area of research must go beyond mere characterization. Models must be developed further to establish the relationships between variables such as spin speed, coating temperature, humidity, solution age, precursor identity, etc, and film properties for dip, spin and other coating processes. Consistent models explaining the observed behavior in terms of fundamental parameters such as coating solution viscosity, solvent evaporation rate, hydrolysis rates, the evolving solution and film mechanical properties during the coating process are required.

Much insight into the spin coating process can be found in the literature on spin coating of polymeric materials such as photoresists (178)-(183). However, the generation of sol-gel ceramic thin films is more complicated than synthesizing photoresist films. In inorganic thin film synthesis, the oxide precursor species in solution are hydrolyzing and condensing during the spinning process, so solution and film properties vary greatly during deposition. As a result, the basics established in early coating investigations using polymers must be modified and extended to describe fully the phenomenology of inorganic coating synthesis.

5.5 Low Temperature Densification of Films

Little work has been done to elucidate the mechanisms responsible for the low temperature densification which occurs in some sol-gel ceramic coatings. As seen in Fig. 16, for example, a 50:50 SiO_2-TiO_2 film densifies to near or full density at between 500 and 600°C (102). The SiO_2-TiO_2 system possesses a stable miscibility gap. Thus one expects that upon heating to a temperature where significant mass flow by diffusion can occur, the films would phase separate and then crystallize. Crystallization is observed in the films but not until they are heated to 700 - 800°C. These results show that substantial film consolidation has occurred at temperatures that are on the order of 100°C lower than those required to facilitate significant material diffusion. The characteristics of this consolidation remain to be clarified. Further study into the low temperature densification of this and other systems may allow these processes to be extended into other multicomponent systems which currently require higher processing temperatures.

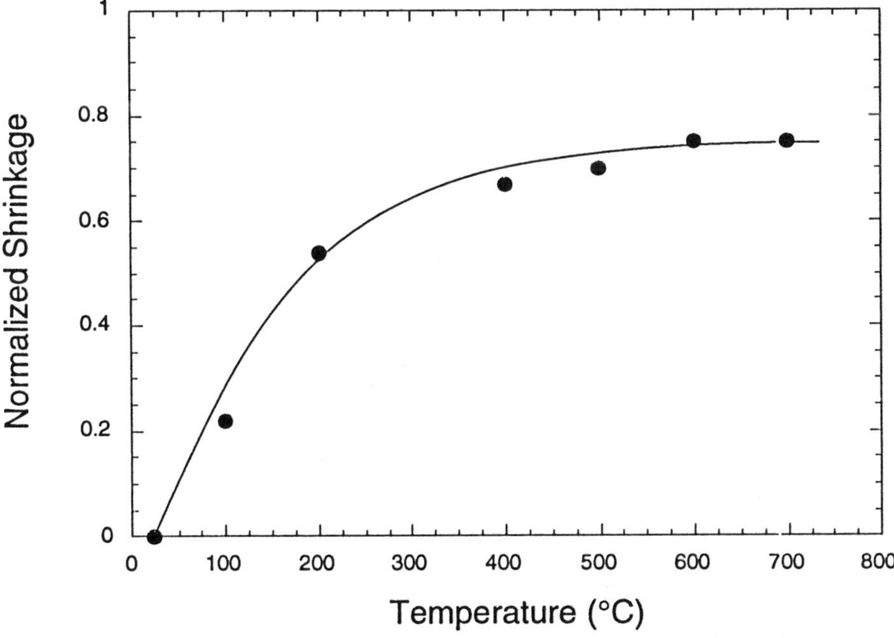

Figure 16. Normalized shrinkage versus processing temperature for SiO_2-TiO_2 films heated for 15 minutes. From Ref. 102.

5.6 Comparison of Thin Film vs. Bulk Ceramics

Measurements of the evolving characteristics of sol-gel derived ceramics in bulk form are often much easier to make than similar measurements on thin films. The very low mass of the sol-gel films in comparison to that of the substrate leads to near zero signal-to-noise ratios for many traditional measurement techniques (including optical absorption, TGA, DTA, porosimetry, etc).

Many factors need to be investigated to establish the link between thin films and bulk samples, since major differences exist between the processing of these two forms. In thin film synthesis, the gelling and drying times are much faster than for bulk materials because the sample thickness is much smaller. In addition, the coating process may lead to orientation of the polymeric species in the film that is absent in bulk samples. Also, because the diffusion distances are shorter, the retention of organic species is expected to be reduced in films.

Moreover, the presence of the substrate may significantly impact thin film development during drying, solvent burnout, densification, etc. For example, thermal expansion mismatching with the substrate can cause stresses to develop in the thin films. The constraints imposed by the substrate may also generate stresses during consolidation, and the substrate may impact the crystallization behavior of amorphous thin films by providing sites for surface nucleation and growth of stable or metastable crystalline phases at lower temperatures than in bulk material. The use of substrates as templates for epitaxial crystallization offers a unique opportunity to generate single-crystal thin films or films with preferred crystalline orientation. Finally, chemical interactions between the substrate and thin film may significantly modify thin film properties and/or performance. Careful exploration of these effects in both thin films and bulk samples of model materials will establish a basis for extrapolation of bulk gel behavior to thin film behavior in a wide variety of ceramic systems.

5.7 The Nature of Sol-Gel Research

While considerable benefit is derived from studies which focus on specific aspects of the sol-gel process, great benefit will continue to be derived from vertically integrated studies as well—i.e., those which focus on establishing chemistry/processing/microstructure/properties relationships. These studies establish the relative importance of various processing steps in the development of a desired coating and often identify previously unknown relationships, effects, and properties.

Also needed are more interactions among chemists, materials scientists, ceramists, and device engineers. This type of plea is not new, and in the past few years a noticeable increase in interdisciplinary work between chemists and materials scientists has resulted in advances in the sol-gel field. For the field to progress to anywhere near its commercial potential, though, the range of interdisciplinary interactions needs to be extended to the other, i.e., device, end of the spectrum.

6.0 FUTURE DIRECTIONS

The unique aspects of sol-gel coatings, discussed in detail above, are the extremely small and controllable pore structures, the ability to prepare organic and inorganic hybrids, and the wide range of mechanical properties and structural states which are accessible during processing. Since these features are difficult or impossible to attain by other available processing routes, much of the future of sol-gel film processing lies with them.

The ability to generate tailored pore structures in the 1 - 10 nm size range has made a myriad of new materials and applications possible including extremely light and efficient insulation (184), nanoparticle filters (185), chemical sensors (186)(187), transpiration cooled windows (188), and highly bioactive ceramic coatings for physiological uses (189), to name a few. As discussed earlier, the use of the pore structure to infiltrate other material into the matrix makes possible the generation of mixed inorganic/organic composites, semiconductor doped glasses, laser dye and non-linear polymer doped matrices. (In some of these cases, the application currently involves bulk materials but extrapolation to coating applications will be obvious to the reader). The majority of these examples currently lie in the developmental stages of research. Clearly the future will see many of these concepts reach maturity and a plethora of new devices and materials which will exploit the porosity of chemically derived films.

The evolving mechanical properties and structure of sol-gel derived films creates new processing opportunities, especially for patterning of coatings by embossing. The feasibility of this approach has been demonstrated for a wide range of applications including surface relief optical grating couplers, lenslets, reflectors and grooved discs for optical data storage. The porous structure of the film which results from drying makes possible the use of lasers to pattern the film as well. These technologies are in their infancy and their full potential has yet to be recognized.

Finally, in a large number of cases, drying and consolidation produces films that are initially amorphous. The ability to generate amorphous structures which may then be controllably crystallized to desired phases and microstructures creates the possibility of synthesizing new materials and the next generation of current materials with substantially improved properties. For example, en route to the synthesis of single-crystal layers for electro-optic applications, oriented crystal films of lithium and tantalum niobates and barium titanates have been synthesized. Huling and Messing (190) have shown that the transformation of the amorphous bulk material can be biased to produce more desirable crystal phases and crystal sizes.

In summary, the areas of research and development which have seen and will continue to see development in the future are those which focus on the unique features of ceramic films made via the sol-gel process. These features include controlled, small scale porosity, the ability to prepare mixed organic/inorganic films, and the wide range of accessible mechanical and structural states during processing.

REFERENCES

1. Eckert, C. and Weatherall, J., *Ceramic Industry,* April 53-57 (1990)
2. Brinker, C. J. and Scherer, G. W., *Sol-Gel Science,* Academic Press, New York (1990)
3. Yamane, M. et al, *J. Non-Cryst. Solids* 48:153 (1982)
4. Dislich, H., *Angew. Chem. Int. Ed. Eng.* 10:363-434 (1971)
5. Schroeder, H., *Phys. Thin Films* 5:87-141 (1969)
6. Schroeder, H. *Optica Acta* 9:249-254 (1962)
7. Johnson, C. E. Hickey, D. K. and Harris, D. C., *SPIE* 683:112 (1986)
8. Guiton, T. A. et al., *Materials Research Society Symposium Proceedings* (Brinker, Clark, and Ulrich, eds.), 121:503, Materials Research Society, Philadelphia (1988)
9. Bagnall C. M. and Zarzycki, J., *SPIE* 1328:108 (1990)
10. Nogami, M., Watanabe, M. and Nagasaka, K., *SPIE* 1328:119 (1990)
11. Tohge, N., Asuka, M. and Minami, M., *SPIE* 1328:125 (1990)
12. Seyferth, D. and Weiseman, G. H., *Science of Ceramic Chemical Processing* (Hench and Ulrich, eds.), p. 354, Wiley, New York (1986)
13. Fabes, B. D., Dale, G. W. and Uhlmann, D. R., *Ultrastructure Processing of Advanced Ceramics* (Mackenzie and Ulrich, eds.), p. 883, Wiley, New York (1988)
14. Pantano, C. G., Glaser, P. M. and Armbrust, D. J., in *Ultrastructure Processing of Ceramics, Glasses, and Composites* (L. L. Hench and D. R. Ulrich, eds.), pp. 161-177, Wiley, New York (1984)
15. Kamiya, K., Nishijima, T. and Tanaka, K., *J. Am. Ceram. Soc.* 73(9):2750-2752 (1990)

16. Scherer, G. W., *J. Non-Cryst. Solids* 87:199-225 (1986)

17. Scherer, G. W., *J. Non-Cryst. Solids* 89:217-238 (1987)

18. Scherer, G. W., *J. Non-Cryst. Solids* 91:83 (1987)

19. Scherer, G. W., *J. Non-Cryst. Solids* 91:101-121 (1987)

20. Scherer, G. W., *J. Non-Cryst. Solids* 92:122-144 (1987)

21. Evans, A. G., Drory, M. D. and Hu, M. S., *J. Mater. Res.* 3:1043-1049 (1988)

22. Garino, T. J., *Mat. Res. Soc. Symp. Proc.* (B. J. J. Zelinski et al., eds.), 180:497-502, Materials Research Society, Philadelphia, (1990)

23. Nogami, M. and Moriya, Y. *J. Non-Cryst. Solids* 48:359-366 (1982)

24. Butts, D. I., LaCourse, W. C. and Kim, S., *J. Non-Cryst. Solids* 100:514-518 (1988)

25. Melpolder, S. M. and Coltrain, B. K., *Mat. Res. Soc. Symp. Proc.* 121:811-816 (1988)

26. Scriven, L. E., *Mat. Res. Soc. Symp. Proc.* 121:717-729 (1988)

27. Elliott, D., *Microlithography Process Technology for IC Fabrication,* McGraw-Hill Book Company, New York (1986)

28. Tuck, E. O. *Phys. Fluids* 26:2352-2358 (1983)

29. Deryagin, B. M. and Levi, S. M., *Film Coating Theory, The Physical Chemistry of Coating Thin Layers on a Moving Support,* The Focal Press, London (1964)

30. Deryagin, B. V., *Zh. Eksp. i. Teoret. Fiz. USSR* 15:9 (1945)

31. Strawbridge, I. and James, P. F., *J. Non-Cryst. Solids* 82:366-372 (1986)

32. Volarovich, M. O. and Levi, S. M., *Koll. Ah.* 18:129 (1956)

33. Hurd, A. J. and Brinker, C. J., *Mat. Res. Soc. Symp. Proc.* 121:731-742 (1988)

34. Yang, C. C., Josefowicz, J. Y. and Alexandru, L., *Thin Solid Films* 74:117-127 (1980)

35. Meyerhofer, D., *J. Appl. Phys.* 49:3993-3997 (1978)

36. Emslie, A. G., Bonner, F. R. and Peck, L. G., *J. Appl. Phys.* 29:858-862 (1958)

37. Pantano, C. G., Glaser, P. M. and Armbrust, D. J., *Ultrastructure Processing of Ceramics, Glasses, and Composites,* pp. 161-177 (1984)

38. Brinker, C. J. and Scherer, G. W., *Sol-Gel Science,* p. 797, Academic Press, New York (1990)

39. Birnie, D. P., III, Zelinski, B. J. J., Marvel, S. P., Melpolder, S. M. and Roncone, R. L., *Film/Substrate/Vacuum-Chuck Interactions During Spin Coating,* submitted to Optical Engineering (1991)

40. Fabes, B. D., and Uhlmann, D. R., *J. Am. Ceram. Soc.* 74:978-988 (1990)

41. Blum, Y. D., Platz, R. M. and Crawford, E. J., *J. Am. Ceram. Soc.* 73:170-172 (1990)

42. Orgaz, F., and Capel. F., *Rivista della Staz* 6:147-152 (1986)

43. Maddalena, A. et al., *J. Non-Cryst. Solids* 100:461 (1988)

44. Yoldas, B. E. and Partlow, D. P., *Thin Solid Films* 129:1-14 (1985)

45. Lipeles, R. A., Coleman, D. T. and Leung, M. S., *Mat. Res. Soc. Symp. Proc.* (Brinker, Clark, and Ulrich, eds.), 73:665, Materials Research Society, Philadelphia, (1986)

46. Budd, K. D., Dey, K. and Payne, D. A., *Br. Ceram. Proc.* 36:107 (1986)

47. Nanao, T. and Eguchi, T. *Japan Kokai Tokkyo Koho,* JP 61/97159 A2 (15 May 1986)

48. Takayama, R. and Tomita, Y., *J. Appl. Phys.* 65:1666 (1989)

49. Croteau, A., Matsubara, S.. Miyasaka, Y. and Shohata, N., *Japan J. Appl. Phys.* 26(Suppl. 2):18 (1987)

50. Adachi, M. et al., *Japan J. Appl. Phys.* 26:550 (1987)

51. Sreenivas, K. and Sayer, M., *J. Appl. Phys.* 64:1484 (1988)

52. A. Okada, *J. Appl. Phys.* 48:2905 (1977)

53. Dey, S. K., Budd, K. D. and Payne, D. A., *IEEE Trans. on Ultra, Ferro and Freq. Control* 35:80 (1988)

54. Yi, B., Wu, Z. and Sayer, M., *J. Appl. Phys.* 64:2717 (1988)

55. Payne, D. A., Presented at the *5th International Conference on Ultrastructure Processing of Ceramics, Glasses, and Composites,* Tucson (1989)

56. Seth, V. K. and Schulze, W. A., *Proc. First Symp. Integ. Ferroelectrics,* Colorado Springs, pp. 175-184 (1989)

57. Chen, C. J., Wu, E. T., Xu, Y. M., Chen, K. C. and Mackenzie, J. D., *Proc. 1st Symp. Integ. Ferroelectrics,* Colorado Springs, pp. 185-188 (1989)

58. Dey, S. K. and Zuleeg, R. *Proc. 1st Symp. Integ. Ferroelectrics,* Colorado Springs, pp. 189-194 (1989)

59. Tuttle, B. A., Garino, T. J., Doughty, D. M., Martinez, S. L. and Yio, J. L., presented at Annual Meeting, American Ceramic Soc., Indianapolis, IN (1989)

60. Hsueh, C. C. and McCartney, M. L., presented at Annual Meeting, American Ceramic Soc., Indianapolis, IN (1989)

61. Lipeles, R. A., Ives, N. A. and Leung, M. S.,in: *Science of Ceramic Chemical Processing* (L. L. Hench and D. R. Ulrich, eds.), pp. 320-326, Wiley, New York (1986)

62. Lipeles, R. A. and Coleman, D. J., *Ultrastructure Processing of Advanced Ceramics* (Mackenzie and Ulrich, eds.), pp. 919-924, Wiley, New York (1988)

63. Haertling, G. M., *Am. Ceram. Soc. Bull.* 43:875 (1964)

64. Shiosaki, T., Adachi, M., Mochizuki, S., and Kawabata, A., *Ferroelectrics* 63:227 (1985)

65. Castellano, R. N. and Feinstein, L. G., *J. Appl. Phys.* 50:4406 (1979)

66. Ishida, M., Matsunami, M. and Tanaka, T., *J. Appl. Phys.* 48:951 (1977)

67. Adachi, H., Mitsuyu, T., Yamazaki, O. and Wasa, K., *J. Appl. Phys.* 60:736 (1986)

68. Adachi, H., Kawaguchi, T., Kitabatake, M. and Wasa, K. *Japan J. Appl. Phys.* 22(Suppl. 2):11 (1983)

69. Volz, H., Koger, K. and Schmitt, H., *Ferroelectrics* 51:87 (1983)

70. Fijima, K., Kawashima, S. and Veda, I., *Japan J. Appl. Phys.* 24(Suppl. 2):482 (1985)

71. Fijima, K., Tomita, Y., Takayama, R. and Veda, I., *J. Appl. Phys.* 60:361 (1986)

72. Okada, M., et al., *Japan J. Appl. Phys.* 28:1030 (1989)

73. Lipeles, R. A., Coleman, D. T. and Leung, M. S., *Mat. Res. Soc. Symp. Proc.* (Brinker, Clark, and Ulrich, eds.), 73:671, Materials Research Society, Philadelphia, (1986)

74. Uhlmann, D. R., et al., in *Mat. Res. Soc. Symp. Proc.* (B. J. J. Zelinski et al., eds.), 180:645, Materials Research Society, Philadelphia, (1990)

75. Suchoff, L. A., U.S. Patent 3,002,861 (Oct. 3, 1961)

76. Mitchell, J. J., U.S. Patent 3,066,048(Nov. 27, 1962)

77. Sakka, S., *Kagaku Sochi* 27:98 (1985)

78. Paulson, B. A., Report IS-T-1336 NTIS (1987)

79. Fukushima, J., Kodaira, K. and Matsushita, T., *Amer. Ceram. Soc. Bull.* 55:1065 (1976)

80. Pebler, A. R., and Charles; R. G., U.S. Patent 4,485,094 (Nov. 27, 1984)

81. Dosch, R. G., *Mat. Res. Soc. Symp. Proc.* (C. J. Brinker, ed.), 32:157, Materials Research Society, Philadelphia, (1984)

82. Mohallem, N. D. S. and Aegerter, M. A., *Mat. Res. Soc. Symp. Proc.* (Brinker, Clark, and Ulrich, eds.), 121:515, Materials Research Society, Philadelphia (1988)

83. Yoko, T., Kamiya, K. and Sakka, S., *Res. Rep. Fae. Eng. Mie Univ.* 49 (1987)

84. Pettit, R. B., et al., *Sol-gel Technology for Thin Films, Fibers, Preforms, Electronics and Specialty Shapes,* (L. C. Klein, ed.), pp. 80-109, Noyes, Park Ridge, NJ, (1988)

85. Yoldas, B. E., U.S. Patent 4,346,131 (Aug. 24, 1982)

86. Rehg, T. J., Ochoa-Tapia, J. A., Knoesen, A. and Miggins, B. G., *Appl. Optics* 28:5215 (1989)

87. Brinker, C. J. and Harrington, M. S., *Sol. Energy Mat.* 5:159 (1981)

88. Ashley, C. S. and Reed, S. T., *Mat. Res. Soc. Symp. Proc.* (Brinker, Clark, and Ulrich, eds.), 73:671, Materials Research Society, Philadelphia (1986)

89. Pettit, R. B., Brinker, C. J. and Ashley, C. S., *Solar Cells,* 15:267 (1985)

90. Dislisch, H., *Sol-gel Technology for Thin Films, Fibers, Preforms, Electronics and Specialty Shapes,* (L. C. Klein, ed.), pp. 50-79, Noyes, Park Ridge, NJ (1988)

91. Hinz, P. and Dislisch, H., *J. Non-Cryst. Solids* 82:411 (1986)

92. Biswas, P. K., Kundu, D. and Ganguli, D., *J. Mat. Sci. Lett.* 8:1436 (1989)

93. Floch, H. G. and Priotton, J. J., *Ceram. Bulletin* 69:1141 (1990)

94. Melpolder, S. M., Hanranhan, M. J. and Musshaven, G. N., Presented at the Presented at the *5th International Conference on Ultrastructure Processing of Ceramics, Glasses, and Composites,* Tucson (1989)

95. Keddie, J. L., and Giannelis, E. P., *Mat. Res. Soc. Symp. Proc.* (B. J. J. Zelinski et al., eds.), 180:383, Materials Research Society, Philadelphia (1990)

96. Partlow, D. P. and O'Keefe, T. W., *Appl. Opt.* 29:1526 (1990)

97. Debsikdar, J. C., *J. Non-Cryst. Solids* 91:262 (1987)

98. Debsikdar, J. C., U.S. Patent 4,830,879 (May 16, 1989)

99. Lukosz, W. and Tiefenthaler, K., *Opt. Lett.* 8:537 (1983)

100. Ulrich, R. and Weber, H. P., *Appl. Opt.* 11:428 (1972)

101. Dale, G. W., et al., *Mat. Res. Soc. Symp. Proc.* (B. J. J. Zelinski et al., eds.), 180:371, Materials Research Society, Philadelphia (1990)

102. Weisenbach, L., et al., *Mat. Res. Soc. Symp. Proc.* (B. J. J. Zelinski et al., eds.), 180:377, Materials Research Society, Philadelphia (1990)

103. Guglielmi, M., Colombo, P., Esposti, L. M. D., Righini, G. C. and Pelli, S., to appear in *SPIE* 1513 (1991)

104. Zelinski, B. J. J., et al., Presented at the *5th International Conference on Ultrastructure Processing of Ceramics, Glasses, and Composites,* Tucson (1989)

105. Weisenbach, L., et al., *SPIE* 1590:50-58 (1991)

106. Martin, A. J. and Green, M., *SPIE* 1328:352 (1990)

107. Roncone, R. L., et al., *SPIE* 1590:14-25 (1991)

108. Roncone, R. L., et al., *J. Non-Cryst. Solids* 128:111 (1991)

109. Matsuda, A., et al., *SPIE* 1328:62 (1990)

110. Matsuda, A., et al., *SPIE* 1328:71 (1990)

111. Krchnavek, R. R., Gilgen, H. H. and Osgood, R. M., *J. Vac. Sci. Technol.* B2:4 641-644 (1984)

112. Taylor, D. J., Fabes, B. D., and Steinthal, M. G., *Mat. Res. Soc. Symp. Proc.* (B. J. J. Zelinski et al., eds.), 180:1047-1052, Materials Research Society, Philadelphia (1990)

113. Fiori, C. and Devine, R. A. B., *Mat. Res. Soc. Symp. Proc.* 61:187-195 (1986)

114. Fabes, B. D., et al., *SPIE* 1328:319-328 (1990)

115. Guglielmi, M., Colombo, P., Esposti, L. M. D., Righini, G. C., Pelli, S. and Rigato, V., presented at the *Sixth International Workshop on Glasses and Ceramics from Gels,* Seville, Spain (October 6-11, 1991)

116. Taylor, D. J. and Fabes, B. D., presented at the *Sixth International Workshop on Glasses and Ceramics from Gels,* Seville, Spain (October 6-11, 1991)

117. Makishima, A., Asami, M. and Wada, K., J. Non-Cryst. Solids 100:321 (1988)

118. Makishima, A., et al., *J. Am. Cer. Soc.* 64:C127 (1986)

119. Mohallem, N. D. S. and Aegerter, M. A., *J. Non-Cryst. Solids* 100:526 (1988)

120. Duran, A., et al., *J. Non-Cryst. Solids* 100:494 (1988)

121. Duran, A., et al., *Riv. Stn. Sper. Vetro* 16:59 (1986)

122. Duran, A., et al., *J. Non-Cryst. Solids* 82:391 (1986)

123. Yamamoto et al., *Yogyo Kyokaishi* 91:222 (1983)

124. Orgaz, F. and Rawson, H., *J. Non-Cryst. Solids* 82: 378 (1986)

125. Kojima, T., *Japan Kokai Tokkyo Koho,* JP 62/32153 (Feb. 12, 1987)

126. Sakka, S., *Sol-Gel Science and Technology,* (M.A. Aegerter et al., eds.), pp. 346-374, World Scientific, Singapore (1989)

127. Matsunami, H., et al., *J. Phys. Soc. Japan* 49 (Suppl. B):194 (1980)

128. Higgshino, H., et al., *Japan J. Appl. Phys.* 24(Suppl. 2):284 (1985)

129. Okada, A., *Ferroelectrics* 14:739 (1976)

130. Vest, R. W. and Xu, J., *Ferroelectrics* 93:21 (1984)

131. Hirano, S., and Kato, K., *J. Non-Cryst. Solids* 100:538 (1988)

132. Hirano, S., and Kato, K., *Adv. Cer. Materials* 3:503 (1988)

133. Hirano, S., and Kato, K., *Solid State Ionics* 32/33:765 (1989)

134. Hirano, S., and Kato, K., *Bull. Chem. Soc.* Japan 62:429 (1989)

135. Hirano, S., and Kato, K., *Mat. Res. Soc. Symp. Proc.* 155:181 (1989)

136. Chen, C. J., Xu, Y., Xu, R., and MacKenzie, J. D., presented at 1st Int'l Cer. Sci. and Tech. Cong., Anaheim (1989)

137. Schwartz, S. L., Bright, S. J ., Busch, J. R., and Shrout, T. R., presented at 1st Int'l Cer. Sci. and Tech. Cong., Anaheim (1989)

138. Uhlmann, D. R., Zelinski, B. J. J., Teowee, G., Boulton, J. M. and Koussa, A., presented at the 4th International Otto-Schott Colloqium, Jena, Germany, July, 1990, accepted for publication in *J. Non-Cryst. Solids*

139. Uhlmann, D. R., et al., Presented at the *5th International Conference on Ultrastructure Processing of Ceramics, Glasses, and Composites,* Tucson (1989)

140. Partlow, D. P. and Greggi, J., *J. Mat. Res.* 2:595 (1987)

141. Eichorst, D. J. and Payne, D. A., *Mat. Res. Soc. Symp. Proc.* (Brinker, Clark, and Ulrich, eds.), 121:773, Materials Research Society, Philadelphia (1988)

142. Kobayashi, Y., Imai, Y. and Kurokawa, Y., *J. Mater. Sci.* Lett. 7:1148 (1988)

143. Kobayashi, Y., Kurokawa, Y., Imai, Y. and Muto, S., *J. Non-Cryst. Solids* 105:198 (1988)

144. Sasaki, M., Kobayashi, Y., Muto, S. and Kurokawa, Y., *J. Am. Ceram. Soc.* 73:453 (1990)

145. Reisfeld, R., Zusman, R., Cohen, Y. and Eyal, M., *Chem. Phys. Lett.* 147:142 (1988)

146. Reisfeld, R., Eyal, M. and Brusilovsky, D., *Chem. Phys. Lett.* 153:210 (1988)

147. Avnir, D., Kaufman, V. R. and Reisfeld, R., *J. Non-Cryst. Solids* 74:395 (1985)

148. Chernyak, V., Reisfeld, R., Gvishi, R. and Venezky, D., *Sens. Mater.* 2:117 (1990)

149. Dunn, B., Knobba, E., McKiernan, J. M., Pouxviel, J. C. and Zink, J. I., *Mat. Res. Soc. Symp. Proc.* (Brinker, Clark, and Ulrich, eds.), 121:331, Materials Research Society, Philadelphia (1988)

150. Tanaka, H., Takahashi, J., Tsuchiya, J., Kobayashi, Y. and Kurokawa, Y., *J. Non-Cryst. Solids* 109:164 (1989)

151. Kaufman, V. R., Levy, D. and Avnir, D., *J. Non-Cryst. Solids* 82:103 (1986)

152. Reisfeld, R., *SPIE,* 1328:29 1990)

153. Covino, J. and McManis, G. E., *Mat. Res. Soc. Symp. Proc.* (Brinker, Clark, and Ulrich, eds.), 121:553, Materials Research Society, Philadelphia (1988)

154. Chemseddine, A., Babonneau, F. and Livage, J., *J. Non-Cryst. Solids* 91:271 (1987)

155. Chemseddine, A., Morineau, R. and Livage, J., *Solid-State Ionics* pp. 9-10 (1983)

156. Chemseddine, A., Henry, M. and Livage, J., *Rev. Chim. Miner.* 21:487 (1984)

157. Kato, I., Ariizumi, A. and Kimura, N., *Japan. Kokai Tokkyo Koho,* JP 61/36292 (Feb. 20, 1986)

158. Yoshino, T., Baba, N. and Yasuda, K., *Nippon Kagaku Kaishi,* 9:1525 (1988)

159. Tompkin, W. R., Coyd, R. W., Hall, D. W., and Tick, P. A., *Opt. Soc. Am.* B4:1030 (1987)

160. Prasad, P. N., *SPIE* 1328:168 (1990)

161. Boulton, J. M., Thompson, J., Fox, H. H., Gorodisher, I., Teowee, G., Calvert, P. D. and Uhlmann, D. R., *Mat. Res. Soc. Symp. Proc.* (B. J. J. Zelinski et al., eds.), 180:987, Materials Research Society, Philadelphia, (1990)

162. Gubina, L. N. and Lygin, V. I., *Russ. J. Phys. Chem.* 54:751 (1980)

163. Hench, L. L., Wang, S. H. and Park, S. C., *SPIE* 505:90 (1984)

164. Hench, L. L., *Science of Ceramic Chemical Processing* (Hench and Ulrich, eds.), pp. 52-54, Wiley, New York (1986)

165. Schmidt, H., *J. Non-Cryst. Solids* 112:419 (1989)

166. Boulton, J. M., Fox, H. H., Neilson, G. F. and Uhlmann, D. R., *Mat. Res. Soc. Symp. Proc.* (B. J. J. Zelinski et al., eds.), 180:773-776, Materials Research Society, Philadelphia, (1990)

167. Schmidt, H., *Sol-Gel Science and Technology,* (M.A. Aegerter et al., eds.), p. 432, World Scientific, Singapore (1989)

168. Pope, E. J. A., Asami, M. and Mackenzie, J. D., *J. Mater. Res.* 4:1018 (1989)

169. Parkhurst, C. S., et al., *Mat. Res. Soc. Symp. Proc.* (Brinker, Clark, and Ulrich, eds.), 73:769-773, Materials Research Society, Philadelphia, (1986)

170. Huang, H. H., Orler, B. and Wilkes, G. L., *Macromol.* 20:1322 (1987)

171. Mark, J. E. and Sun, C. C., *Polym. Bull.* 18:259 (1987)

172. Spinu, M., Arnold, C. and McGrath, J. E., *Polym. Preprints* 30:125 (1989)

173. Yoldas, B. E., *J. Mater. Sci.* 10:1856 (1975)

174. Shoup, R. D., *Ultrastructure Processing of Advanced Ceramics* (Mackenzie and Ulrich eds.), pp. 347-354, Wiley, New York (1988)

175. Garino, T. J. and Bowen, H. K., *J. Am. Ceram. Soc.* 70:C315 (1987)

176. Hubert-Pfalzgraf, L. G., *New J. Chem.* 11:663 (1987)

177. Brinker, C. J., Hurd, A. J. and Ward, K. J., *Ultrastructure Processing of Advanced Ceramics* (Mackenzie and Ulrich, eds.), p. 223, Wiley, New York (1988)

178. Lawrence, C. J., *Phys. Fluids* 31:2786 (1988)

179. Higgins, B. G., *Phys. Fluids* 29:3522 (1986)

180. Matsuba, I. and Matsumoto, K., *IEEE Trans. Elect. Dev.* ED-33: 1263 (1986)

181. Weill, A. and Dechenaux, E., *Poly. Eng. Sci.* 28:945 (1988)

182. Skrobis, K. J., Denton, D. D. and Scrobis, A. M., *Poly. Eng. Sci.* 30:193 (1990)

183. Daniels, B. K., Szmanda, C. R., Templeton, M. K. and Tretonas, P., III, *SPIE* 631:192 (1986)

184. LeMay, J. D., Hopper, R. W., Hrubesh, L. W. and Pekala, R. W. *Mat. Res. Soc. Symp. Proc.* 15:30 (1990)

185. Klein, L. C., *Sol-gel Technology for Thin Films, Fibers, Preforms, Electronics and Specialty Shapes,* (L. C. Klein, ed.), p. 382, Noyes, Park Ridge, NJ (1988)

186. Lukosz, W., and Tiefenthaler, K. *Sensors and Actuators* 15:273 (1988)

187. Frye, G. C., Brinker, C. J., Ricco, A. J., Martin, S. J., Hilliard, J. and Doughty, D. H., *Mat. Res. Soc. Symp. Proc.* (B. J. J. Zelinski et al., eds.), 180:583, Materials Research Society, Philadelphia (1990)

188. Fosmoe, A. G., II, and Hench, L. L., *Mat. Res. Soc. Symp. Proc.* (B. J. J. Zelinski et al., eds.), 180:843, Materials Research Society, Philadelphia (1990)

189. Hench, L. L. and Wilson, J., *Mat. Res. Soc. Symp. Proc.* (B. J. J. Zelinski et al., eds.), 180:1061, Materials Research Society, Philadelphia (1990)

190. Huling, J. C. and Messing, G. L., *Mat. Res. Soc. Symp. Proc.* (B. J. J. Zelinski et al., eds.), 180:515, Materials Research Society, Philadelphia (1990)

8

Electronic Thick Film Technology

Daniel J. Shanefield

1.0 INTRODUCTION

The type of electronic circuit usually referred to as a *thick film* consists of a cermet coating approximately 25 µm (0.001 inch) thick, on top of a supporting sheet of ceramic called a *substrate*. (By contrast, another type of circuit which is referred to as a *thin film* is approximately 1 µm thick.). Thick films are often used where a moderate degree of reliability is required, while the more expensive thin films are most often reserved for higher reliability applications. Thick films find large scale use in consumer applications such as television receivers, personal computers, and automobiles. However, thick films are not excluded from all high performance applications, and they are reliable enough for certain designs used in satellites, telephones, military applications, and supercomputers, especially where the particular device characteristics can vary somewhat during normal service without harming system performance.

The total U.S. market (1) for thick films in 1989 was estimated at about 5 billion dollars worth of product, with an annual growth rate of about 11%. (The thin film market was estimated at about two billion dollars, with a growth rate of about 13%.)

Thick films are an offshoot of other aspects of ceramic technology, since they involve the firing of ceramic powders and glasses, in combination with the some of the technologically similar processes of powder metallurgy. Historically, these materials (as well as electronic thin films) were derived from the strictly decorative gold or silver circular lines that had previously been printed onto dishes and drinking glasses. During World War II, the highly miniaturized circuitry needed for electronic artillery fuses made use of printed and fired-on films of silver, on top of insulating ceramic substrates. However, when using pure silver, the particular resistors which required high resistance values were difficult to produce, because the metal is too conductive. A major improvement was made by James D'Andrea of DuPont, by blending silver metal powder with glass powder and an organic chemical paint vehicle, which could then be printed in a complex pattern and fired to form a high resistance cermet (2). Eventually, many other material combinations were added to the technology.

Present day thick films are almost always screen printed onto previously fired aluminum oxide substrates, dried to evaporate the temporary solvents, and then fired at about 850°C to sinter the powders to a dense solid film. Screen printing, or *serigraphic printing*, makes use of a wire or Nylon mesh, some areas of which are plugged with *emulsion* polymer (usually photoresist). A cross section of the screen is shown schematically in Fig. 1 (not drawn to scale). The screen is placed close to the previously fired ceramic substrate, and ink (usually called *paste* in thick film technology) is placed on top of the screen. A soft rubber wiper or "squeegee" is moved across the screen, driving the paste in front of it and down through the unplugged areas onto the substrate. (The screen usually deflects in a complex manner, but that is not shown in this diagram.) The screen is then removed, leaving a patterned deposit of paste on the substrate, which is then dried and fired to become the thick film.

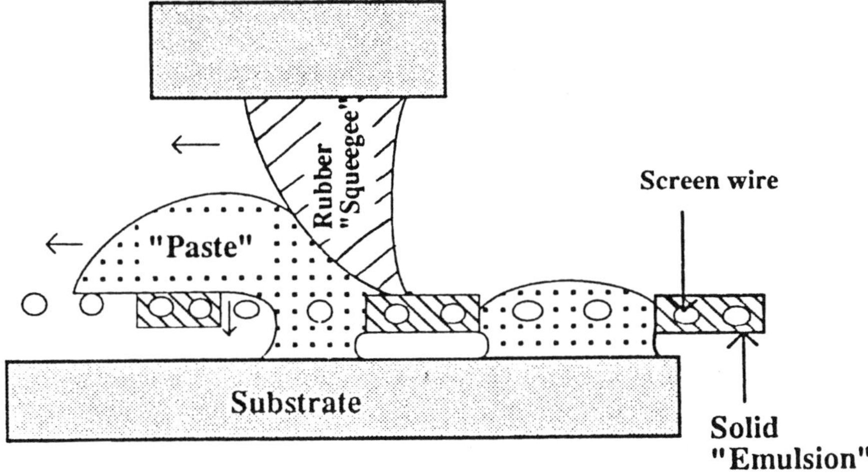

Figure 1. Schematic drawing of screen printer operation.

This patterning process is essentially *additive.* In other words, film material is deposited onto the substrate only where it is needed, but it is not deposited elsewhere and then etched away as it is in the *subtractive* patterning of thin films. This yields an important cost advantage because it is an inherently simple process, and it does not involve high vacuum or a great degree of cleanliness. The wet printing process tends to bridge over small defects, dirt particles, and substrate roughness.

Thick films can incorporate conductive lines, *crossovers* where two lines cross but do not make contact (including rather complex multilayer planes), resistor areas, capacitor areas, and the addition of silicon integrated circuit chips. The combinations of the thick films plus silicon chips are called *hybrid integrated circuits*, or simply "HICs." (Thin films can also be used in hybrid ICs.)

Photographs of a thick film circuit in various stages of manufacture are shown in Fig. 2.

1.1 Comparisons to Competing Technologies

Printed wiring boards (sometimes referred to as *printed circuits*) are in some ways a similar technology. These are made with copper conductors on polymer substrates, usually with separate resistors and capacitors, etc., soldered onto the copper. They are not referred to as *integrated circuits* because the resistors and capacitors are not made integral to the rest of the printed circuit, as they are in thick films and thin films. A major advantage of thick and thin film HICs is the fact that conductors, insulators, resistors, and capacitors can all be made in the same general type of processing, with thousands of devices being made at the same time. This increases the reliability and, at the same time, lowers the cost of the product. (However, a new type of polymer-substrate thick film is now being used for ultra-low cost, fairly low reliability applications, where some of the integration advantages of thick films are merged with the lower cost of polymeric printed circuits.)

Table 1 shows some comparisons between typical thick and thin films. An important difference is in the means by which adhesion to the ceramic substrate is obtained. In thick films, the adhesion is mostly mechanical. That is, a controlled degree of roughness in the surface of the substrate interlocks with a conformal roughness in the thick film. Also, a small amount of easily melted glass in the substrate composition intermingles with glass in the thick film cermet during the firing step. This provides additional adhesion strength. As indicated in Table 1, the substrate contains several percent impurities and is relatively rough, in order to provide the necessary adhesion.

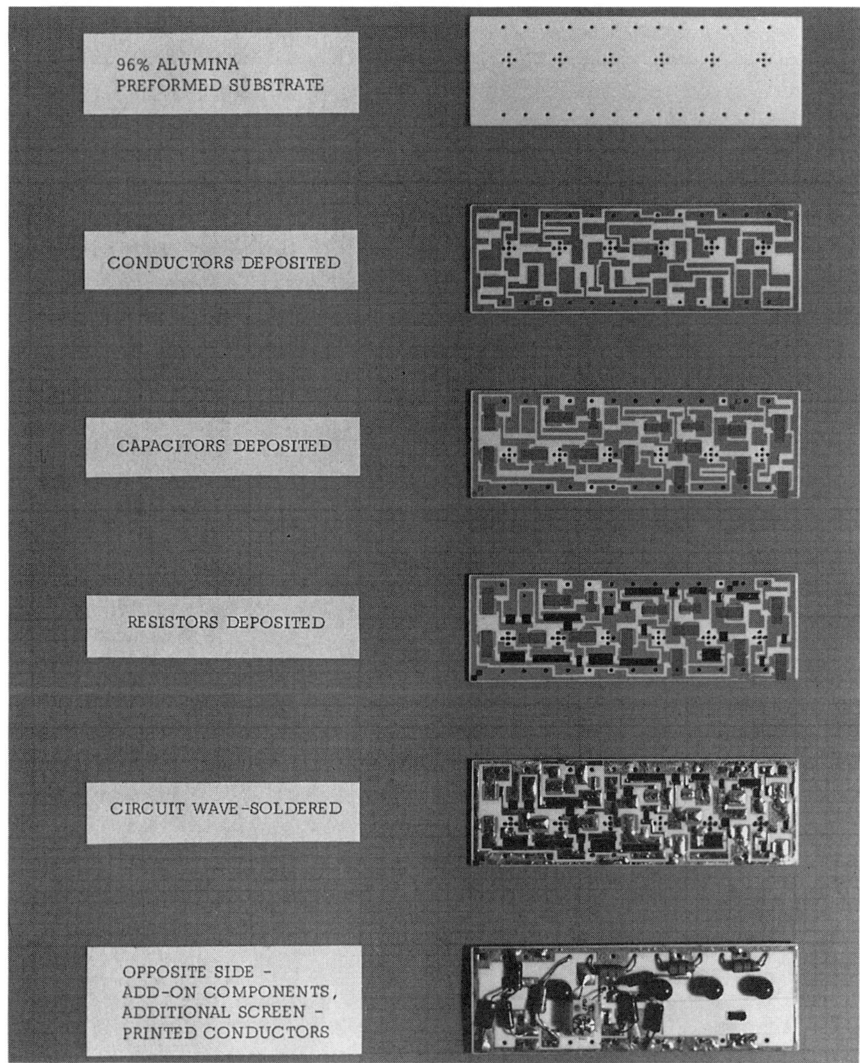

Figure 2. Successively printed and fired thick film conductors and devices, with later addition of separate devices.

Table 1. Comparison of Hybrid IC Films

	Thin Films	Thick Films
Thickness	1 μm	1 mil (25 μm)
Line Width	≥ 3 μm	≥ 3 mils
Conductors	Ti, Pd, Au Cr, Ni, Cu	AgPd Au Cu* (Usually with glass)
Resistors	NiCr Ta_2N	PdO_x RuO_2 LaB_6*
Dielectrics	SiO_2 Ta_2O_4	Glass Glass Ceramic
Adhesion	Chemical	Mechanical (Chemical if Bi, Cd)
Reproducibility	Excellent	Good
Reliability	Excellent	Good
Cost	High	Low
Deposition	Evaporation Sputtering	Screen Print and Fire
Substrate	99 ± 0.5% Alumina Smooth (0.1 μm)	95 ± 5% Alumina Rough (0.5 μm)

* Fired in nitrogen

Several general sources of information on the overall thick film technology are available (3)-(5) although they do not cover recent developments or provide detailed formulations, since these are likely to be proprietary.

2.0 MATERIALS

2.1 Substrates

Thick film circuits are mostly deposited onto alumina substrates, which contain about 5% glassy additives to enhance adhesion of the film, as discussed briefly above. These substrates are ordinarily made by ball milling the materials shown in Table 2 for about 24 hours, and then casting the resulting material to form a thin sheet *(tape)*, using a *doctor blade* to control the thickness (6). The milling is best done in two stages, the first few hours without the binder and plasticizer, and the remaining time with those two last materials added into the mill. The sheet of cast and dried tape is then cut to a rectangular shape (typically about 15% larger than the desired final size, to compensate for firing shrinkage), and this is fired at about 1500°C for approximately one hour.

Table 2. Alumina Substrate Starting Materials

Material	Function	Amount
Alumina Powder, 2 μm median size	Ceramic Precursor	69 wt.%
Clay, Kaolinite	Adhesion Promoter	3
Talc, 80 m^2/gm	Adhesion Promoter	3
Toluene	Solvent	15
Ethanol	Solvent	5
Menhaden Fish Oil	Dispersant	0.5
Polyvinyl Butyral, Molecular Wt: Approx. 10,000	Binder	2.5
Polyethylene Glycol, Molecular Wt: Approx. 400	Plasticizer, Release Agent	2.0

One of the reasons for mounting the chips on ceramic substrates to form HICs is to provide a good heat sink for the chip, through the large area ceramic substrate. There is a current trend toward a larger scale of integration in silicon chip technology which is leading to greater heat dissipation of the chips. Therefore materials with better heat conductivity than alumina are being sought as substrates for the newer chip designs. Table 3 (assembled from data in Ref. 7) summarizes some important properties of alternative materials such as aluminum nitride, which are being actively considered for substrates of the future. Increased heat conductivity would also allow improved design of the thick film power resistors, especially for highly miniaturized power supplies and other resistor applications.

Another factor in choosing materials for improved substrates is the thermal expansion coefficient. As silicon chips mounted on HICs get larger, the mismatch in expansion between the silicon and the ceramic substrate causes stresses to build up in the chip. This can change transistor characteristics or even break the chip. Table 3 shows that aluminum nitride, silicon carbide, and a few other materials have better matches to silicon than alumina.

Some materials such as cordierite or boron nitride might appear to be good candidates; however, most materials other than AlN and SiC have some sort of disadvantage such as relatively poor mechanical strength (usually because of intergranular stresses from highly anisotropic thermal expansion). Some candidates exhibit poor resistance to chemical corrosion in processing etchants (cordierite), high cost (SiC), or poor adhesion of films to the material in question (BN).

In fact, the adhesion of films to aluminum nitride has been highly unreproducible in large scale HIC manufacturing processes, in spite of the fact that smaller scale laboratory results have been promising. Possibly current developments such as the growth of optimal-thickness oxide surface layers will overcome this problem (8).

Another development area with newer substrate materials is the matching of thermal expansion coefficients in the interface between the substrate and thick film resistor materials. The current large scale production resistor materials are designed to match alumina substrates, but when AlN or SiC are used, unpredictable variations in resistance occurs as the temperature changes. Further development work may solve the problem (9).

2.2 Conductors

Although pure silver was historically the first thick film conductor material, it has two serious disadvantages. One is that it dissolves readily

Table 3. Alternative Substrate Materials

	Coefficient of Thermal Expansion ppm/°C	Thermal Conductivity W/km	Dielectric Constant (Usually 1 MHz)	Bending Strength (MOR) ksi
Silica, Amorph.	0.5 - 0.7	0.5 - 2	3 - 4	1 - 10
Pyrex, Corning 7740	3.3	1.2	4.7 - 5.1	1
Glass, Corning 7059	4.6	1.2	5	1 - 12
Porcelain, Electrical	4 - 9	1.7	5.5	12 - 24
Mullite	4 - 5	4 - 7	5.4 - 6.6	18 - 40
Cordierite	1 - 3(A)	4	4 - 5	17
Alumina ≥ 99%	6.5 a5.2*,c6.7*	25 - 40, 50*	8 - 9	30 - 40
Alumina 96%	6.3 - 9.1	12 - 26	8 - 10	30 - 50
AlN	$\underline{3.1}$ - 4.6	60 - 230, 320*	8 - 10	40 - 50
Silicon Nitride Hot pressed, alpha	3.0 - 4.0	10 - 33, 80*	5 - 7	45 - 100
SiC Hot pressed, alpha	$\underline{2.5}$ - 4.7	33 - 270, 330*	40 - 100+	40 - 71
Boron Carbide	4.5	300		
BN	0.4 - 3.8(A)	12 - 250, 1300*	3.4 - 5.1	12
BN, Cubic	4	950 - 1300	3 - 6	
BeO 99.5%	5 - 7	200 - 280, 320*	5.8 - 6.7	20 - 40
Si*	$\underline{2.7}$ - 3.6	130 - 1240	12 - 100+	2

* Single Crystal
(A) Highly anisotropic
Underlined value is at 50°C

in molten tin-lead solder, making the solder bonding of termination wires difficult. Another problem is that continuous electric fields, in the presence of high humidity, tend to cause the slow growth of silver filaments across bare substrate areas which should be electrical insulators (10). These silver (or semiconductive silver oxide) filaments eventually lead to short circuits in the system.

Both of these problems are solved to some degree by alloying the silver with 20% palladium, making this 80/20 alloy the most popular thick film conductor metal in current use.

To manufacture a conductor ink *(paste)* for screen printing, a small amount of glass powder *(frit)* is typically ball milled or three-roll milled with the pre-alloyed metal powder, together with the organic chemicals shown in Table 4, for about 15 minutes. The metal powder (3) should have a B.E.T. specific surface area of about 10 m^2/gm. Various glass frits have been used successfully, but the composition of Table 5 is popular. Frit compositions have been reported in the literature (3)(10), but a more reliable indication of what is actually used is available from published analyses of commercial thick film pastes (11)(12). The size of the glass powder particles is not critical, since the milling step tends to break down the particles to a nearly constant size.

Table 4. Thick Film Conductor Composition

Material	Function	Amount
80 Ag, 20 Pd Alloy Powder, 4 m^2/gm	Conductor	71 wt.%
Glass Powder (See Table 5)	Adhesion Promoter	4
Alpha Terpineol	Solvent	18
Ethyl Cellulose	Binder	3.5
Menhaden Fish Oil	Dispersant	1
Palmitic Acid	Thixotropy Agent	2.5

Table 5. Adhesion Promoting Glass Composition

Material	Amount
Silica	66 wt.%
Boron Trioxide	15
Alumina	13
Barium Oxide	6

The oxides of sodium, potassium, and lead were incorporated into the glass compositions of early thick films, in order to lower the firing temperature (3). Modern electronic industry formulators, however, tend to avoid these materials, because of a tendency toward degraded electrical resistance on nearby insulating areas of the substrate, and environmental problems. Barium and boron oxides are used as rather benign substitutes.

Instead of the glass additive shown in the table, a few percent of bismuth or cadmium or copper can be added to the metal alloy in order to provide adhesion. During the firing step, either of these non-noble elements tends to migrate to the ceramic-metal interface and oxidize slightly, thus intermingling with the glass phase of the ceramic substrate and giving the system adhesion. The electrical conductivity of the system can be somewhat improved by eliminating the glass frit, as can the thermosonic bondability for wire bonding. However, in order to achieve good results with this technique, the firing must be quite precisely adjusted to avoid underfiring or overfiring and a consequent loss of adhesion.

The solvent should be a liquid which does not evaporate rapidly at room temperature. Screen clogging would occur if fast evaporating solvents were used. Therefore, toluene and alcohol, which are excellent for fast drying casting formulations (Table 2), are not satisfactory for screen printing. However, alpha terpineol or butyl carbitol acetate are suitably nonvolatile, but they can be evaporated in a controlled manner after printing by gentle heating.

Palmitic acid (or other organics such as 2-furoic acid) serve to provide a low viscosity during screen printing, but a high viscosity immediately after printing, so that freshly printed lines do not become distorted by gravity or capillary wetting forces. This viscosity characteristic is commonly called *thixotropy* in the hybrid IC industry, although a more scientific description is *psuedoplasticity*. A good viscosity value for a screen printable mixture is 600 poise, measured at 100 sec^{-1} with a cone and plate viscometer.

A disadvantage of the most common type of thick film is the need for expensive noble metals, which can withstand firing in air without complete oxidation. However, newer thick film technologies tend to use copper metal, fired in nitrogen, for both lower cost and better electrical conductivity. W. R. Glave et al, have reported that 10 ppm of oxygen in otherwise highly purified nitrogen gas is an optimized atmosphere for firing copper thick films (13). Too much oxygen degrades the electrical conductivity of the copper, and too little prevents binder burnout and interferes with the development of adhesion to the ceramic substrate. The polymeric binder for firing in an essentially nitrogen atmosphere should be modified in the direction of evaporation removal, rather than burning. D. Whitman (14) has reported that acrylic binders are therefore advantageous in such applications, and commercial copper thick films often use such modified binder systems.

Instead of copper powder, a solid sheet of copper foil can be bonded to alumina ceramic, by using a layer of copper oxide as the adhesion promoting agent. The adhesion mechanism operates by the melting of a mixed oxide eutectic, followed by the formation of copper aluminate spinel at the interface. The process is known as *direct bonded copper*, or DBC. This process for bonding copper foil to alumina, followed by chemical etching for pattern generation, was invented at General Electric Company (15) and is in small scale use there. It is only used at a few other companies, probably because it requires extremely tight control of firing temperatures, oxygen concentrations, etc. Research at Rutgers University has recently shown that a high yield of strongly adhering foils can be made by optimizing the substrate composition, with alumina that contains less than 1% silica (16). In this variation of the process, the copper foil is pre-oxidized by heating it in air. Direct bonded copper has also been used on aluminum nitride, for mounting large power transistors, where a great deal of heat must be removed from the operating device (8).

Another way to coat a ceramic substrate with a smooth, thick layer of pure copper is to print and fire a conventional thick film, followed by electroplating of copper. The mismatch in thermal expansion coefficients between the relatively thick copper and the ceramic could cause failure of adhesion during temperature cycling. However, sufficient enhancement of the adhesion can be obtained by depositing electroless palladium onto the ceramic before printing the thick film (17). An additional advantage to this modified thick film process is that substractive or semi-additive processing with photoresist can be used to yield extremely well defined line edges. This is useful with microwave circuits and waveguides, where roughness can cause excessive losses. Other semi-additive or subtractive processing techniques have also been used with thick films to improve fine line resolution.

Adhesion mechanisms correlated to various microstructures of the films and substrates have been reviewed by T. T. Hitch (18). In some of the analytical experiments by Hitch, the glassy phase was removed with hydrofluoric acid, and in other work he removed the noble metal by extraction in mercury. Generally, thick films were found to consist of intertwined spongy networks of the metal and the glass.

For special purposes, various other metals are used in thick film conductors. For example, gold is utilized in cases where thermosonic or thermocompression bonding must be done in order to connect to transistors, to hermetic package leads, etc. Platinum or platinum-gold alloys are used where dissolution of the film by molten solder must be minimized.

In addition to metallic conductors, ceramic superconductors of yttrium barium cuprate can be made using screen printing and firing techniques (19). Although alumina substrates have been used here, sintered magnesia or strontium titanate substrates have also been used.

3.0 RESISTORS

Metal powder sintered together with glass powder can form a cermet, with moderately high to very high resistance values. These are used to make thick film (as well as other) resistors. The conduction mechanism is considered to be quantum mechanical tunneling (20) based on studies of conductivity versus temperature and versus metal concentration, etc.

A disadvantage of simply diluting metal conductors in making resistive devices is that the resistance rises strongly with temperature. In other words, the temperature coefficient of resistance *(TCR)* is positive. However, in many circuit designs, the TCR should be negative or close to zero.

If palladium is used as the metal in thick film resistors, it will oxidize slightly during normal firing, and the oxide is a semiconductor with the usual activation energy type of negative TCR. Therefore, by controlling the ratio of metal to oxide, the TCR can be controlled. In a strongly oxidizing atmosphere, at high temperature, or by premixing palladium oxide into the starting materials, the TCR can be made negative or positive or nearly zero (about 1 ppm/°C). This is the basis of many thick film resistor compositions.

In general, higher resistance values can be obtained by incorporating less metal into the composition. This can be fine tuned by adjusting the oxidation during firing *(trimming).* In addition, part of the resistor structure can be removed with a laser beam or mechanical abrasive device, in order to adjust the resistance value after firing. This is commonly done at the same time as measurements are made, allowing accurate resistance values to be obtained.

An unusually stable, easily controlled material for thick film resistors is ruthenium oxide. This has a cubic pyrochlore structure, and it is a semiconductor of widely variable but very repeatable resistivity. Control of the firing temperature and time can yield TCR values close to zero. The ruthenium oxide powder is used as a replacement for the palladium-silver alloy in the composition of Table 4. Microstructures and other properties of these materials have been studied extensively (21)(22).

The above materials do not behave predictably when fired in nitrogen, but advanced types of thick films such as copper must be fired in that atmosphere. A commonly used substitute for use with nitrogen firing is lanthanum boride (23), although control of the TCR is not as readily achieved as in palladium oxide.

4.0 DIELECTRICS

If the metal powder is eliminated from the composition of Table 4, and extra glass is added, an insulating thick film can be made. This can be used as a patch over a conductor line, and another conductor can then be printed and fired on top of it, producing a crossover.

A disadvantage of using glass as the insulating or dielectric layer is that the top conductor tends to diffuse or blend into the glass, since they are both fired at the same temperature (typically 850°C), and the glass usually melts at about 700°C. This intermingling can cause short circuits in crossovers.

Improved dielectric layers can be made from glass ceramics. These are materials which start as glasses and fire to full density, but upon annealing at 800°C they can crystallize to become ceramics. The melting point of the ceramic is higher than 850°C, and the top conductor is therefore not likely to blend into the layer. A material which is often used for this purpose is one part (by weight) barium oxide, one part aluminum oxide, and two parts silicon dioxide (24).

An increasing number of thick film applications involve multilayers, where several layers of dielectric completely cover the substrate, with appropriate holes for metal lines to purposely make contact from layer to layer. The holes filled with printed metal are termed *vias.*

Some multilayer structures involve unfired *green* ceramic sheets, with via holes punched into them mechanically before assembly. Thick films can be printed onto the sheets, and the paste is encouraged to fill the via holes with a vacuum tool underneath. The green sheets are then pressed together *(laminated)* and the whole composite is fired once, to sinter the ceramic and sinter the metal all in one step. Since the alumina ceramic requires firing at a higher temperature than the usual noble metal pastes can tolerate, a refractory metal such as tungsten or molybdenum must be used. These must be fired in hydrogen, to prevent oxidation. Water vapor is added to the hydrogen *(wet hydrogen),* and at these high temperatures the water decomposes to some extent into hydrogen and oxygen. The oxygen partial pressure is sufficient to burn out the organic polymer binder. This technology has been carried to a high level of complexity at IBM Corporation, where thirty-three layers have been laminated together in mass production (25).

A recent development in thick film multilayers involves the green sheet idea, except that the ceramic substrate is a low temperature firing material (850°C), which is compatible with noble metals and resistors (26). Thus the multilayer can be fired in air, and resistors can be included in the design.

5.0 CAPACITORS

Dielectric layers of glass ceramic can be used to make capacitors, with metal layers printed above and below them, all on an inert alumina substrate. These capacitors have been incorporated into many different thick film circuits (see Fig. 2). To increase the available capacitance values, barium titanate powder can be substituted for the usual dielectric material. However, the restraints of inexpensive firing atmospheres, the need to add metal contacts, etc., limit the practical capacitance values to less than one microfarad.

6.0 FUTURE DIRECTIONS

In order to use insulating layers with lower dielectric constants, polymeric layers can be combined with ceramic layers. This is done in the NEC Corporation supercomputer, where alumina, polyimide, and copper metal layers are all used to optimize heat conductivity, low capacitance between signal lines, and other properties (25).

In the future, the boundaries of material science applied to film circuits, thick and thin, are likely to be opened even wider. Substrate materials such as silicon are already being incorporated into advanced structures for film circuits. Relatively thick electroplated copper can be patterned onto silicon, with ceramic heat sinks underneath. The silicon IC chips are inherently well matched to this interconnection environment. AT&T has disclosed a multilayer structure with plated copper and silicon substrates (27) and this has been chosen for future designs by the Semiconductor Research Corporation cooperative agency (28). In this *wafer scale integration* technology, rectangular holes are etched in a large silicon wafer (see Fig. 3). The wafer is coated with a layer of insulating material having a low dielectric constant, and small round via holes are etched in that layer. The wafer is then metallized and patterned by standard integrated circuit techniques. Silicon IC chips are made, mounted on *tape automated bonding* strips (*TAB technology*), and tested. The TAB strips are a well known means of transporting, testing, and later interconnecting chips in highly automated equipment. They utilize electroformed, patterned copper foil, usually gold plated. (A long plastic strip is also involved in transporting the chips, but it is discarded after bonding. The plastic strip is not shown in Fig. 3.)

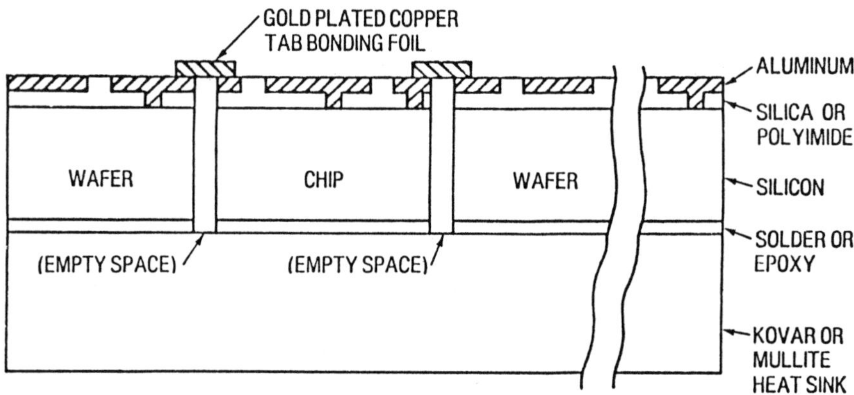

Figure 3. Advanced hybrid integrated circuit using silicon substrate *(wafer)* with rectangular holes for chips.

Those silicon IC chips which pass the *burn-in* tests are then placed in the rectangular holes of the large wafer, using the TAB transport system. The short strips of gold plated copper foil are used to interconnect the chips to the aluminum metallization on the wafer, with either reflow soldering or thermosonic bonding. (The relatively thick copper foil can actually be a major part of the interconnection system, although for simplicity in this diagram it is shown as being only a short pair of interconnection strips.)

The wafer assembly is mounted on a thick metal or ceramic heat sink whose thermal expansion coefficient matches that of silicon. Thus the main components are well matched in expansion properties, the signal paths are adjacent to low dielectric constant materials (which can aid in achieving high speed operation), and the chips can be tested before assembly. Although the system is somewhat complex, it uses proven technologies and could provide high overall performance.

Hybrid integrated circuit designs will probably make increased use of ceramic-metal-polymer composites, in addition to the simpler systems of the past. As has happened with silicon ICs, the market is likely to increase drastically, along with improvements in the technical capabilities.

REFERENCES

1. Allen, R. V. and Weiland, D. A., "Hybrid Market Forecast," *Hybrid Circuit Tech.* 16 (October 1989)

2. D'Andrea, J. B., "Ceramic Composition and Article," U.S. Patent 2,924,540 (1960)

3. Hamer, D. W. and Biggers, J. V., *Thick Film Hybrid Microcircuit Technology,* J. Wiley & Sons, New York (1972)

4. Miller, L. F., *Thick Film Technology,* Gordon and Breach, New York (1972)

5. Jones, R. D., *Hybrid Circuit Design and Manufacture,* Marcel Dekker, New York (1982)

6. Shanefield, D. J., "Casting Ceramic-Polymer Sheets," *Proc. Matls. Res. Soc.* 40:69 (1985)

7. Shanefield, D. J., "Comparison of Ceramics for Multichip Hybrids," *Proc. Natl. Electronic Packaging Conf.-West* 937 (1989)

8. Takahashi, T. et al., "Properties and Reliability of AlN Ceramics for Power Devices," *Advances in Ceramics* 26:159 (1989)

9. Ishigame, I. et al., "Thick Film Resistors for AlN Substrates," *Proc. Intl. Soc. Hybrid Microelectronics* 349 (1988)

10. Hornung, A., "Diffusion of Silver in Borosilicate Glass," *Proc. Electronic Components Conference* 250 (1968)

11. Hitch, T. T., "Analysis of Thick Film Materials," *Trans. IEEE,* PHP-11 248 (1975)

12. Creter, P., "Analyzing Thick Film Inks," *Circuits Manufacturing* 56 (September 1976)

13. Glave, W. R. et al., "Nitrogen Furnace Atmosphere in Copper Thick Film Manufacturing," *Proc. Intl. Soc. for Hybrid Microelectronics* 414 (1988)

14. Whitman, D., "Mechanisms of Char Formation in Nitrogen Fired Thick Film Materials," *Proc. Intl. Soc. for Hybrid Microelectronics* 421 (1988)

15. Burgess, J. F., Neugebaur, C. A. and Flanagan, G., "The Direct Bonding of Metals to Ceramics by the Gas-Metal Eutectic Method," *J. Electrochem. Soc.* 40 (1974)

16. Holowczak, J. E., Greenhut, V. A. and Shanefield, D. J., "Effect of Alumina Composition on Direct Bonded Copper," *Ceram. Eng. & Sci. Proc.* 10:1283 (1989)

17. Shanefield, D. J. and Crosby, G. E., "Method for Bonding a Metal Pattern to a Substrate," U.S. Patent 3,679,472 (1972)

18. Hitch, T. T., "Adhesion, Phase Morphology, and Bondability of Thick Film Conductors," *J. Electronic Matls.* 3:553 (1974)

19. Giacobbe, F. W., "Resistance Measurements in Thick Films of Yttrium Barium Cuprate," *J. European Ceramic Soc.* 5:87 (1989)

20. Springett, B. E., "Conductivity of Metal Parts Dispersed in an Insulating Medium," *J. Appl. Phys.* 44:2925 (1973)

21. Iles, G. S., "Ruthenium Resistor Glazes for Thick Film Circuits," *Proc. Intl. Soc. for Hybrid Microelectronics* 161 (1968)

22. Iizuka, K. and Yamaguchi, T., "Mechanism of Glass-Ruthenium Oxide Interaction in Thick Film Resistors," *Intl. Soc. for Hybrid Microelectronics* 345 (1988)

23. Sayers, P. and Petsis, W., "Nitrogen Firable Materials for Hybrids," *Proc. Intl. Soc. for Hybrid Microelectronics* 11 (1987)

24. Larry, J. R., Rosenberg, R. M. and Uhler, R. D., "Thick Film Technology," *Trans. IEEE,* CHMT-3, 211 (1980)

25. Tummala, R. R., "Ceramics in Microelectronic Packaging," *Advances in Ceramics* 26:3 (1989)

26. Sawhill, H. T., "Low Temperature Cofired Ceramic Packages," *Advances in Ceramics* 26:307 (1989)

27. Shanefield, D. J., "Wafer Scale Integration," U.S. Patent 4,866,501 (1989)

28. Johnson, R. W. et al., "Planar Hybrid Interconnection Technology," *Intl. J. Hybrid Microelectronics* 10:28 (1987)

9

Electronic Films From Metallo-Organic Precursors

Robert W. Vest

1.0 INTRODUCTION

1.1 Overview of MOD Technology

The metallo-organic decomposition (MOD) process is a technique for producing inorganic films without processing in vacuum or going through a gel or powder step. Figure 1 is a flow chart which shows the essential steps in MOD processing. The processing starts with metallo-organic compounds of the desired elements dissolved in an appropriate solvent. A metallo-organic compound is one in which a metal atom is bonded to a hetero-atom (e.g., O, N, S, or P) which in turn is bonded to an organic radical. These solutions of individual metallo-organic compounds are then mixed in the appropriate ratio to give the desired cation stoichiometry for the final film to produce a formulation, which is itself a true solution. This formulation is deposited on a substrate by any of a variety of techniques to produce a wet film, which is then heated, first to remove any solvent that did not evaporate during the deposition step, then to decompose the metallo-organic compounds to produce an inorganic film. A significant volume change occurs in going from the wet film to the inorganic film; if the inorganic film produced by a single pass through the process is not as thick as desired, the deposition and pyrolysis steps can be repeated as many times as necessary to produce a multilayer film of the required thickness. After desired film thicknesses are achieved, the films are often subjected to a further heat treatment to control features such as oxygen stoichiometry, grain size or preferred orientation.

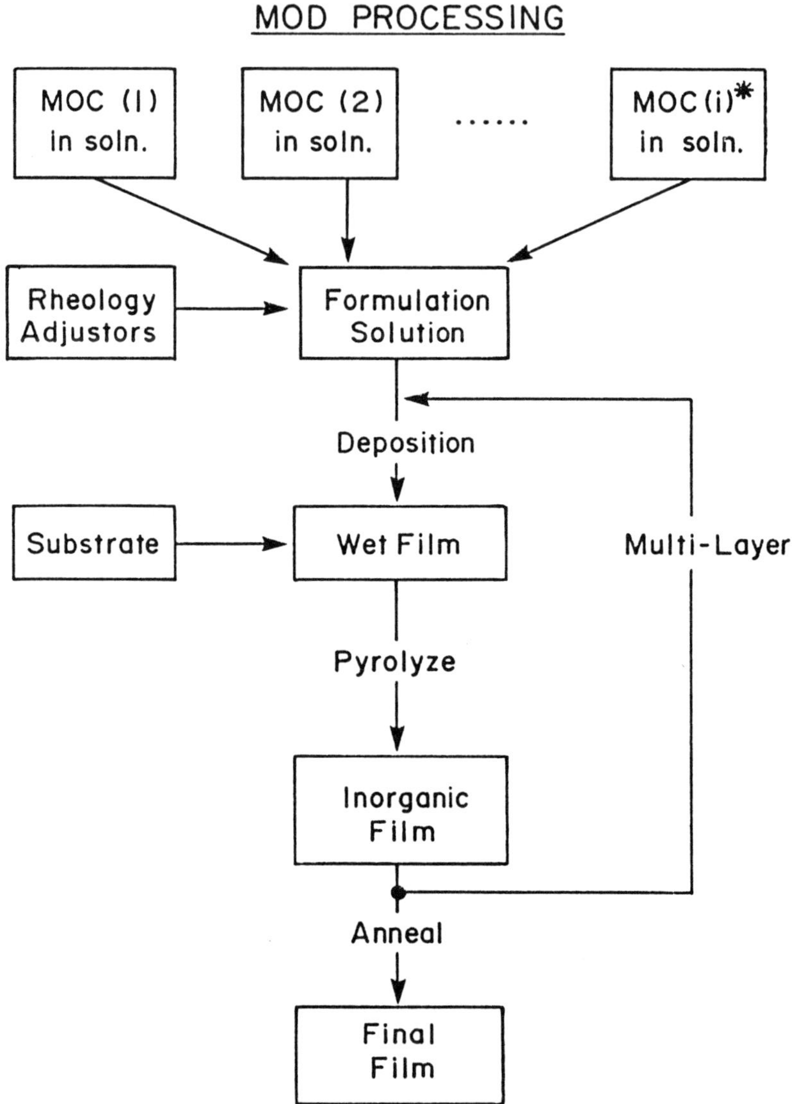

Figure 1. Flow diagram for MOD processing.

The ambient used for MOD processing depends on the particular films being prepared. In almost all cases the deposition step can be carried out in air, although dust free conditions are often desirable. For most oxides and the noble metals, the pyrolysis step can also be carried out in air. The pyrolysis step must be carried out in a low oxygen partial pressure atmosphere for producing base metal films, and in some cases the atmosphere during pyrolysis should have a controlled oxygen partial pressure in order to fix the oxygen stoichiometry of the inorganic film. However, the oxygen stoichiometry is most commonly controlled during a final annealing step. The MOD process has wide applicability to a variety of metal and ceramic films, discussed in later sections of this chapter.

1.2 Historical Review

Thin metallic films prepared by the decomposition of metallo-organic solutions have been used for many hundreds of years, and the first publications dealing with this technology appeared more than a hundred years ago. Although the technique has been known for a long time, its primary application has been in the decorative trade, with the main emphasis on producing films of precious metals on ceramic and glass articles. The earliest known reference is one describing the manufacture of bright gold for decorating porcelain (1), but this 1861 paper states that the process had been used since 1830. References 2 - 9 are review articles published between 1911 and 1964, which describe the development of MOD technology for producing thin films of precious metals on glass or ceramic.

The earlier references to technical applications of MOD technology are almost all in the patent literature. The use of MOD films for resistors was described in an 1895 patent (10), and gold or platinum film electrodes for capacitors were discussed in a 1934 patent (11). Another of the early technical applications of MOD technology was to produce noble metal films on glass for optical purposes (12)-(14). A very extensive review of the literature on MOD processing for both decorative and technical applications from its origins through 1965 was given by Langley (15), but this government report is somewhat difficult to obtain. The term "MOD" was coined by C. Y. Kuo in his 1974 paper (16) which discussed numerous applications of the metallo-organic decomposition process in electronics.

Almost all of the work prior to 1980 used metallo-organic compounds derived from resins or other natural products, and were commonly called metal resinates. These resinates were and still remain suitable for most applications in the decorative trade, but the variability of chemistry, which is an inescapable result of the preparation from natural products, was a major impediment to the extensive development of technical applications of the MOD process. It has only been in recent years that MOD films have been produced from pure, well characterized compounds. It is the preparation and utilization of pure synthetic compounds in MOD processing that forms the thrust of this chapter.

1.3 Advantages and Limitations

There are many advantages of MOD processing compared to alternate techniques for producing metal and ceramic films. The MOD process yields the equilibrium phases of the desired systems at relatively low temperatures, which circumvents the problem of selective volatility of different species. In general, the low temperature processing yields extremely fine grain size polycrystalline films; in many cases, the initial inorganic films are amorphous to x-rays. This allows for precise control of grain size by annealing after preparation of the films. The low temperature processing and the achievement of equilibrium phases is primarily due to the fact that the formulation deposited on the substrate is a true solution, hence the mixing of the various ingredients is on the molecular (or micelle) level. This means that the diffusion distances in the inorganic film after pyrolysis required to achieve thermodynamically stable phases are very short. This ultimate mixing and high reactivity also can be used to advantage in preparing very dense films. In most cases, films with near theoretical density can be achieved. Starting from solution also leads to films with extremely uniform composition over large areas, and allows for uniform doping in the ppm or ppb ranges. High purity can be maintained during MOD processing by appropriate care in the various processing steps. The MOD films can be patterned by a wide variety of techniques as discussed in Sec. 3.4, and MOD processing is cost effective because it does not require processing in vacuum.

There are some intrinsic limitations to MOD processing, however. The volume change in going from the deposited wet film to the fired inorganic film is always large, with typical ratios of deposited to fired thicknesses of 6 to 30. In addition to requiring care during thermal processing, this large volume change means that the fired films will always be thin (less than 1 μm in almost all cases). This limitation of film thickness can be overcome by the multilayer approach as indicated in Fig. 1, but there may be technical or economic limits to the number of layers that can be produced. A second intrinsic limitation is one of the advantages cited in the previous paragraph, namely, that thermodynamic equilibrium is achieved very rapidly because of the extremely high reactivity upon thermal decomposition. Many of the electronic films in use today (e.g., thick film resistors and high K capacitor dielectrics with low temperature coefficients) have their desirable properties because of a non-equilibrium microstructure. These non-equilibrium microstructures cannot be duplicated by MOD technology, but alternate approaches to achieve equivalent electrical properties with an equilibrium microstructure can often be utilized. Another limitation due to the thermodynamic equilibrium achieved is that only oxides or only metals of certain elements can be produced. This limitation is best understood with reference to a phase stability diagram such as that shown in Fig. 2. The metal is the thermodynamically stable phase in the T-P_{O_2} region below the line indicated for each element, while the lowest oxide is the thermodynamically stable phase for that element above its line. Also shown in Fig. 2 is the curve for the C-CO-CO_2 equilibrium.

Since MOD processing must always be carried out under conditions that are oxidizing to carbon, the pyrolysis step must be carried out in the T-P_{O_2} region above the carbon curve. In some cases, any thermodynamically stable compound in a metal-oxygen system can be obtained; copper is an example of this situation. Firing in the T-P_{O_2} region above the carbon curve but below the copper line will yield copper metal films, whereas firing in the region above the copper line will produce Cu_2O films. While not shown on Fig. 2, the line corresponding to the equilibrium between Cu_2O and CuO also lies within the possible regime for MOD processing, so films of Cu, Cu_2O or CuO can be produced by controlling the oxygen partial pressure during pyrolysis. Only oxide films can be produced for those elements whose lines lie well below the carbon curve on Fig. 2 (e.g., B, Si, Al or any of the alkaline earth metals). The lines for the more noble metals (e.g., Pt, Au or Ag) lie at oxygen partial pressures greater than one atmosphere for the temperature range shown in Fig. 2, and only metal films of these elements can be produced.

2.0 METALLO-ORGANIC SYSTEMS

2.1 Selection of Compounds

The successful application of the MOD process to produce films is totally dependent upon the metallo-organic compounds used as precursors for the various elements. The ideal compounds should satisfy the following ten requirements:

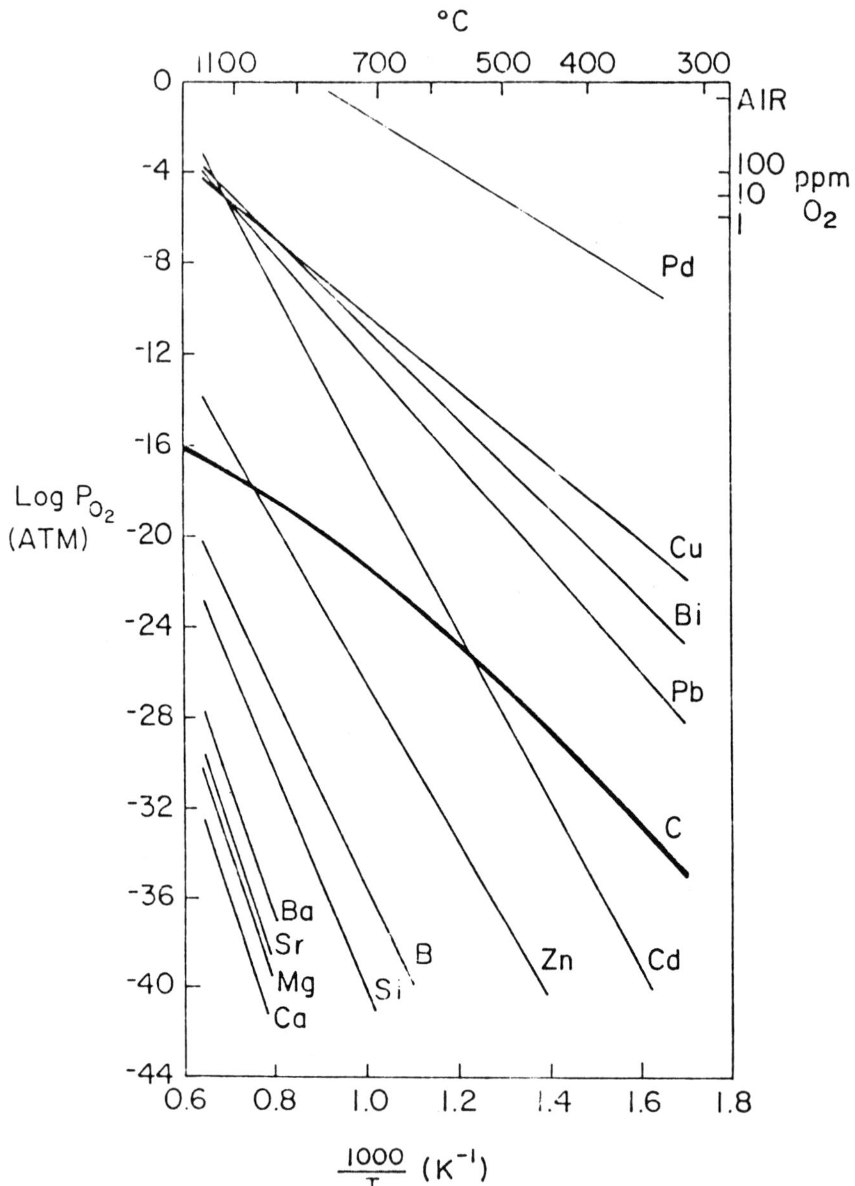

Figure 2. Phase stability diagram for selected metal-oxygen systems along with the C-CO-CO_2 equilibria.

1. Pure compounds with well defined formulas and structures. This characteristic is necessary so that reproducible films with well defined characteristics can be produced. The great bulk of the early work utilizing the MOD process involved compounds derived from natural resins, which are useful in the decorative trade but are not satisfactory for electronic applications where much better control over thin film quality is a requirement.
2. Easy synthesis and purification. Commercial sources for most of the pure compounds required to produce MOD electronic films are not presently available, and so the compounds must be capable of being synthesized in pure form using standard chemical procedures.
3. Thermally decompose without evaporating, melting or leaving a carbon deposit. This is the heart of the MOD process and distinguishes it from alternate technologies, such as chemical vapor deposition (CVD) where the requirements for the compounds are that they must evaporate without decomposing at some low temperature and subsequently decompose on the substrate at a higher temperature. This requirement is sometimes difficult to realize and compromises must be made. The decomposition without evaporation is an absolute requirement unless all constituents in the formulation evaporate to the same extent so that the stoichiometry is preserved, which is a most unlikely situation. If the compound does melt to some extent before decomposition it can adversely affect film quality or line definition, but these problems can be overcome in some cases by altering other parts of the processing. Similarly, if a compound leaves some carbon deposit on decomposition it may be possible to burn this off during subsequent higher temperature processing without adversely affecting the film quality.
4. High metal content. This requirement reflects the desire to minimize the volume change in going from the metallo-organic compounds deposited on the substrate to the inorganic film.
5. High solubility in a common solvent. This requirement is also related to the desire to achieve the minimum volume change in going from the deposited organic film to the inorganic film, but in most cases is not compatible with requirement 4. To achieve the highest metal content the compounds should have the minimum number of carbon atoms, but solubility in organic solvents normally increases with increasing chain length of the organic radicals. Obviously, a compromise is necessary between requirements 4 and 5.
6. Stable under ambient conditions. It is very desirable to carry out the MOD processing in normal air, which means that the compounds should not be sensitive to normal constituents of air such as water vapor or CO_2. This

requirement provides a distinction between compounds suitable for the MOD process and those used in sol-gel processing. For the MOD process, the compounds should not gel when exposed to moisture, which presents a particular problem for compounds of silicon, boron and aluminum. If a compound meets the other primary requirements, the gelling tendency can be overcome by keeping the compound under a controlled atmosphere until it is deposited on the substrate; gelling after deposition does not cause significant problems.

7. Compatible with other compounds in the formulation. One of the major virtues of MOD processing is the ability to control the composition of multi-component systems. Therefore, the various compounds must be somewhat similar chemically so that they do not react with each other in the formulation.

8. Proper decomposition temperature. The metallo-organic compounds for all of the elements in a given formulation should have similar decomposition temperatures. This requirement is alleviated to some extent by the decomposition mechanisms which involve the formation of free radicals and the subsequent chain reaction associated with them.

9. Non-toxic and produce benign off-gasses on thermal decomposition. This requirement is desirable from a processing standpoint since a significant volume of off-gases is produced by the MOD process, and removal of these should be simple and straightforward.

10. Cost effective to produce. The MOD process is inherently cost effective because capital equipment requirements are minimal. Synthesis of the metallo-organic compounds should not negate this advantage in processing.

From the above discussion of the compound requirements it is apparent that all of these cannot be optimized simultaneously, and that compromises are required. The desire to have non-toxic compounds which produce benign off-gases on thermal decomposition led to the selection of the class of metallo-organic compounds in which the central metal atom was linked to the organic ligands through hetero-atoms of oxygen or nitrogen. In order to minimize carbon residues on thermal decomposition, only saturated organic ligands were considered and these ligands should contain only carbon, hydrogen and oxygen. The alkoxides are not generally suitable compounds because most tend to evaporate to some extent prior to thermal decomposition.

These general requirements and considerations led to a preference for compounds with the metal atom linked to carboxylate ligands, and with or without additional alkoxide or amine groups. For a metal of valence z, the three types of compounds found most suitable were $(RCOO)_zM$, $(RCOO)_{z-a}M$

$(OR')_a$ or $(RCOO)_zM(NR')_n$ with the carboxylate groups containing eight to ten carbon atoms. Further, the carboxylate groups should have secondary or tertiary hydrocarbon chains in order to increase the solubility in organic solvents for the same number of carbon atoms. Thus, the choice was narrowed to metal salts (soaps) of 2-ethylhexanoic acid or neodecanoic acid.

2.2 Synthesis of Compounds

A few metallo-organic compounds suitable for MOD processing are commercially available, but compounds for most of the elements of interest are not. The synthetic methods developed for various metallo-organic compounds to produce metal or metal oxide films by the MOD process can be divided into six categories and are discussed below. Many of the individual procedures are described in a series of reports and journal articles from the Turner Laboratory at Purdue University (17)-(21).

Neutralization. This method involves the direct neutralization of an alcohol solution of a metal hydroxide by 2-ethylhexanoic acid or neodecanoic acid. The resulting metal soap is separated by filtration if it is a solid, or extracted in solvents such as xylene or toluene after vacuum distillation if it is an oily liquid. This procedure can be used to make soaps of K, Na and Li, but the yield is generally low because these soaps have appreciable solubility in water and alcohols. The general reaction is:

$$M(OH)_z + zRCOOH \longrightarrow M(RCOO)_z + zH_2O$$

An example synthesis is given for potassium neodecanoate. In a 500 ml flask, 11.5 g (0.205 moles) of KOH is dissolved in 25 ml of methanol and stirred until a clear solution is formed. To this solution, 39.9 ml (0.205 moles) of neodecanoic acid is added and the mixture refluxed with stirring for 18 hours. Most of the water and methanol are removed by vacuum distillation at 40 - 50°C using a rotary vacuum unit. Xylene (25 ml) is added to adjust the viscosity, and the last traces of water are removed by drying over molecular sieves.The reaction is:

$$KOH + C_9H_{19}COOH \longrightarrow C_9H_{19}COOK + H_2O$$

Double Decomposition From Ammonium Soap. This method involves the preparation of the ammonium soap of 2-ethylhexanoic acid or neodecanoic acid in the first stage, then mixing the soap with a metal salt (e.g., chloride or nitrate) in aqueous solution. The resulting metal soap is separated by filtration if it is a solid, or extracted in solvents such as xylene or toluene if it is an oily liquid. The general reactions are:

$$NH_4OH + RCOOH \longrightarrow RCOONH_4 + H_2O$$

$$MX_z + zRCOONH_4 \longrightarrow M(RCOO)_z + z(NH_4)X$$

This method can be used to produce soaps of Mg, Ca, Sr, Ba, Y, Nd, Yb, Zr, Hf, Mn, Ag, In, Sn, Pb and Bi. In some cases (e.g., Bi or Sn), the metal salt must be dissolved in an acidic solution instead of a neutral solution to avoid the formation of insoluble precipitates.

An example synthesis is given for barium neodecanoate. One hundred fifty grams (0.61 moles) of $BaCl_2 \bullet 2H_2O$ is dissolved in about 300 ml of water. In a 1000 ml beaker, 245 ml (1.2 moles) of neodecanoic acid is neutralized with 75 ml (1.2 moles) of NH_4OH (NH_3 assay: 30%). The solution is stirred for 20 - 25 minutes. To this ammonium soap solution, the $BaCl_2$ solution is added with vigorous stirring. White gummy barium soap forms as a top layer which is dissolved in xylene (200 ml), washed with water and separated from the aqueous layer with the help of a separating funnel. The clear xylene solution is then filtered and subjected to vacuum distillation at 40°C using a rotatory vacuum unit to get a concentrated solution. The final traces of water are removed by drying over molecular sieves. The reactions are:

$$NH_4OH + C_9H_{19}COOH \longrightarrow C_9H_{19}COONH_4 + H_2O$$

$$2C_9H_{19}COONH_4 + BaCl_2 \longrightarrow 2NH_4Cl + (C_9H_{19}COO)_2Ba$$

Double Decomposition From Amine Soap Some metal salts form insoluble complexes when treated with ammonium soap. In those cases, triethylamine is used instead of ammonium hydroxide to neutralize the organic acid. The amine is mixed with 2-ethylhexanoic acid or neodecanoic acid in aqueous media, then an aqueous solution of the metal salt is added to it. The reaction mixture is warmed on a water-bath and the oily product is extracted in an organic solvent.

The general reactions are:

$$RCOOH + (C_2H_5)_3N \longrightarrow RCOO^-(C_2H_5)_3NH^+$$

$$MX_z + zRCOO^-(C_2H_5)_3NH^+ \longrightarrow M(RCOO)_z + z((C_2H_5)_3N \bullet HX)$$

This method has been found useful to produce soaps of Fe, Ru, Rh, Ir and Pt. An example synthesis is given for ruthenium 2-ethylhexanoate. A mixture of 5 g (0.096 moles) of 2-ethylhexanoic acid, 9.71 g (0.096 moles) of triethylamine and 15 ml of water are stirred together in a 250 ml round bottom flask, and a solution containing 0.024 moles of ruthenium chloride in 30 ml of water is slowly added. The mixture is stirred at room temperature for 30 minutes, then heated and stirred on a water bath at 60 - 70°C for 45 minutes. A black oil separates at the bottom and sides of the flask. It is further stirred for 1 hr, then left at room temperature overnight. The clear aqueous solution is removed and the black oil washed six times with 50 ml of warm water (40 - 50°C), and the product extracted in xylene (~10 ml). The solution is then dried over molecular sieves overnight. The reactions are:

$$C_7H_{15}COOH + (C_2H_5)_3N \longrightarrow C_7H_{15}COO^-(C_2H_5)_3NH^+$$

$$RuCl_3 + 3C_7H_{15}COO^-(C_2H_5)_3NH^+ \longrightarrow Ru(C_7H_{15}COO)_3 + 3(C_2H_5)_3N \bullet HCl$$

Metathesis Reaction From Metal Acetate. A solution of metal acetate in alcohol or hydrocarbon solvent is mixed with 2-ethylhexanoic acid or neodecanoic acid in an evaporating dish and warmed on a steam-bath for 2 - 3 hours. The solvent and the by-product acetic acid are removed and more solvent added until the reaction is complete. The product is then extracted in a hydrocarbon solvent:

$$(CH_3COO)_zM + zRCOOH \xrightarrow[R''OH]{R'H \text{ or}} (RCOO)_zM + zCH_3COOH$$

This method has proven useful in preparing soaps of La, Cr, Ni, Pd, Cu and Zn. A sample synthesis is given for copper 2-ethylhexanoate. Fifty ml (0.1 mole) of ethanol, 10 g (0.05 moles) of cupric acetate and 14.42 g (0.1 mole) of 2-ethylhexanoic acid are added to a porcelain evaporating dish that is placed on top of a steam bath and stirred. A clear green solution is obtained as the mixture heats. Most of the ethanol and the acetic acid by-product evaporate slowly leaving behind a thick, green oily residue. Fifty ml of ethanol is added and the solution stirred and evaporated once more. The green residue is poured into a beaker containing 300 ml of water and mixed thoroughly to remove any acetic acid as well as unreacted cupric acetate.

The green semisolid is repeatedly washed with water and finally extracted in 100 ml of toluene. The toluene layer is washed three times with water and dried over molecular sieves. When the solvent is evaporated in a rotary vacuum unit at 40°C, a green dry solid is left. The reaction is:

$$Cu(CH_3COO)_2 + 2C_7H_{15}COOH \longrightarrow Cu(C_7H_{15}COO)_2 + 2CH_3COOH$$

Metathesis Reaction From Metal Alkoxide. The metal alkoxide is mixed with 2-ethylhexanoic acid or neodecanoic acid in the presence of an alcohol or a hydrocarbon solvent, refluxed for 4 - 5 hours, and the volatile alcohol by-product and the solvents are removed under reduced pressure. The general reaction is:

$$M(OR')_z + aRCOOH \xrightarrow[R'''OH]{R''H \text{ or}} M(OR')_{z-a}(RCOO)_a + aR'OH$$

This method has been used to prepare satisfactory compounds of Li, Na, K, Mg, Ba, Ti, Nb, Ta and Al. For example, titanium di-methoxy-di-neodecanoate can be prepared by putting 86 g (0.5 moles) of titanium methoxide in a 1000 ml flask under an inert atmosphere (N_2). One mole of neodecanoic acid (172 g) and about 20 ml of methanol are added to the flask without disturbing the inert atmosphere. The reaction mixture is refluxed under the inert atmosphere for 5 hours at 65°C to produce a pale yellow viscous liquid. Methanol is removed from that liquid by vacuum distillation, and the liquid dissolved in xylene (~300 ml). The solution is filtered to obtain a clear pale yellow solution of titanium di-methoxy-di-neodecanoate in xylene, and dried over molecular sieves. The reaction is:

$$(CH_3O)_4Ti + 2C_9H_{19}COOH \longrightarrow Ti(OCH_3)_2(C_9H_{19}COO)_2 + 2CH_3OH$$

Metal Amine Carboxylates. Gold, and to a lesser extent platinum, present a particular problem in synthesizing suitable metallo-organic compounds for use in the MOD process. Compounds having both amine and carboxylate ligands have proven to be the most satisfactory solution to this problem. An example is gold amine 2-ethylhexanoate, which is synthesized following closely U.S. Patent No. 4,201,719.

An aqueous solution of 2-ethyl-4-methyl imidazole [$C_6H_{10}N_2$] (13.75 g = 0.124 moles) in water (125 ml) is added slowly to a stirred solution of gold chloride [$HAuCl_4 \cdot H_2O$] (10.82 g = 0.0274 moles) in water (50 ml). The buff colored precipitate which forms is stirred for an additional 30 minutes and kept at room temperature for 18 hours. The precipitate is filtered, washed with water (8 x 10 ml), and then suspended in water (15.2 ml) and stirred. In a separate flask, a mixture of 2-ethylhexanoic acid (19 ml) and 2-ethyl-4-methyl imidazole (13.30 g) in water (125 ml) is prepared, and added slowly to the stirred suspension of gold complex in water. After most of the amine soap is added, the buff colored precipitate disappears and a thick reddish oil separates. The mixture is further stirred for 30 minutes and set aside for 3 - 4 hours. The supernatent water solution is removed and the oil which remains is repeatedly washed with warm (40°C) water (4 x 40 ml). The honey colored residue is dissolved in xylene and dried over molecular sieves overnight. The solution can be further concentrated by using rotary vacuum evaporation at 35°C. The reactions are:

$$HAuCl_4 \bullet 3H_2O + 2C_6H_{10}N_2 \longrightarrow AuCl_3 \bullet C_6H_{10}N_2 + C_6H_{10}N_2 \bullet HCl + 3H_2O$$

$$C_6H_{10}N_2 + C_7H_{15}COOH \longrightarrow C_6H_{10}N_2H^+C_7H_{15}COO^-$$

$$AuCl_3 \bullet C_6H_{10}N_2 + 3C_6H_{10}N_2H^+C_7H_{15}COO^- \longrightarrow Au(C_7H_{15}COO)_3 \bullet C_6H_{10}N_2 + 3C_6H_{10}N_2 \bullet HCl$$

2.3 Solvent Considerations

Requirements. The solvent(s) used in the MOD process must satisfy a number of requirements. First and foremost is the desire to provide high solubility for the individual metallo-organic compounds without chemically interacting with them. Secondly, the solvent should be such that the formulation solution can be adjusted to the proper rheology for whatever deposition method is to be used. This consideration may require solvents with high or low viscosity, high or low surface tension, or compatibility with various rheology adjustors. A third requirement concerns the vapor pressure of the solvent, with the desired value ranging from high for deposition by spinning or ink jet printing to very low for deposition by screen printing. The rate of solvent vaporization is also important, because the solvent should be removed from the system as soon as possible after satisfying the above three requirements. While related to vapor pressure, the rate of vaporization is also influenced by chemical and electrostatic interactions between solvent and solute. Finally, the solvent should be low cost, recoverable, non-toxic, and non-corrosive. As was the case with the metallo-organic compounds, compromises must be made because a single solvent cannot satisfy all of the diverse requirements.

Solvency. Solvency is the ability of a solvent to form a stable solution with one or more solutes, but this quantity cannot be given in exact terms because the solvency of a given solvent is different for each solute. The science of solubility is based on statistical thermodynamics, and while significant progress has been made in developing a theory of non-electrolyte solutions (e.g., see Ref. 22), the theory still falls short of being able to predict solubility relationships with sufficient accuracy for practical applications. The theory is, however, useful in identifying properties of solvents related to their solvency, and these will be discussed below.

The driving force for a solution process is described in terms of the free energy of solvation, ΔG_s, expressed as:

Eq. (1) $$\Delta G_s = \Delta H_s - T\Delta S_s$$

where ΔH_s and ΔS_s are the enthalpy and entropy of solvation, respectively. Solvation will occur only if ΔG_s is negative. For an ideal solution $\Delta H_s = 0$ and the entropy of solvation is given by:

Eq. (2) $$\Delta S_s = -k\,(n_1 \ln x_1 + n_2 \ln x_2)$$

where n_1 and n_2 are the number of moles of solvent and solute, respectively, x_1 and x_2 are their mole fractions, and k is Boltzmann's constant. Of course, most real solutions deviate from ideality, and the next level of sophistication is to consider a *regular solution* in which ΔS_s has the ideal value but ΔH_s is not equal to zero. However, the concept of a regular solution also has limited applicability unless combined with consideration of the molecular properties of solvent and solute. Consideration of the intermolecular forces acting between solvent and solute molecules leads to the division of systems into simple and complex solutions. The interactions in simple solutions result exclusively from dispersion (London) forces, which are sometimes called *nonspecific interactions.* In complex solutions the molecules have a permanent nonuniform distribution of charge so they interact through electrostatic forces (primarily dipole-dipole interactions) in addition to the dispersion forces; this leads to some degree of specific orientation of one molecule with respect to an adjacent molecule, and these are sometimes called *specific interactions.*

For a simple solution containing molecules of similar size, the enthalpy of solvation can be written as (23):

Eq. (3) $$\Delta H_s = (x_1V_1 + x_2V_2)\,\phi_1\phi_2\,(\delta_1 - \delta_2)$$

where V_1 and V_2 are the molar volumes of solvent and solute, respectively, ϕ_1 and ϕ_2 are their volume fractions, x_1 and x_2 are their mole fractions, and δ_1 and δ_2 are their solubility parameters defined as:

Eq. (4) $$\delta_i = \left(\frac{E_i}{V_i}\right)^{1/2}$$

where E_i is the energy change accompanying the isothermal vaporization of the saturated liquid of species i to the ideal gas state. In this treatment, ΔH_s is always positive, which is usually the case for solutions of nonpolar components. The solubility parameter concept combined with the regular solution assumption has the appeal of simplicity because it relates solubility to the difference between δ's of the two components. For example, if the solvent and solute both have a molar volume of 100 ml/mole, this theory predicts that there will be some solubility at room temperature until $(\delta_1 - \delta_2)$ exceeds 3.5.

Solubility parameters have been calculated for most of the common organic solvents by rewriting Eq. 4 as

Eq. (5) $$\delta = \left(\frac{\Delta H_v - RT}{V}\right)^{1/2}$$

and using measured values for the heat of vaporization (ΔH_v). The solubility parameter can also be calculated from surface tension (γ) according to the relation (24):

Eq. (6) $$\delta = 4.1\left(\frac{\gamma}{V^{1/3}}\right)^{0.43}$$

The agreement between the solubility parameters for petroleum solvents calculated by the two methods is very good (25). Values for some selected solvents are given in Table 1 in order of increasing δ. Unfortunately, solubility parameters are generally not available for the types of metallo-organic compounds discussed in Sec. 2.1 because they are supposed to decompose without vaporizing (hence, ΔH_v data are not available) and most of them are solids at 25°C (hence, γ data are not available). However, the δ values given in Table 1 can serve as an empirical guide for solvent selection; lacking any other information the first three solvents tried should be ones with high, low and intermediate solubility parameters. This is as much guidance as the simple solution concept can give.

Table 1. Solubility parameters (δ) and dielectric constants (K) at 25°C and 1 atm pressure, and boiling points for selected solvents.

Solvent	δ $(cal/cm^3)^{1/2}$	K	B.P. (°C)
pentane	7.02	1.84	36
hexane	7.27	1.89	69
cyclohexane	8.19	2.02	81
α-terpineol	8.58		215
xylene	8.88	2.37	139
ethyl acetate	8.91	6.02	77
butyl carbitol (diethylene glycol monobutyl ether)	8.91		231
toluene	8.91	2.38	111
2-pentanone (methylpropylketone)	8.99	15.4	102
2-butanone (methyl ethylketone)	9.04	18.5	80
benzene	9.16	2.28	80
tetrahydrofuran	9.32	7.58	66
acetone	9.62	20.7	56
carbon disulfide	9.92	2.64	46
m-dichlorobenzene	10.04	5.04	173
o-dichlorobenzene	10.04	9.93	181
3-pentanol (diethyl carbinol)	10.16	13.9	
nitrobenzene	10.40	35.74	211
pyridine	10.62	12.3	116
butyl alcohol (1-butanol)	11.60	17.8	117
ethanol (ethyl alcohol)	12.78	24.30	79
methanol (methyl alcohol)	14.50	33.62	65
glycerol	15.53	42.5	290

In order to include the specific interactions which occur in complex solutions, the polarity of the solvent and solute must be considered in order to evaluate the importance of the dipole-dipole interactions. In general, a highly polar solute will have higher solubility in a highly polar solvent than in one with low polarity. A measure of the polarity of the molecules of a liquid is the dielectric constant (K), and K values are given for the selected solvents in Table 1. Neither the dielectric constants nor the dipole moments of metallo-organic compounds of the type discussed in Sec. 2.1 are available, but relative polarity can be assessed by considering their structure.

The carboxylate group, which is present in all of the types of compounds selected for the MOD process, is a resonance structure in which the carboxyl carbon is joined to each oxygen by a "one-and-one-half" bond to give the ion:

O
R —— C ⊖
O

Further, the carboxyl carbon is joined to the 3 atoms by σ bonds which lie in a plane 120° apart. For metallo-organic compounds of the monovalent metals the following single molecule planar structure is possible:

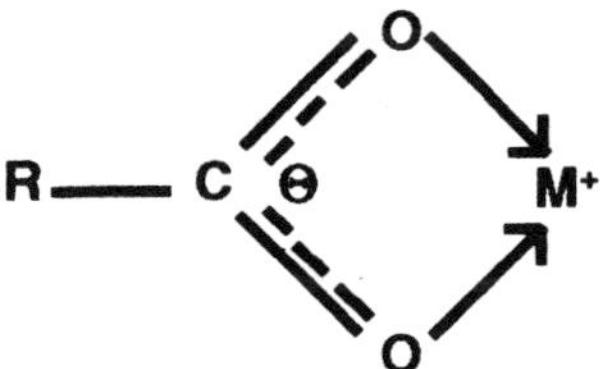

This is a highly polar structure, which can help explain why the alkali metal carboxylates have an appreciable solubility in water (K = 80.4) as discussed in Sec. 2.2. However, the physical properties of soap solutions are best interpreted in terms of the micellar theory (26), where a micelle is an aggregate of three or more soap molecules existing in the liquid in thermodynamically stable equilibrium. A proposed (27) structure for alkali metal soap micelles is a puckered chain:

R R R
C C C
O O O O O O O
M+ M+ M+ M+

This is still a polar structure but its dipole moment is less than that of the single molecule. Micelles are certainly the rule for soaps of the divalent, trivalent, tetravalent and pentavalent metals. It has been shown (27) that carboxylates of strongly coordinated small ions form compact, closed configurations containing approximately 4 cations rather than linear chains. This produces a non-polar structure, and these soaps (e.g., Be, Zn or Al) give low viscosity solutions in low polarity solvents such as benzene.

A consequence of the micellar nature of soap solutions in organic solvents is their ability to solubilize certain compounds by enclosing their molecules in the cores of the micelles (28). This enclosure causes the micelles to swell, and the change in size is frequently accompanied by a change in shape. This phenomenon can sometimes be advantageous; for example, lanthanum acetate has very low solubility in xylene but it is readily soluble in a xylene solution of titanium di-methoxy-di-neodecanoate. However, the micelle swelling can also cause problems; for example, a benzene solution of aluminum soap is a clear, low viscosity liquid, but it exhibits a marked increase in viscosity when a small amount of water is added and eventually develops a gel structure (29). This is due to a change in shape of the soap micelles from small spheres to long flexible rods or threads, which can then interlink to form a three dimensional network.

It is well established that solubility is influenced by the nature of the organic radical bonded to the carboxyl carbon. Spherically symmetrical molecules with high molecular weight will have higher dispersion (London) forces which lead to higher solubility. This is why secondary and tertiary carboxylates were selected as the metallo-organic compounds of choice. However, there is little reliable information available on the part played in solubility by the metallic portion of the molecule. Nickel and manganese dodecanoates appear to be more soluble in toluene than calcium, cadmium, lead and silver dodecanoates (30), and magnesium carboxylates appear to be more soluble in benzene than the corresponding cadmium soaps (31). The effect of the metal on solubility cannot be predicted theoretically, and insufficient experimental data are available to establish empirical correlations.

Selection Procedure. The main conclusion that can be drawn from the preceding section on solvency is that there is not a solid theoretical basis for selecting the optimum solvent for the type of metallo-organic compounds suitable for MOD processing. However, the theories can provide a guide for empirical selection. The approach which is used at the Turner Laboratory at Purdue University, and which has proven rather successful, is based on the solvent properties given in Table 1. We first try a non-polar solvent with a low solubility parameter (e.g., xylene). If this does not prove successful, the next solvent tried is one that is moderately polar and has an intermediate solubility parameter (e.g., tetrahydrofuran). For stubborn cases, solvents with a high solubility parameter and low polarity (e.g., m-dichlorobenzene) or a low solubility parameter and high polarity (e.g., methyl ethyl ketone) can be tried. Optimization sometimes requires using mixtures of high and low polarity solvents (e.g., xylene plus pyridine). In preparing formulations containing metallo-organic compounds of several different elements, the order of addition is often important because of the micellar nature of the soap solutions. Before discarding a particular solvent, all permutations for combining the individual compounds should be evaluated.

If the deposition method requires a low vapor pressure solvent, as is the case for screen printing, it is usually desirable first to dissolve the appropriate compounds in a low viscosity, high vapor pressure solvent such as xylene or tetrahydrofuran, and then affect a solvent exchange. This is accomplished by adding a low vapor pressure solvent, such as α-terpineol, to the solution and removing the high vapor pressure solvent at reduced pressure and evaluated temperature. Rotation or agitation is required during the solvent exchange process in order to effectively remove the high vapor pressure solvent.

3.0 PROCESSING

3.1 Film Deposition

In principle, the formulation solution (see Fig. 1) can have any desired viscosity and surface tension by use of appropriate solvents and additives, which means that any technique that has ever been used to deposit a liquid on a solid surface can be used to deposit the formulation solution on the substrate. A number of printing methods for patterning during the deposition step are discussed in Sec. 3.4. This section concentrates on general considerations applicable to the deposition process and on methods of depositing a uniform film over the entire substrate.

Several of the advantages of MOD processing discussed in Sec. 1.3 require that the formulation deposited on the substrate be a true solution, so the deposition method must not cause any segregation of the metallo-organic compounds. Regardless of the solvent system used, the different metallo-organic compounds in a given formulation solution will have different solubilities, so if there is any solvent evaporation during the deposition step it should be very rapid so as to minimize segregation. A desirable feature of any deposition method is the ability to control both the magnitude and uniformity of film thickness because many physical and structural properties of MOD films are related to the single layer thickness.

The methods that have been used to deposit a uniform coating of the formulation solution on a substrate are generally those which have been developed for depositing photoresist in the microelectronics industry and include spinning, dipping or spraying techniques. For substrates with holes, the spraying technique gives good results, and machine spraying is the preferred method to achieve a uniform film over a large area. Dipping with a controlled rate of withdrawal is the deposition method recommended by Nippon Soda* for depositing solutions of their metallo-organic compounds of Si, Ti, In, Zr and Ta. They recommend deposition by submerging the substrate in the metallo-organic formulation and then withdrawing at speeds from 10 - 40 cm/min. For example, in their trade literature they show that starting with a formulation solution that will produce 8 wt.% SiO_2, the fired film thickness can be varied from 150 to 250 nm by varying the withdrawal speed from 10 to 40 cm/min.

By far the most common technique used to deposit a uniform film of the formulation solution on a substrate has been spin coating. The substrate is placed on a turntable and a pre-measured amount of formulation solution is dispensed onto the substrate. The spinning is initially done at a slow rate to assure complete coverage of the substrate, followed by fast spinning for a longer duration. In order to avoid dust streaks in the film the spinning should be carried out in a clean environment, preferably class 100 or better. A fluid dynamics analysis of the spin coating process derived from the Navier-Stokes equation in cylindrical coordinates was reported (32) for a simple system of a non-volatile Newtonian fluid, and included the contribution of interface slip between the liquid film and the rotating disk. This resulted in the following equation for the time rate of change of film thickness h:

Eq. (7) $$\frac{dh}{dt} = \frac{-2\rho_l\omega^2h^2}{3\eta}(h + 3\lambda)$$

where ρ_l is the liquid density, ω is the angular velocity, η is the viscosity of the liquid and λ is the slip coefficient. The second term in Eq. 7, which represents an increased flow due to interfacial slip, is generally negligible because the slip coefficient is very small for most liquid-solid couples. If the slip term is neglected, Eq. 7 can be integrated to give:

* Nippon Soda Co., Ltd., Shin Ohtemachi Bldg., 2-2-1 Ohtemachi, Chiyoda-ku, Tokyo, Japan.

Eq. (8)
$$h = \frac{h_o}{\left[1 + \frac{4\rho_l \omega^2 h_o^2 t}{3\eta}\right]^{1/2}}$$

where h_o is the liquid film thickness at t = 0, which represents the time at which fast spinning is initiated. Some typical values for spin coating of formulation solutions are:

ρ_l = 1 g/cm^3
η = 5 mPa.s
ω = 209 radians/s (2000 RPM)
h_o = 10 μm
t = 30 s

Using these values, the second term in the denominator in Eq. 8 is 35, which is sufficiently greater than 1 that Eq. 8 reduces to

Eq. (9)
$$h = \frac{1}{2\omega}\left(\frac{3\eta}{\rho_l t}\right)^{1/2}$$

One consequence of Eq. 9 is that the final liquid film thickness is independent of the initial film thickness or the amount of formulation solution transferred to the substrate. The thickness of the solid film after pyrolysis (h_s) can be calculated from:

Eq. (10)
$$h_s = h \frac{\rho_l}{\rho_s} C$$

where ρ_s is the density of the solid film and C is the concentration of the formulation solution (g solid film/g solution). Combining Eqs. 9 and 10 gives

Eq. (11)
$$h_s = \frac{C}{2\omega\rho_s}\left(\frac{3\eta\rho_l}{t}\right)^{1/2}$$

An equimolar formulation solution of lead neodecanoate and titanium dimethoxy-di-neodecanoate in xylene solvent was used to collect experimental data (33) to compare with Eq. 11. The thicknesses of the fired $PbTiO_3$ films were determined for various formulation solution concentrations and viscosities, and for the spinning parameters of speed and time. Figure 3a shows the experimental and calculated fired film thickness as a function of spinning speed at fixed spinning time and solution viscosity. At the lower spinning speed the calculated thickness agrees very well with the experimental values. With increasing spinning speed the shift of the values calculated by Eq. 11 from the experimental results was larger, and at a speed of 3,000 RPM the discrepancy reached 30%. This may indicate that the assumption of Newtonian behavior of the solution at high shear stress (high spinning speed) is not a good one. The calculated and experimental results of the change of fired film thickness with solution viscosity at fixed spinning speed and time are shown in Fig. 3b. At all points in Fig. 3b, the film thickness calculated from Eq. 11 is less than the experimental value, which may be due to some solvent evaporation; both C and η will increase with evaporation of the xylene solvent, and this will tend to increase the fired film thickness over that calculated by Eq. 11. However, the agreement between the predictions of Eq. 11 and the experimental results shown in Fig. 3 is reasonably good, and Eq. 11 has proven to be very useful for predicting thickness of the fired MOD films.

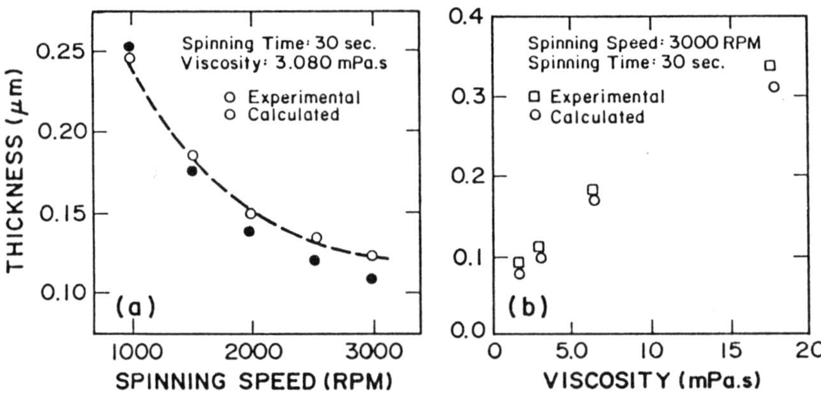

Figure 3. The calculated and experimental layer thickness of $PbTiO_3$ film as a function of (a) spinning speed at fixed spinning time and viscosity and (b) viscosity at fixed spinning time and speed.

3.2 Pyrolysis

After deposition of the formulation solution onto a substrate, the next step in MOD processing is to increase the temperature to remove any remaining solvent not evaporated during the deposition step and to convert the metallo-organic compounds into an inorganic film. As discussed in Sec. 1.3, the pyrolysis must always be carried out under temperature and oxygen partial pressure conditions that are oxidizing to carbon. In most cases, the pyrolysis step is the most critical one in all of MOD processing because this is where the microstructure of the film is developed.

A large volume change occurs during pyrolysis and this may lead to cracks in the fired film. The smallest volume change reported (34) during pyrolysis of a MOD film is shown in Fig. 4, which gives the surface profile across a line of silver MOD ink deposited on a glass substrate (Fig. 4a), and a profile of the line after firing (Fig. 4b). The change in volume after firing is about a factor of 6, and this small (for MOD processing) volume change is due to the deposition method. The film in Fig. 4a was deposited by ink jet printing using a 12 µm nozzle diameter, which led to almost complete evaporation of the toluene solvent by the time the droplets impacted the substrate. Therefore, the metallo-organic film in Fig. 4a was almost pure silver neodecanoate. The volume change associated with films deposited by spin coating are invariably greater than 10, and sometimes as high as 30.

Thermogravimetric analyses can be used to determine the minimum temperature required to remove all of the carbon from the film, and to suggest appropriate heating rates in different temperature ranges. Figure 5 shows a thermogram of a formulation solution to produce $PbTiO_3$ (33), and the decomposition temperature (T_d) is seen to be slightly above 300°C. Therefore, the pyrolysis step can be carried out at any temperature greater than 300°C.

The rate of heating from room temperature to $T > T_d$ is always very important in producing good quality films, but unfortunately no general rules can be given. Looking at the thermogram in Fig. 5 would suggest that the heating rate should be quite low during the solvent evaporation phase below 100°C, and also that rather slow heating rates seem in order between 100 and 300°C where the compound decompositions are occurring, and for most systems studied a slow heating rate the order of 10°C/min to the decomposition temperature is preferable. However, in the case of indium-tin oxide (ITO) films just the opposite was observed (35), and nonuniformities in the films were always present if slow heating was used. An extreme example is shown in Fig. 6 where the ITO film has segregated into nearly circular

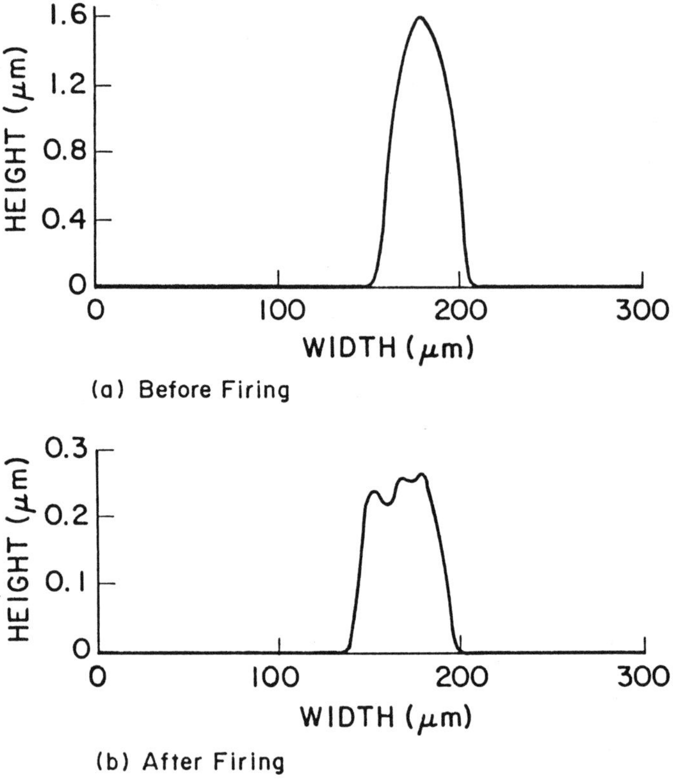

Figure 4. Thickness profile of a MOD silver line on a glass substrate before and after pyrolysis.

regions on the SiO_2 glass substrate. It is believed that this effect is due to a wetting phenomena because both the indium and tin 2-ethylhexanoates used in the formulation solution are very viscous liquids at room temperature. One of the requirements for an ideal metallo-organic compound for MOD processing is that it decomposes without melting, but ideal compounds cannot always be found.

The viscosity of both compounds used in the ITO formulation decrease with increasing temperature until thermal decomposition initiates, and in the temperature range where the compounds are still liquids the viscosity becomes sufficiently low that they will assume their equilibrium contact angle with the substrate; in the case of SiO_2 glass substrates this required the film breaking up into discrete droplets. An identical film processed under the same conditions on a silicon substrate did not show the individual droplets as in Fig. 6 but rather showed interconnected regions of thicker ITO.

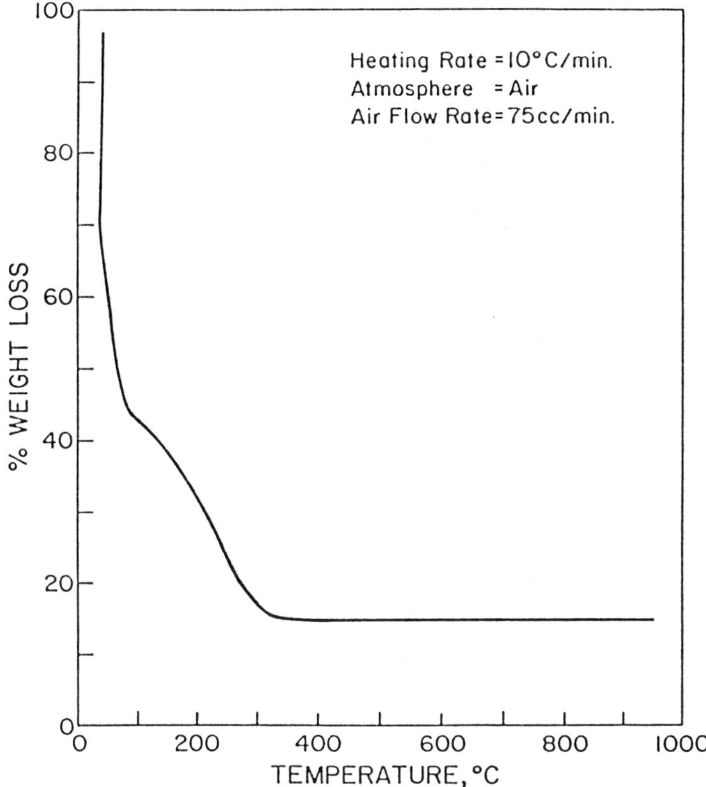

Figure 5. Thermogram of a xylene solution of an equimolar mixture of lead 2-ethylhexanoate and titanium di-methoxy-di-neodecanoate.

Nonuniformities having still different appearances were observed for films on single crystal quartz and sapphire substrates. Since the contact angle of a liquid on a solid surface depends on the liquid-solid and solid-vapor interfacial energies in addition to the surface tension of the liquid, different behaviors would be expected on different substrate materials. It was determined that the degree of nonuniformity in the ITO films decreased as the heating rate increased, and that the best quality films were obtained by inserting the substrates directly into a muffle furnace at 550°C. Using this very rapid heating rate, films such as those shown in Fig. 7 were obtained. It was feared that the very rapid heating rates required to keep the films from segregating prior to decomposition would lead to rough films due to the rapid release of the organic materials. This was not the case, however, and the films fired by placing them directly into a 550°C muffle furnace had a surface roughness equivalent to that of the substrate.

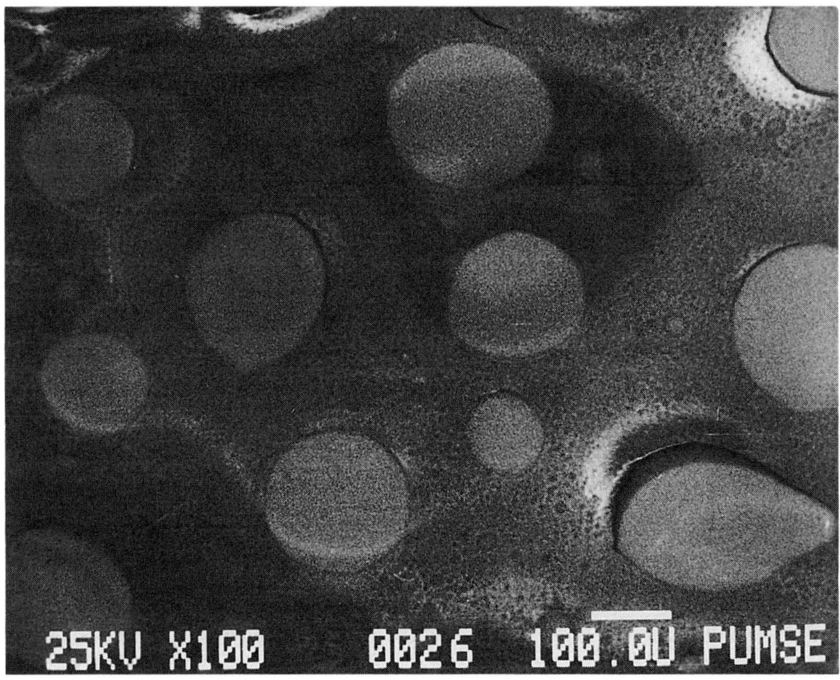

Figure 6. An ITO film fired at 20°C/minute to 500°C on an SiO_2 glass substrate.

The thermochemistry which applies during the pyrolysis step is very complex. Figure 8 (33) is a thermogram of lead 2-ethylhexanoate which shows that T_d is about 380°C, and Fig. 9 shows a T_d of about 375°C for titanium di-methoxy-di-neodecanoate although a small additional weight loss is observed between 375° and 500°C. When these two compounds dissolved in xylene are mixed to give a $PbTiO_3$ formulation solution, the thermogram of Fig. 5 is obtained, which shows that T_d is lower than for either of the individual compounds. This result indicates that some type of "domino effect" is operative in the decomposition of a mixture of compounds. The most likely pyrolysis mechanism for carboxylates $M(RCOO)_z$ involves a rate-determining free radical generation by thermal fission, followed by a fast fragmentation of the radical R and a very fast oxidative chain reaction.

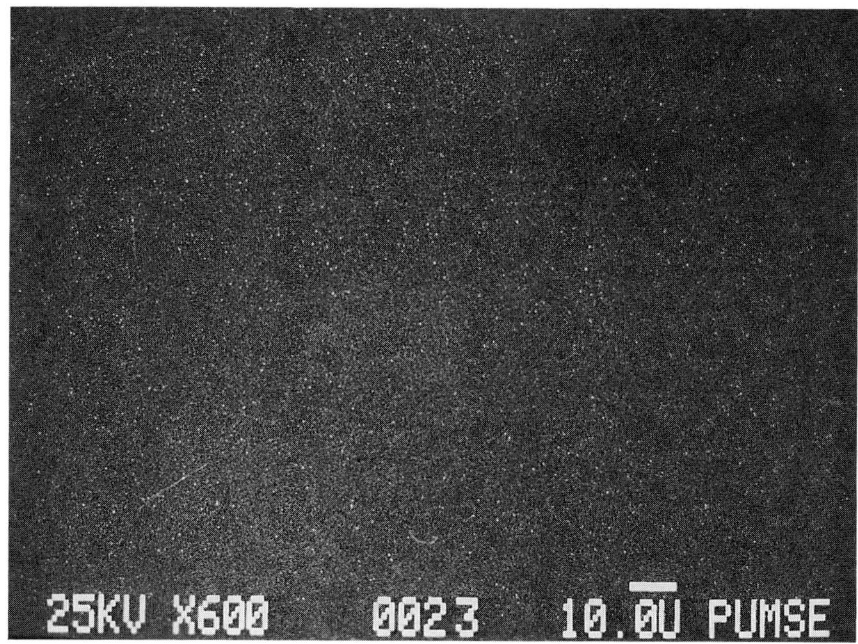

Figure 7. An ITO film fired directly at 550°C on an SiO_{22} glass substrate.

The free radical mechanism would account for the observed "domino effect." If this is the mechanism, then the decomposition temperature should decrease as the chain length of R increases, as the oxygen partial pressure increases, and as the degree of branching of R increases. These predictions have been found to be valid in many, but not all cases. A study (19) of the decomposition temperature for five different silver carboxylates with R containing 3 to 9 carbon atoms, and with branching varying from primary to secondary to tertiary, showed that the decomposition temperatures of all compounds were within 5°C of each other; this results indicates that it is the silver-oxygen bond that fractures first to initiate the decomposition, and that the nature of the organic radical is immaterial in affecting the decomposition temperature. The thermochemistry involved in pyrolysis of this class of metallo-organic compounds is a fertile area for basic research.

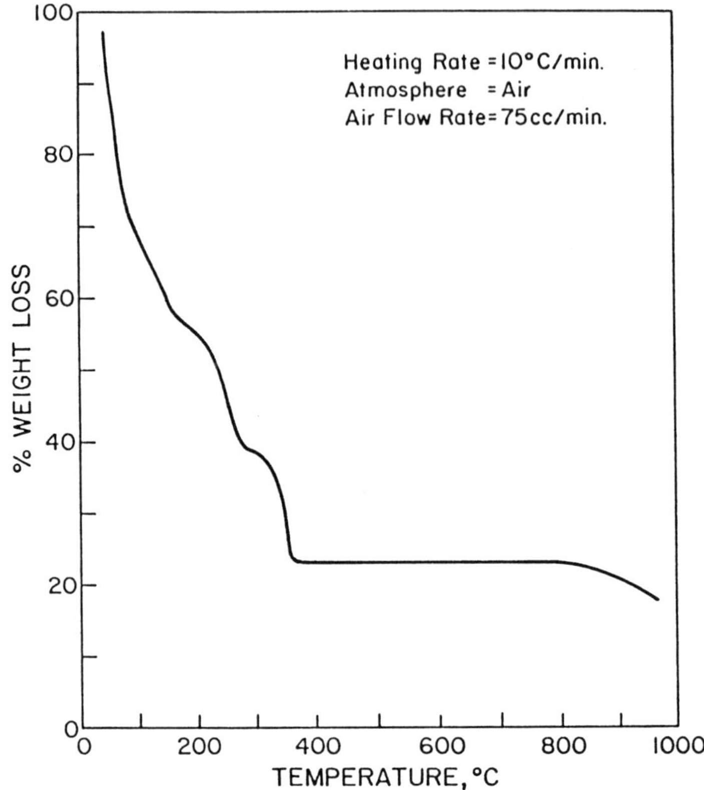

Figure 8. Thermogram of lead 2-ethylhexanoate solution in xylene.

One of the advantages of MOD processing is illustrated by comparing Figs. 8 and 5. The decrease in weight above 800°C in Fig. 8 is due to vaporization of PbO, which is a common problem during processing of lead-containing ceramics. However, when the lead 2-ethlyhexanoate is mixed with titanium di-methoxy-di-neodecanoate the reactivity during pyrolysis is so high that crystalline $PbTiO_3$ is formed below 500°C (36) and no lead loss is observed at higher temperatures (Fig. 5).

3.3 Annealing

If MOD films are fired to temperatures only slightly above the decomposition temperature during the pyrolysis they usually show an amorphous x-ray diffraction pattern, as shown in Fig. 10 for the $PbTiO_3$ film fired at 435°C for one hour (36). Annealing at higher temperatures develops

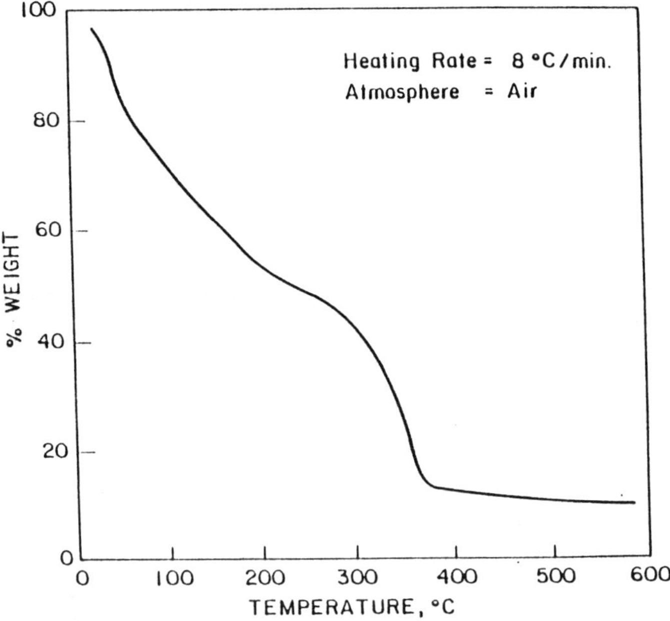

Figure 9. Thermogram of $(CH_3O)_2Ti(C_9H_{19}COO)_2$ solution in xylene.

the crystallinity of the film as indicated for the 475 and 494°C anneals in Fig. 10. The increase in grain size with increased annealing temperatures can be followed by using x-ray line broadening techniques, and an example of such results (37) for $BaTiO_3$ films is shown in Fig. 11 for one hour anneals at temperatures from 780° to 1200°C.

Annealing is sometimes necessary to control the oxygen stoichiometry in the MOD films. This is particularly true for the superconducting oxides such as $YBa_2Cu_3O_{7-\delta}$ The extent of grain size control and oxygen stoichiometry control by annealing is often limited by substrate-film interactions. For example, the grain size data in Fig. 11 for temperatures of 1100°C and below were taken for films deposited on ITO coated silicon wafers. When these films were annealed at temperatures above 1100°C, the x-ray diffraction patterns showed some new peaks which were not characteristic of $BaTiO_3$, ITO or silicon, which indicated that a new phase or phases had formed due to interactions in the film-electrode-substrate system.

While MOD films are always polycrystalline, it is sometimes possible to achieve preferred orientation during the annealing step. The x-ray

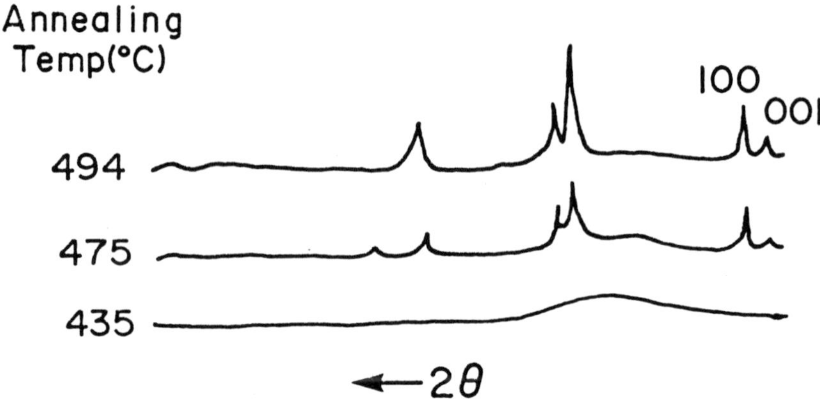

Figure 10. X-ray diffraction patterns (with CuK_{α}) for $PbTiO_3$ films (1 μm) fired on Pt foil at various temperatures.

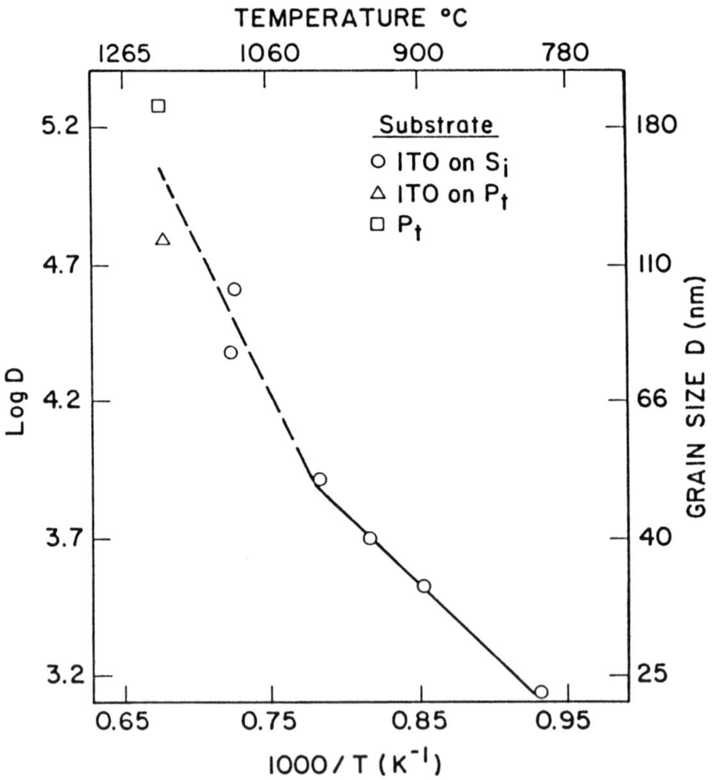

Figure 11. Grain growth kinetics for 1 hour anneal of $BaTiO_3$ films.

diffraction pattern in Fig. 12 (33) for a MOD platinum film on a (111) silicon substrate shows a very strong degree of (111) preferred orientation, which is probably due to an epitaxial effect. Figure 13 shows x-ray diffraction patterns for PLZT, $Pb_{0.92}La_{0.08}(Zr_{0.65}Ti_{0.35})_{0.98}O_3$, films on sapphire substrates annealed at two different temperatures compared to the x-ray diffraction pattern of powder having the same composition. The pattern of the film annealed at 650°C was identical to the powder pattern, indicating random orientation of the grains in the film, but the pattern of the film annealed at 750°C shows that the grains were oriented with (001) planes parallel to the substrate surface. This preferred orientation cannot be due to an epitaxial effect because there is a large lattice mismatch between (1010) sapphire and (001) PLZT.

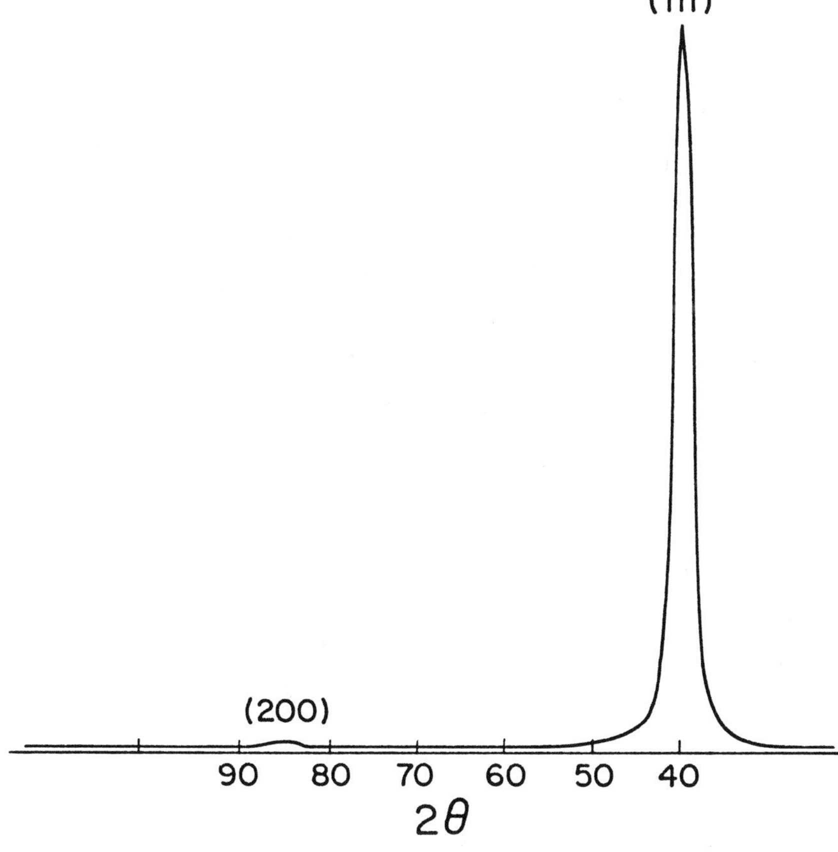

Figure 12. X-ray diffraction pattern of thin (60 nm) MOD Pt film on (111) Si wafer.

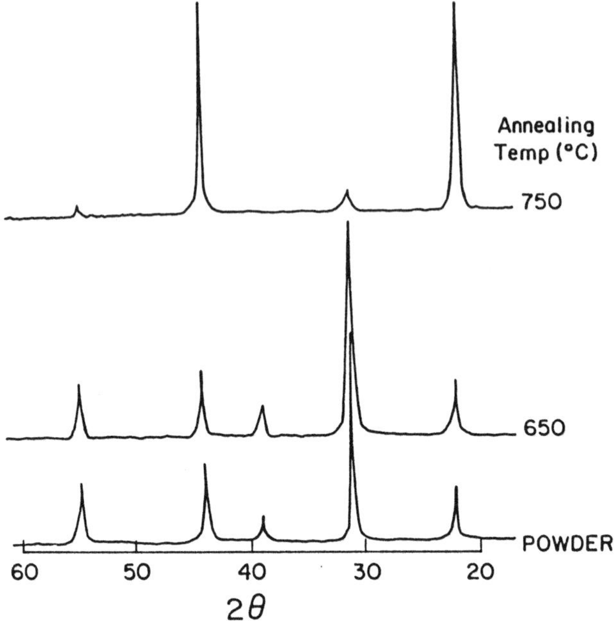

Figure 13. Comparison of x-ray diffraction patterns of PLZT films annealed at 650 and 750°C on (1010) sapphire with powder of the same composition.

3.4 Patterning

The MOD films can be patterned at four different points during the processing shown in Fig. 1: during the deposition step, after deposition and prior to pyrolysis, during pyrolysis, or after pyrolysis. Studies of patterning MOD films during deposition by screen printing (20)(38) and by ink jet printing (39) have been reported, but, in principal, any process from spraying through a stencil to offset printing could be used to pattern during the deposition step. For screen printing, an ink must have rather high viscosity (the order of 100 Pa.s) and the solvent system used must have very low volatility so that the viscosity does not change when the ink is spread out on the screen during the printing process. The synthesis procedures for most of the metallo-organic compounds of choice result in a solution of the compounds in a high vapor pressure solvent, such as xylene or tetrahydrofuran. It is therefore necessary to affect a solvent exchange during which the high vapor pressure solvent is removed and replaced by a low vapor pressure solvent. A screen printable silver ink was obtained (20) by using either benzene or tetrahydrofuran as the high vapor pressure solvent and replacing it with a mixture of butyl carbitol acetate and neodecanoic acid. A screen printable gold ink was prepared (38) by using xylene as the high vapor pressure solvent and replacing it with Penzoil mineral jelly #20. While screen printing of MOD inks has been successful, it is not the method of choice because the formulations' solutions typically have low viscosity.

A technique for patterning during deposition that requires a low viscosity ink is computer controlled ink jet printing, and this technique has been successfully applied for patterning MOD inks. For the drop on demand technique used in our studies (39)(40) for ink jet printing of MOD inks, both the viscosity and surface tension of the inks are important in the printing characteristics. A mathematical model was developed (41) for relating the quantity of ink deposited by ink jet printing to the surface tension and viscosity of the ink and the various parameters of the ink jet printing system. For ink jet printing, the most suitable values for surface tension were 30 - 50 mN/m and viscosities from 1 - 10 mPa·s. These values can easily be achieved by proper selection of low vapor pressure solvent and concentration of the metallo-organic compounds in solution.

The MOD films can also be patterned after film deposition and prior to the pyrolysis step. The wet films can be given a soft bake to remove most of the solvent and then standard photolithographic techniques used in silicon technology can be applied. A photoresist can be deposited on top of the soft baked MOD film by spinning, and then exposed through a mask. After exposure the photoresist is developed, and the developing step usually removes the MOD film along with the photoresist. The patterned film can then be fired with photoresist still on top because it burns off during the pyrolysis step along with the carbonaceous material in the MOD film.

One of the more interesting patterning methods for MOD films involves patterning during the pyrolysis step by using a laser, electron or ion beam as a local heat source. A scanning laser was used successfully to pattern silver MOD films for electrodes on photovoltaic cells (42). Silver is probably the simplest system for laser patterning because it has the feature of self limiting power absorption. When the laser beam raises the film temperature to the decomposition temperature of the silver neodecanoate used in this study, the formation of the metallic silver film provides almost complete reflection of the laser energy and further heating of the film does not occur. Using laser pyrolysis for films that do not become totally reflective after decomposition is a more challenging problem, but some success has been reported (43) for patterning Y-Ba-Cu-O superconducting films. The selective pyrolysis approach has also been demonstrated using scanning electron (44) or ion (45) beams.

The fourth option to patterning MOD films is after the pyrolysis step. Any of the subtractive techniques (e.g., plasma etching or laser ablation) developed for microelectronic fabrication can be used at this point.

4.0 EXAMPLES OF MOD FILMS

4.1 Conductor Films

Metals. As discussed in Sec. 1.3, metal films can be produced by the MOD process only for those elements whose lines on a phase stability diagram (Fig. 2) lie above the carbon curve. This includes all of the noble metals and a few base metals. Thus, all of the conductor films typically prepared by thick film technology (e.g., Au, Au/Pt, Ag/Pd, Cu) can also be prepared by MOD technology, but some metal films prepared by thin film technology (e.g., Al, Ti, Ni/Cr) cannot be prepared by MOD.

Gold conductor films were prepared (38) by screen printing a MOD formulation onto alumina substrates. A single print followed by firing to a maximum of 850°C produced gold films approximately 0.5 μm thick and having near theoretical density. The gold metallo-organic precursor compound was gold 2-ethyl-4-methyl imidazole-tri-2-ethylhexanoate, which was synthesized as described in Sec. 2.2. The formulation also contained metallo-organic compounds of bismuth and copper to enhance adhesion to the alumina substrates, rhodium to enhance the formation of continuous films, probably due to the precipitation of rhodium oxide in the grain boundaries of the gold film which inhibits grain growth, and palladium to improve aged wire bond strength and solder leach resistance. The adhesion strength reported for the MOD gold films was higher than that of conventional thick film gold conductors, and aluminum wire bond tests showed that pull strengths measured for the MOD gold films were comparable to conventional thick film gold conductors.

Platinum films for conductors on silicon wafers were produced (36) by spin coating the wafers with a MOD formulation containing platinum di-1,2-diaminopropane-tetra-2-ethylhexanoate dissolved in tetrahydrofuran. The formulation spun onto the silicon wafer also contained bismuth 2-ethylhexanoate in a concentration which produced a fired film with composition 5 wt.% Bi_2O_3 and 95 wt.% platinum after firing at 350°C in order to improve adhesion. Subsequent layers were spun on and fired using a formulation containing only the platinum metallo-organic compound. A four layer Pt film with total thickness of 50 nm had a sheet resistance of 12 ohms/sq and the surface roughness was 4 to 6 nm, which was almost the same as that of the silicon substrate.

The preparation of MOD silver films using silver neodecanoate as the metallo-organic compound have been reported by both screen printing (20) and by computer controlled ink jet printing (46). For both patterning methods, a small amount of bismuth was added to the formulation in the form of bismuth 2-ethylhexanoate in order to enhance adhesion to the substrates. The silver films were fired to a maximum of only 300°C, and the screen printed films produced a 1 μm thick conductor that was 100% solderable, 100% solder leach resistant and had good line definition and excellent long term adhesion. The resistivities were very close to that of bulk silver, indicating that the films were very dense.

Fine line palladium films were prepared (47) by spinning a chloroform solution of palladium acetate onto fused quartz or silicon substrates, and using a focused Ar ion laser as a local heat source. The films were in air during the laser writing, so the local temperature must have been above 800°C in order to produce Pd instead of PdO (See Fig. 2).

Copper MOD films were prepared (48) from an ink containing copper 2-ethylhexanoate screen printed onto alumina substrates and fired in a CO/CO_2 atmosphere such that the oxygen partial pressure was always between the copper line and the carbon curve in Fig. 2. The major problem encountered in this study was the poor adhesion of the copper films to the aluminum substrates. However, good adhesion was achieved if the alumina substrates were pre-glazed with a lead borosilicate glass.

Oxides. One of the more common electronically conducting oxide films used in a variety of electronic devices is tin doped indium oxide (ITO). MOD films with composition $In_{1.91}Sn_{0.09}O_3$ were prepared (35) by spinning a solution of indium and tin 2-ethylhexanoates onto various substrates, firing the wet films to 550°C in air, then annealing at various temperatures in air. These films had resistivities as low as 4 mΩ-cm, and the transmittance of the films was greater than 95% in the 450 - 1000 nm wavelength range.

The report of high T_c oxide superconductors (49) in 1986 has done more than any other event to stimulate research in MOD technology. A number of studies on the preparation of $YBa_2Cu_3O_{7-\delta}$ films by MOD technology have been reported (50)-(54); most of these studies (50)-(53) used the 2-ethylhexanoates or the neodecanoates as precursor compounds but the use of stearates and naphthenates (54) has also been reported. MOD processing of the high T_c superconductor films on $SrTiO_3$, MgO or yttrium stabilized zirconia substrates seem to give films with the highest transition temperature and narrowest transition range, although films on sapphire substrates with a superconducting onset temperature greater than 90°K have been reported (50). Research in this area was progressing very rapidly in a number of laboratories at the time of the writing of this chapter, and it is too early to predict the extent of applicability of MOD superconducting films to various proposed devices.

4.2 Resistor Films

Very little research has been reported on the preparation of resistor films by MOD technology. Most of the studies have been attempts to duplicate compositions successfully used from thin film or thick film technology. The metal alloys typically used in thin film technology (e.g., Ni-Cr) cannot be prepared by MOD technology because the nickel and chromium lines are below the carbon curve on the phase stability diagram, and thick film resistors made with a metallic conducting oxide such as RuO_2 and a glass have non-equilibrium microstructures which cannot be reproduced by MOD technology.

A study (55) using a formulation containing metallo-organic compounds of ruthenium, lead, boron and silicon showed that the inorganic films after firing did indeed contain crystalline RuO_2 in a lead borosilicate glass matrix, but that the RuO_2 was confined to isolated regions and did not form a continuous network throughout the glassy matrix as is the case in conventional thick film resistors. Some success in preparing MOD resistor films was reported (56) using the 2-ethylhexanoates of Ru, Ir or Rh to generate the conductive metal oxide after pyrolysis in air, along with compounds of glassy elements (Si, Bi, Pb, Zr, Sn and Al). The films were screen printed onto glazed alumina substrates and fired in air at a peak temperature of 800°C. It was found that the sheet resistance could be varied from 30 Ω/sq to 20 kΩ/sq by changing thickness and composition. The most stable resistors were prepared using a formulation that contained a molar ratio of 0.5 for Ru, Ir or Rh, 0.25 for Si and 0.25 for Bi.

It is apparent from the research that satisfactory MOD resistor films having a wide range of sheet resistance values cannot be produced by copying the materials systems used successfully in thin or thick film technology. A more logical approach may be to produce suitable metal alloy (e.g., Ag/Pd) films for the low sheet resistance range, and for the high sheet resistance range to go to doped semiconductor films such as the MOD SnO_2 films reported (57) for gas sensing applications.

4.3 Dielectric Films

Linear Dielectrics. The MOD process should have wide applicability in the fabrication of linear dielectric films since the classical ingredients in linear dielectrics (e.g., Al_2O_3, SiO_2, B_2O_3, and the alkaline earth oxides) are produced by decomposition of the precursor compounds in air. However, very little research has been reported on the preparation of linear dielectric films by MOD technology, probably because few applications in electronic devices have been identified. It has been shown (55) that lead borosilicate glass films on alumina substrates can be prepared starting with formulation solutions containing lead 2-ethylhexanoate, boron di-methoxy-2-ethylhexanoate or borane pyridine, and silicon di-ethoxy-di-2-ethylhexanoate or silicon tri-ethoxy-2-ethylhexanoate. The precursor compounds were present in concentrations to produce a fired film with composition 63 wt.% PbO - 25 wt.% B_2O_3 - 12 wt.% SiO_2. Continuous glass films were formed in the temperature range of 600 to 700°C, whereas a firing temperature of 800°C is required to produce a pin hole free glaze when the same glass composition is deposited on an alumina substrate as a -325 mesh frit.

Dense, crack free films of ZrO_2 and yttria stabilized ZrO_2 on Si substrates have been prepared (58) using the 2-ethylhexanoates as precursors. The films fired at 500°C for two hours were amorphous and had insulation resistances in the range of 10^{12} to 10^{13} ohm-cm over the temperature range of -55 to +150°C. The yttria stabilized zirconia films showed promise for use as oxygen electrolytes in micro-ionic devices.

Ferroelectrics. Films of a variety of ferroelectric compositions have been prepared by MOD processing, and this remains an active area of research at laboratories in the United States and in Japan. Films of $PbTiO_3$ with thickness 0.5 µm to 2.0 µm were prepared (36)(59) by the MOD process using a multilayer spinning technique. The precursor compounds were lead neodecanoate and titanium di-methoxy-di-neodecanoate. The xylene solution of these compounds was deposited on platinum coated silicon wafers, pyrolyzed at 370°C, and annealed at various temperatures to a maximum of 600°C. The processing produced dense, crack free films with easy control of crystal structure and composition. The dielectric constant of the films was 100 at room temperature, and reached 3 x 10^4 at the Curie temperature of 493 - 495°C. The temperature dependence of the dielectric constant was found to be a function of the c/a ratio, which could be modified by either control of the single layer thickness or the strength of the applied DC field during film preparation near the Curie temperature. The P-E hysteresis loops were measured and the spontaneous and remnant polarizations were found to be 5.2 $\mu C/cm^2$ and 3.8 $\mu C/cm^2$, respectively. The coercive field was 33.2 kV/cm and the dielectric strength was greater than 100 kV/cm. The grain orientation in the films was random at annealing temperatures below 550°C, but a strong tendency for preferred orientation with the c axis perpendicular to the substrate surface was observed at annealing temperatures above 600°C.

Crack free and dense $BaTiO_3$ films with 4 - 8 µm thickness were prepared (37) by the multilayer spinning technique using barium neodecanoate and titanium di-methoxy-di-neodecanoate as the precursor compounds. After pyrolysis, the films were annealed at temperatures from 800 to 1200°C in order to control the grain size. Films with an average grain size of 0.2 µm had room temperature spontaneous polarization (3.1 $\mu C/cm^2$) and bias field dependence of dielectric constant similar to bulk polycrystalline $BaTiO_3$. Spontaneous polarization decreased with decreasing grain size, and films with average grain size 34 nm were not ferroelectric. The dielectric constants for films with three different grain sizes are shown as a function of temperature in Fig. 14. For the very fine grain size film the dielectric constant was 200 and was constant over the temperature range -15°C to 160°C, whereas the film with 0.2 µm grain size showed behavior more like bulk $BaTiO_3$. Figure 14 demonstrates the wide range of properties that can be achieved by controlling grain size of dielectric films during MOD processing.

Transparent PZT ($PbZr_{0.5}Ti_{0.5}O_3$) films were prepared (60) by the MOD process using lead 2-ethylhexanoate, zirconium acetylacetonate and titanium tetrabutoxide as precursor compounds with butanol as the solvent. After pyrolysis, the films were annealed at 500 to 800°C for 30 minutes. The films deposited on platinum substrates were smooth and uniform, but microcracking was observed for films deposited on fused silica substrates. Thin layer films deposited on platinum substrates and annealed at 700°C had a spontaneous polarization and remnant polarization of 35.7 $\mu C/cm^2$ and 30.6 $\mu C/cm^2$ respectively, and the coercive field was 45 kV/cm. The dielectric constant and dielectric loss angle were about 300 and 0.05, respectively.

Very dense films 0.4 to 2.5 µm thick of $Pb_{0.92}La_{0.08}(Zr_{0.65}Ti_{0.35})_{0.98}O_3$, described in a short notation as PLZT (8/65/35), were prepared (61) by spinning a formulation solution onto sapphire substrates, pyrolyzing at 500°C and annealing the films in air in the range 600 to 850°C. The precursor compounds were lead neodecanoate, lanthanium 2-ethylhexanoate, zirconium n-propoxide and titanium di-methoxy-di-neodecanoate. The annealing temperature range gave films with grain sizes from 0.2 to 1.0 µm; films with grain size 0.3 µm had a dielectric constant of 1800, a dissipation factor of

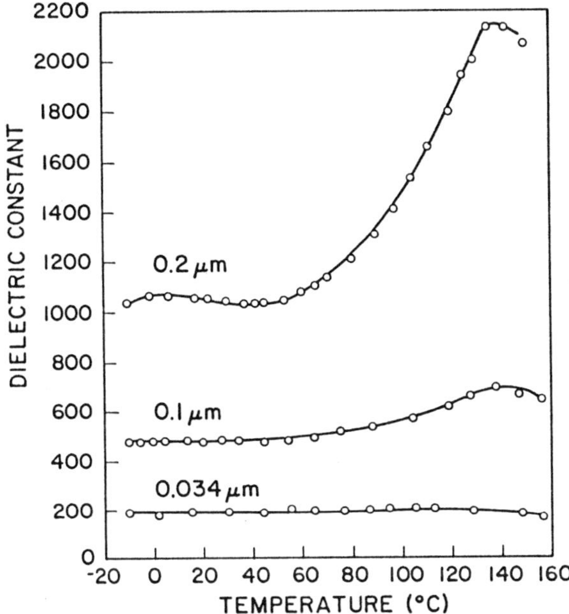

Figure 14. Variation of dielectric constant of $BaTiO_3$ films of varying grain sizes with temperature.

2%, a spontaneous polarization of 5 μC/cm, and a coercive field of 8 kV/cm. The change in birefringence with electric field for a PLZT film with 0.3 μm grain size is shown in Fig. 15. The linear and quadratic electro-optic coefficients calculated from the data in Fig. 15 were 30 x 10^{-12} m/V and 50 x 10^{-18} $(m/V)^2$, respectively. These coefficients are smaller than those of the same composition ceramic with average grain size of 3.0 μm, but are large enough for a number of potential device applications.

Films of lead strontium titanate, lead magnesium niobate, lead iron niobate and lead nickel niobate have also been prepared by the MOD process in the Turner Laboratory at Purdue University, but the results of these recent studies have only appeared in a report to the Office of Naval Research (62).

5.0 SUMMARY

It has been shown that MOD processing has very wide applicability to the preparation of electronic films. There are a number of advantages of

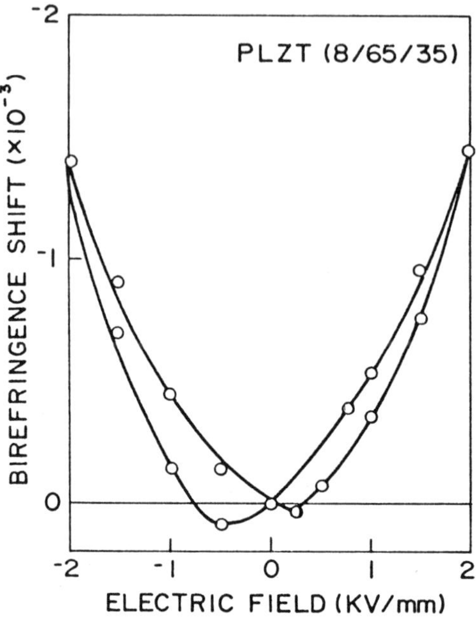

Figure 15. Birefringence shift with electric field for a PLZT (8/65/35) film.

MOD processing compared to alternate methods of electronic film preparation, but there has been very little basic research on MOD processing compared to these alternate methods. This is probably due to the fact that MOD processing is a multi-disciplinary area including organic synthesis, solution chemistry, surface chemistry, thermochemistry, film formation, grain size effects and film-substrate interactions, to name a few. Because of the shortage of fundamental research, MOD processing in many respects is still an art rather than a science. As a better understanding of the fundamental aspects is obtained, MOD processing can be brought under tighter control and the degree of acceptance and applicability should increase markedly.

ACKNOWLEDGEMENTS

The large majority of the results presented in this chapter were from various research programs at the Turner Laboratory for Electroceramics at Purdue University over the period 1980 to 1988. These research programs were sponsored by the Naval Avionics Center, the Jet Propulsion Laboratory and the Office of Naval Research, and their support is gratefully acknowledged.

REFERENCES

1. Dutertre Bros., *US Bulletin de la Societe d'Encouragement pour l'Industrie Nationale,* 129-134 (1861)

2. Falk, C., *Glasind.* 22:20-21 (1911)

3. Anon, *Glasind.* 25 (1914)

4. Marmsworth, W., *Sands, Clays, Minerals,* 3:49 (1936)

5. Morgan, J. E. and Short, O. A., *Am. Ceram. Soc. Bull.* 40:496-497 (1961)

6. Shaw, K., *Pottery and Glass,* 39:48 (1961)

7. Allison, G. D., *J. Canad. Ceram. Soc.* 31:35 (1962)

8. Mopper, R. T., *Ceram. Ind.* 80:74-76 (1963)

9. Gualandi, A., *Ceramica,* 19:60-62 (1964)

10. German Pat. No. 85262 (February 1895)

11. Siemens and Malske Aktiengesellschaft, British Pat. No. 420,774 (1934)

12. Taylor, G. F., *J. Optical Soc. Am.* 18:138-142 (1929)

13. Shulz, M., *Glasshutte,* 66:685-686 (1936)

14. Anon, *Platinum Metals Review,* 2:128 (1958)

15. Langley, R. C., "Improved Coatings from Organometallic Solutions." AFML TR 65-262, 30 pp. Wright Patterson A. F. B., Ohio (Aug. 1965)

16. Kuo, C. Y., *Solid State Technology,* (February 1974)

17. Vest, R. W., Final Technical Report on Contract No. N00163-79-C-0352, Purdue University, W. Lafayette, IN, (November, 1980)

18. Vest, R. W. and Vest, G. M., Final Technical Report on Contract No. N00163-83-C-0167, Purdue University, W. Lafayette, IN (April 1985)

19. Vest, G. M. and Vest, R. W., Final Technical Report, 7/1/85, JPL Flat Plate Solar Array Project, Pasadena, CA, DOE/JPL-956679-84

20. Sabo, C. J., Vest, G. M., Singaram, S. and Mis, D., *Proc. Intl. Soc. Hybrid Microelectronics Symp.*, pp. 59-65, Anaheim, CA (Nov. 11-14, 1985)

21. Vest, G. M. and Singaram, S., *Mat. Res. Soc. Proc.*, Vol. 60, pp. 35-42 (1986)

22. Acree, W. E., *Thermodynamic Properties of Non-Electrolyte Solutions*, Academic Press, Orlando (1984)

23. Reynolds, W. W., *Physical Chemistry of Petroleum Solvents*, 40, Reinhold, New York (1963)

24. Hildebrand, J. H. and Wood, S. E., *J. Chem. Phys.* 1:817 (1933)

25. Reference 23, p. 43

26. Pilpel, N., *Chem. Rev.* 63:221-234 (1963)

27. Honig, J. G. and Singleterry, C. R., *J. Chem. Soc.* 1114-1119 (August, 1956)

28. McBain, M. E. L. and Hutchinson, E., *Solubilisation and Related Phenomena*, Academic Press, New York (1955)

29. Pilpel, N., *Trans. Faraday Soc.* 56:893 (1960)

30. Nelson, S. M. and Pink, R. C., *J. Chem. Soc.* 1744 (1952)

31. Von Ostwald, W., and Riedel, R., *Koll. Zeit.* 69:185 (1934)

32. Yanagisawa, M., *J. Appl. Phys.* 61:1034-37 (1987)

33. Xu, J., "Preparation and Properties of Electroceramics Films using the Metallo-Orgnaic Decomposition Process." Ph. D. Thesis, Purdue University (1988)

34. Teng, K. F., "Ink Jet Printing in Thick Film Hybrid Microelectronics." Ph. D. Thesis, Purdue University (1986)

35. Xu, J., Shaikh, A. S. and Vest, R. W., *Thin Solid Films,* 161:273-80 (1988)

36. Vest, R. W. and Xu, J., *IEEE Trans. UFFC,* 35:711-717 (1988)

37. Xu, J., Shaikh, A. S. and Vest, R. W., *IEEE Trans. UFFC,* 36:307-312 (1989)

38. Vest, G. M. and Vest, R. W., *Intl. J. Hybrid Microelectronics,* 2:62-68 (1982)

39. Vest, R. W., Tweedell, E. P., and Buchanan, R. C., *Intl. J. Hybrid Microelectronics,* 6:261-267 (1983)

40. Teng, K. F. and Vest, R. W., *IEEE Trans. CHMT,* 12:545-549 (1987)

41. Teng, K. F. and Vest, R. W., *Appl. Math. Modelling,* 12:182-188 (1988)

42. Rohatgi, R., *IEEE Proc. 18th Photovoltaic Specialist Conf.,* Las Vegas, NV, p. 782 (Oct 21-25, 1985)

43. Mantese, J. V., Catalan, A. B., Mance, A. M., Hamdi, A. H., Micheli, A. L., Sell, J. A. and Meyer, M. S., *Appl. Phys. Lett.* 53:1335-1337 (1988)

44. Craighead, H. G. and Schiavone, L. M., *Appl. Phys. Lett.* 48:1748-1750 (1986)

45. Ohmura, Y., Shiokawa, T., Toyoda, K. and Namba, S., *Appl. Phys. Lett.* 51:1500-1502 (1987)

46. Teng, K. F., and Vest, R. W., *IEEE Trans. CHMT,* 11:291-297 (1988)

47. Gross, M. E., Appelbaum, A., and Gallagher, P. K., *J. Appl. Phys.* 61:1628-1632 (1987)

48. Sparks, D. R., "Thick Film Copper Conductors from Solutions of Copper Compounds." M. S. Dissertation, Purdue University (1982)

49. Bednorz, J. G. and Muller, K. A., *Z. Phys.* B64:189-193 (1986)

50. Vest, R. W., Fitzsimmons, T. J., Xu, J., Shaikh, A., Liedl, G. L., Schindler, A. I. and Honig, J. M., *J. Solid State Chem.* 73:283-285 (1988)

51. Hamdi, A. H., Mantese, J. V., Micheli, A. L., Laugal, R. C. O., Dungan, D. F., Zhang, Z. H. and Padmanabhan, K. R., *Appl. Phys. Lett.* 51:2152-2154 (1987)

52. Gross, M. E., Hong, M., Liou, S. H., Gallagher, P. K. and Kevo, J., *Appl. Phys. Lett.* 52:160-162 (1988)

53. Davison, W. W., Shyu, S. G., and Buchanan, R. C., *Mat. Res. Soc. Symp. Proc.,* Vol. 99, pp. 289-292 (1988)

54. Kumagai, T., Yokota, H., Kawaguchi, K., Kondo, W. and Mizuta, S., *Chem. Lett.,* Chem. Soc. Japan 1645-1646 (1987)

55. Herzfeld, C. J., "Application of Metallo-Organics to Ruthenium Dioxide/ Lead Borosilicate Glass Thick-Film Resistor Inks." M. S. Thesis, Purdue University (1985)

56. Baba, K., Takahashi, K., Shiratsuki, Y. and Katoh, R., *Proc. Intl. Soc. Hybrid Microelectronics Symp.,* Seattle, WA, pp. 381-386 (Oct. 17-19, 1988)

57. Micheli, A. L., Chaug, S-C, and Hicks, D. B., *Cer. Eng. Sci. Proc.,* pp. 1095-1105 (Sept.-Oct. 1987)

58. Davison, W. W. and Buchanan, R. C., *Mat. Res. Soc. Proc.,* Vol. 108 (1988)

59. Vest, R. W. and Xu, J., *Proc. Sixth IEEE Intl. Symp. Appl. of Ferroelectrics,* Bethlehem, PA, pp. 374-380 (June, 1986)

60. Fukushima, J., Kodaira, K. and Matsushita, T., *J. Mat. Sci.* 19:595-598 (1984)

61. Vest, R. W. and Xu, J., *Ferroelectrics,* 93:21-29 (1989)

62. Vest, R. W., Vest, G. M., Shaikh, A. S. and Liedl, G. L., "Metallo-Organic Decomposition Process for Dielectric Films." Annual Report on Contract No. N00014-83-K-0321, Purdue University, W. Lafayette, IN (June, 1988)

10

Chemical Characterization Techniques for Thin Films

Robert Caracciolo

1.0 INTRODUCTION

Surface phenomena represent one of the last frontiers of the physical sciences, because, while it is generally agreed that bulk properties of gases, liquids, and solids are well understood, at least in principle, this can in no sense be said for surfaces. Due to the increased availability of surface analytical instruments, modern technology has been able to examine and utilize the surface and near-surface regions of materials. Such processes as ion implantation, pulsed electron beams and lasers are used to modify composition and structure. Fabrication of thin films by various techniques, i.e., CVD, MOCVD, sputtering, laser ablation, yield different properties. Furthermore as devices grow smaller, and consequently thin films thinner, the role of surface and near-surface regions grow more and more significant.

This chapter is designed to describe some of the more popular techniques for surface and thin film characterization such as XPS, AES, RBS, SIMS, etc. In the limited space of this chapter, it is not possible to go into each technique in depth, however the basic principles are outlined and some illustrative examples are given. See Refs. 1 - 4 for more in-depth discussions of these characterization techniques and others.

Chemical characterization of materials is generally based on the measurement of the resultant emission of radiation, electrons or ions from a material that has been irradiated by the same. Among the processes that occur are coulomb scattering (RBS and ISS) to determine mass and depth of an atom in a material, photoionization (XPS) to determine the electronic structure of atoms near the surface, and mass analysis (SIMS) to determine elements and clusters of elements at the surface. Table 1 gives a summary of incident beams, measured responses, and information obtained.

Table 1. Comparison of Chemical Characterization Techniques.

Technique	Excitation	Response	Sampling Depth	Sensitivity at.%	Information
Auger Electron Spectroscopy (AES)	Electrons (2 - 10 kV) or x-rays	Auger Electrons via e^- Energy Measurement	20 - 30 Å	10^{-1}	Elements Li-U
X-ray Photoelectron Spectroscopy (XPS or ESCA)	X-rays typically MgKα or AlKα (1253 eV or 1486 eV)	Photoelectrons via e^- Energy Measurement	20 - 30 Å	10^{-1}	Elements Li-U -chemical state
Rutherford Backscattering Spectrometry (RBS)	MeV He^+ (RBS)	Scattered He^+ via Energy Measurement	50,000 Å (RBS)	1	Depth Profiles (RBS) Element Be-U
or Ion Scattering Spectroscopy (ISS)	KeV He^+ (ISS)		Monolayer (ISS)	10^{-3} ML	Element Be-U
Secondary Ion Mass Spectrometry	Ions Ar^+, Ne^+, Kr^+, O^+ (1 - 10 kV)	Sputtered Ions via Mass Analysis	10 Å	10^{-7}	Depth Profile Elements H-U

Many important engineering problems center on surface aspects. As examples, corrosion limits the usefulness and lifetime of manufactured goods more than any other consideration, and heterogeneous catalytic processes are used more than all others for refining energy sources, making chemicals, and reducing air and water pollution. The surface composition of ceramic whiskers and fibers used in composite materials can affect the mechanical properties. The recent development of high T_c $YBa_2Cu_3O_{7-x}$ superconducting thin films has lead to the extensive use of the above characterization techniques to determine surface and bulk structure of these films.

2.0 X-RAY PHOTOELECTRON SPECTROSCOPY (XPS)

Although the earliest measurements of the electron kinetic energy distribution induced by x-ray irradiation of solid materials were reported in the early part of this century, the energy resolution attainable at that time was insufficient to observe actual peaks in the photoelectron spectra, therefore the technique was inadequate for real surface analysis. In 1954, a group of Swedish scientists headed by Siegbahn (5) operated for the first time a high resolution electron spectrometer for low energy electrons produced by x-ray irradition. It was with this instrument that the phenomenon of the photoelectron was first observed. XPS quickly became a technique for studying atomic orbital energy systematics as it was far more accurate than the techniques used previously. In a classic paper published in 1958 (6), Nordling et al. showed that the difference between copper and its oxide (Fig. 1) could be clearly distinguished with XPS. The chemical shift effect was elaborated upon by the Swedish laboratories and it was shown that the chemical states of non-metallic atoms as well as the oxidation states of metals could be distinguished in many cases. Thus, because of its potential applications, the Swedish grouped named this technique "Electron Spectroscopy for Chemical Analysis." The corresponding acronym ESCA is widely used synonymously with the generic name "X-ray Photoelectron Spectroscopy" or XPS. The technique of XPS has become one of the more popular and commonly used surface analytical techniques of the last two decades. This is in part due to commercial availability of high resolution spectrometers.

2.1 Experimental Considerations

In photoelectron spectroscopy, the basic process is the adsorption of a quantum of energy or photon, h , and the ejection of an electron, the photoelectron. The photoelectron process is illustrated in Fig. 2. The photoionization process is

Eq. (1) $$A + h\nu \rightarrow A^{+} + e^{-}$$

Alternatively, in the case where the photon energy is less than the binding energy BE, a photoexcitation process takes place

Eq. (2) $$A + h\nu \rightarrow A^{*}$$

The excited atom A* can relax with the production of an Auger electron (Auger process) or by emission of a photon (fluorescence), explained in a later section.

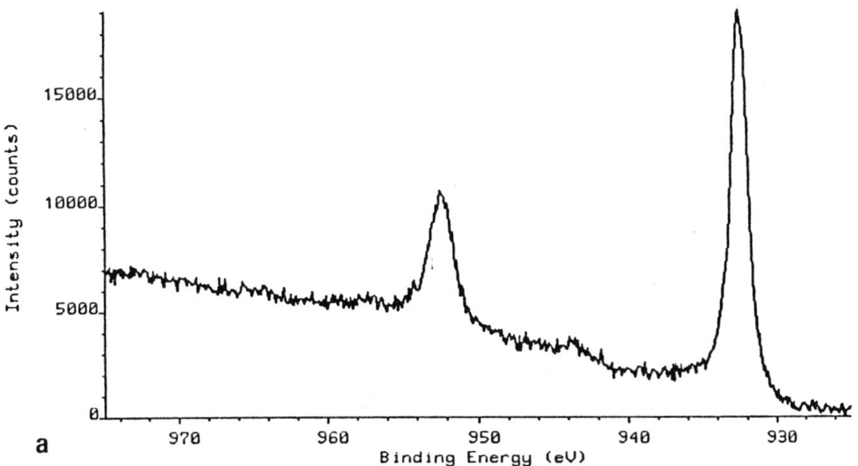

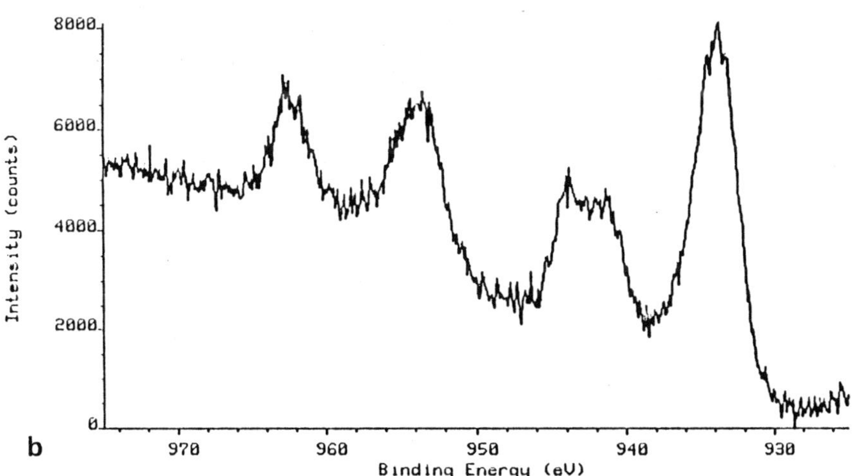

Figure 1. High resolution Cu2p spectra of a) copper metal, and b) copper oxide.

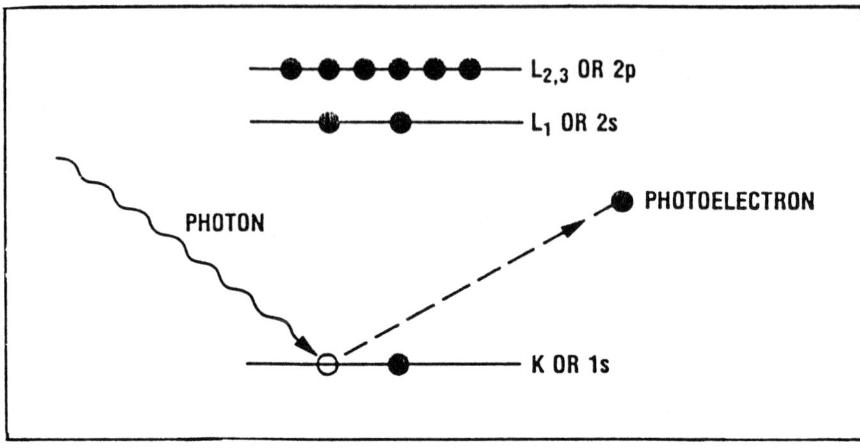

Figure 2. Energy level diagram depicting the photoelectron process.

As a photon interacts with atoms in a material, the photon is absorbed with a probability proportional to the photoelectric cross section. The entire photon energy is absorbed by electrons bound to the atom. If the photon energy hν is greater than the electron binding energy, then the electron will escape from the atom with a kinetic energy KE given by Eq. 3:

Eq. (3) $$KE = h\nu - BE - \theta_s$$

The binding energy is defined as the electron orbital energy level with respect to the Fermi level, or the energy with which the electron is bound to the atom. Since the kinetic energy of photoelectrons is typically ≤1200 eV, they can only escape from shallow depths (≤30 Å) without inelastically scattering off another atom.

As can be seen from Eq. 3, in order to accurately determine BE, hv must be known and KE must be measured. Both a source of monochromatic radiation and an electron spectrometer are required. As is common to all the electron spectroscopies where the escape depth is 20 - 30 Å, careful sample preparation and clean vacuum systems are required.

The spectrometer work function θ_s is dealt with as a calibration parameter. Its value is determined by calibrating the system with standard specimens such as Au, Cu, or Ag.

2.2 Radiation Sources

Typically in XPS, x-rays produced by the electron bombardment of Mg or Al targets are employed. These x-rays, *soft x-rays* (~1 keV), are characterized by a relatively low intensity of *bremsstrahlung* or x-ray continuum with respect to the Kα lines, as opposed to *hard x-rays*, i.e., Cu. In the case of Mg and Al, about one-half of the x-rays produced by electron bombardment are the Kα x-rays. The contribution from the continuum spectrum is not significant since it is spread over a range of several keV. The widths of the exciting x-radiation is normally the major limitation on the resolving power of the instrument. Table 2 gives widths $E_{1/2}$ at half maximum for some characteristic x-ray lines. A 1 eV FWHM is sufficient for most applications. In cases where better resolution is required, an x-ray monochrometer can be used to achieve FWHM's of 0.2 - 0.3 eV, but with a corresponding decrease in x-ray intensity.

Table 2. Characteristic X-ray Line Widths

Target Material	Characteristic Line Energy (eV)		Approx Width Half Height (eV)
Mg	K$\alpha_{1,2}$	1253.6	0.7
Al	K$\alpha_{1,2}$	1486.6	1.0
Ti	Kα	4510.9	1.2
Ag	Kα	2984.3	2.6

Ultraviolet photoemission spectroscopy (UPS) typically uses resonance light sources such as an He discharge lamp (He I: 21.2 eV and He II: 40.2 eV). The intensities of the light sources are high and the energy widths are sharp. The energy resolution in these experiments is generally limited by the electron energy analyzer. This spectroscopy is used mainly to analyze valence electrons, as opposed to XPS which is used to study the core energy levels.

Synchrotron radiation provides a continuous spectrum with intensities far in excess of conventional x-ray sources. This radiation is used in specialized experimental investigations rather than routine analysis due to their limited access.

2.3 Electron Energy Analyzers

The energy of photoelectrons is determined by their deflection in electrostatic fields. Three types of electron energy analyzers are typically used: the hemispherical analyzer (HSA), Fig. 3a, cylindrical mirror analyzer (CMA), Fig. 3b, and 127° sectors, not shown. The principle of operation of these analyzers will not be discussed here but can be found in Refs. 2, 3, and 4. An electron energy analyzer measures the number of emitted electrons as function of their kinetic energy.

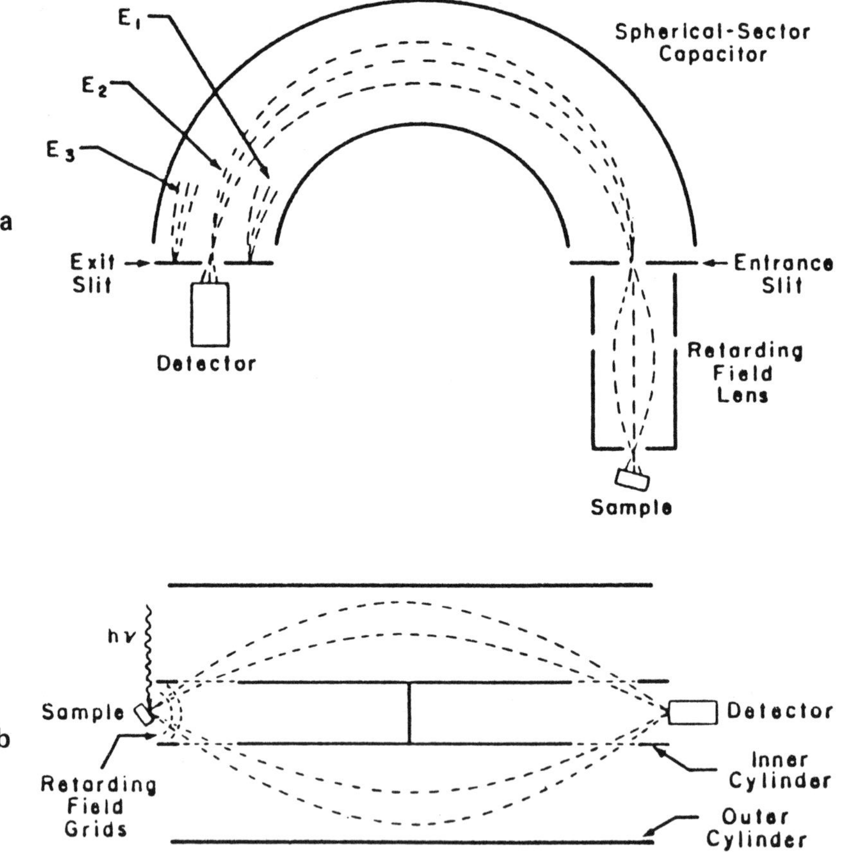

Figure 3. Typical electrostatic electron energy analyzers: a) Hemispherical Analyzer (HSA), and b) Cylindrical Mirror Analyzer (CMA)

The HSA is commonly used today for most techniques, XPS, angle resolved XPS, ISS, and AES. The CMA is used more commonly for collecting Auger spectra in differential mode dN/dE. Data acquisition modes are described in detail in Refs. 3 and 4. Energy resolution for these spectrometers is on the order of 1 eV. The 127° sector analyzers have specialized usage, typically for High Resolution Electron Energy Loss Spectroscopy (HREELS). Energy resolution for this spectrometer is on the order of meV.

2.4 The XPS Spectrum

Chemical identification is accomplished by measuring the KE, hence the BE of the various photoelectrons. These values are illustrated (7) and tabulated (4) in several manuals.

An XPS spectrum of K-doped silica is illustrated in Fig. 4a. The O1s, K2s, K2p, Si2s, Si2p photoelectrons along with the O_{KLL}, and K_{KLL} Auger transitions are labelled. A high resolution spectrum of the O1s photoelectron is given in Fig. 4b. As can be seen, there are three chemical states identifiable, two under the main peak at 533 eV and the small shoulder on the low energy side of the main peak. The main peak is due to oxygen atoms bonded to two silicon atoms in a bridge-bonded configuration (8)(9). The second state is oxygen atoms bonded to one silicon and one potassium atom. The lower binding energy is due to the ability of electropositive K atoms to donate electrons to oxygens, hence these oxygens have more electrons and the consequential "screening effect" lowers the binding energy.

2.5 Insulators

When the sample is not grounded, the photoelectrons and secondary electrons leaving the sample will tend to make the sample acquire a positive charge. However with the normal dual anode type of x-ray gun there is a fair amount of white radiation or "bremsstrahlung" accompanying the characteristic line, and this, when scattered about the source chamber, causes secondary electrons to be released from the walls of the chamber; the electrons move back to the specimen and neutralize some of the positive charge and a charge equilibrium is attained. The process is almost instantaneous and the equilibrium is very steady. Under such conditions Eq. (3) can be written as

Eq. (4) $$KE = h\nu - BE - C - \theta_s$$

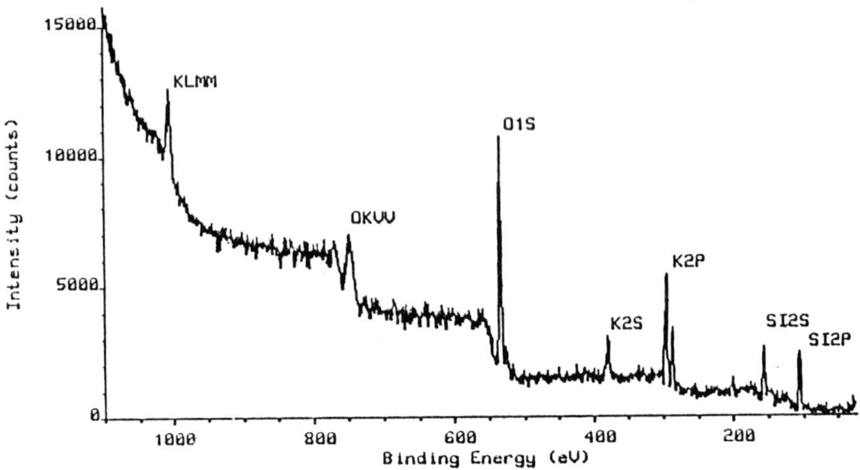

Figure 4a. XPS spectrum of a silica (SiO_2) surface with one monolayer of potassium adsorbed onto it.

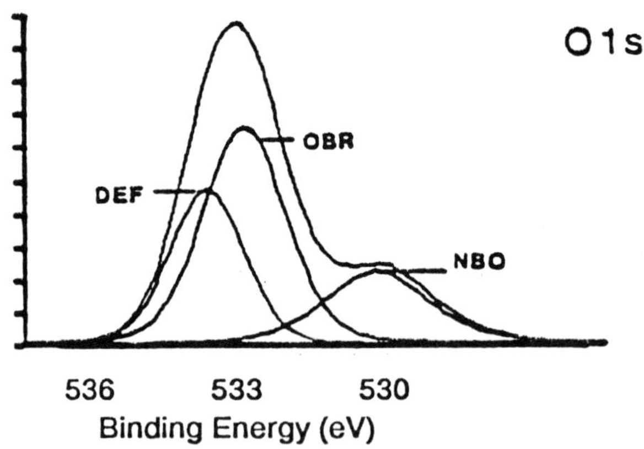

Figure 4b. High resolution XPS spectrum of the O1s peak. Two chemicals are identifiable, the main peak being bridging oxygens, and the low energy shoulder being non-bridging oxygens.

where C is the charge acquired by the sample. If accurate binding energies are to be acquired, some reference will have to be added to the sample so that C can be determined. Possible methods include the following:

(i) coating the sample with a very thin (<5 - 10 Å) conductive layer of Au or similar material

(ii) intimately mixing the sample with a fine conductor, e.g., graphite powder, etc.

(iii) using an internal standard such as a known species with a known BE, or the Auger parameter

When a monochromator is being used for XPS, the action of the monochromator removes the "bremsstrahlung" and this severely cuts down the number of secondaries released from the walls of the sample analysis chamber, and this means that a charge equilibrium is only very slowly set up and the final shift, which is only a few volts in the normal case, can become fifty volts or greater in the monochromator case. Consequently some additional means of charge neutralization is desirable. Suitable means are a low energy electron gun or a UV lamp which liberates copious numbers of secondary electrons from the sample analysis chamber walls.

The Auger parameter is defined as the difference of the Auger line and the photoelectron line plus the energy of the exciting radiation, or more simply,

Eq. (5) $$\alpha' = \text{KE(Auger)} - \text{KE(photoelectron)} + h\nu, \text{ or}$$

$$\alpha' = \text{KE(Auger)} + \text{BE(photoelectron)}$$

where the zero reference for both the Auger and photoelectrons is the Fermi edge. This quantity is useful because it is not subject to problems with determination of steady state charge. Figure 5 is a plot of Auger energy versus photoelectron energy, termed *chemical state plots.*

2.6 Sampling Depth in XPS

In principle, as the angle the electron exits relative to the sample surface is decreased, the effective escape depth appropriately is decreased. This can be utilized to enhance the surface sensitivity of XPS and AES simply by increasing the angle between surface normal and the spectrometer axis. As the angle is increased from 0° to almost 90°, the escape depth of electrons detectable by the spectrometer approaches zero as illustrated in

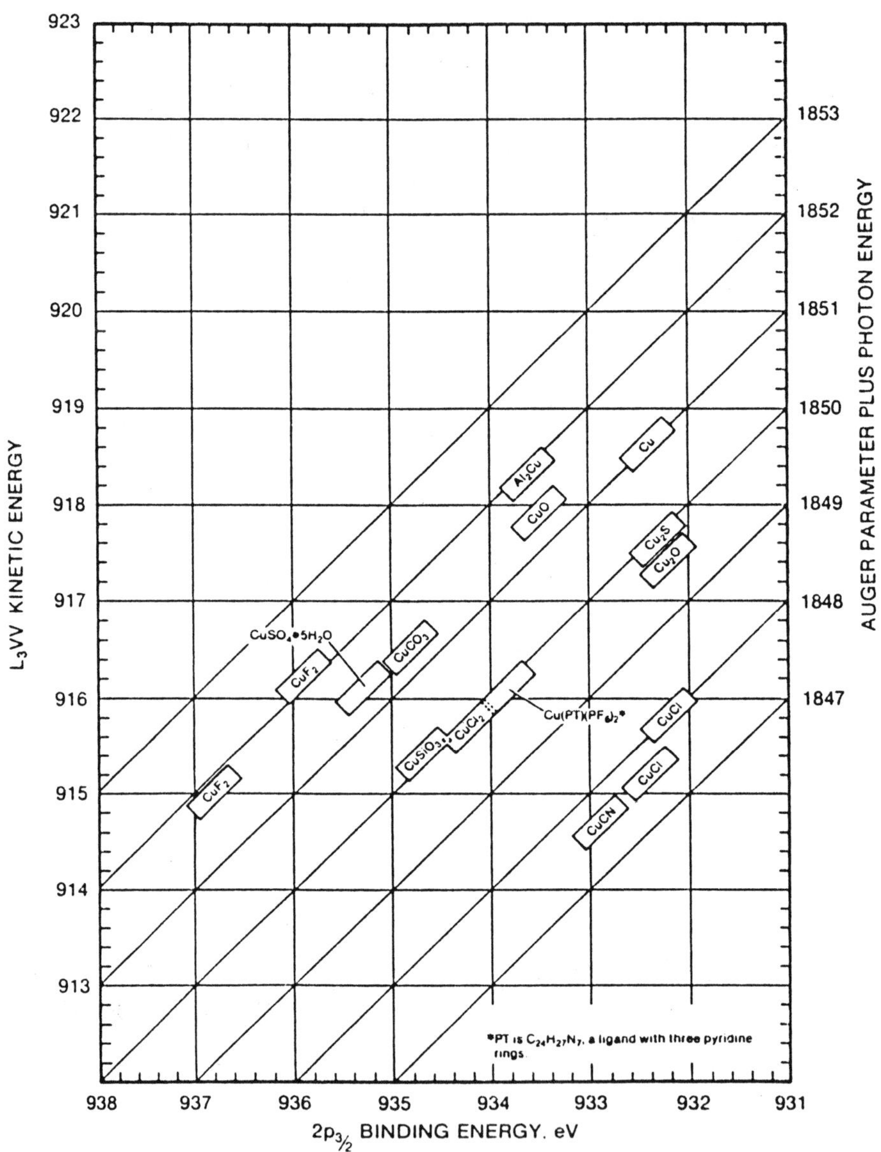

Figure 5. An example of the chemical state plot indicating the Auger parameter from the various compounds of copper.

Fig. 6. The probability of electrons escaping from a depth z with no occurrence of inelastic scattering is proportional to:

Eq. (6) Escape Probability ~ $\exp[-z/(\lambda \sin \theta)]$

where

z = depth
λ = mean free path
θ = angle the electron exits parallel to the surface.

The two characteristic x-ray lines most used in XPS are MgKα and AlKα with energies of 1253.6 eV and 1486.6 eV respectively, and obviously photoelectrons of energy greater than these energies are not going to be released. Because AlKα and MgKα from the standard dual anode x-ray gun are accompanied by some bremsstrahlung, there may be some low intensity Auger lines of greater than 1500 eV. Thus sometimes the Si_{KLL} series is observed in high silicon content samples. The energy of the electrons lies in the range 0 - 1500 eV and it is known that the mean free path (MFP) of electrons of such energy (i.e., the average distance an electron travels without undergoing a collision event and generally losing energy) is not long. A typical graph of mean free path versus energy is shown in Fig. 7. The graph is constructed from a whole series of results of different materials (10). The graph shows that for an electron of 1000 eV the typical mean free path is about 20 Å. The sampling depth is taken to be approximately three times the mean free path (or escape depth) and at 1000 eV is of the order of 60 Å, thus XPS samples the surface but it can also look at the bulk. The use of *angle-resolved* XPS allows differential between the surface and the 'near surface' (bulk) chemistry.

Figure 7 shows how the Escape Depth (Y-axis) varies with the kinetic energy of the ejected electron. It is about 5 Å at around 50 to 100 eV increasing to about 20 Å at 1000 eV. Because of this, low energy peaks will be more rapidly attenuated by a contamination overlayer.

Alternatively, the surface sensitivity may be tuned by changing the angle at which electrons leaving the surface are collected (take-off angle). This is easily done when the sample is mounted on a rotatable manipulator. A variation of up to a factor of ten in surface sensitivity may be obtained by this method. Since the volume sample decreases, the signal decreases and the signal to noise ratio decreases. Therefore to enhance surface sensitivity the proper compromise must be found such that the S/N ratio is adequate.

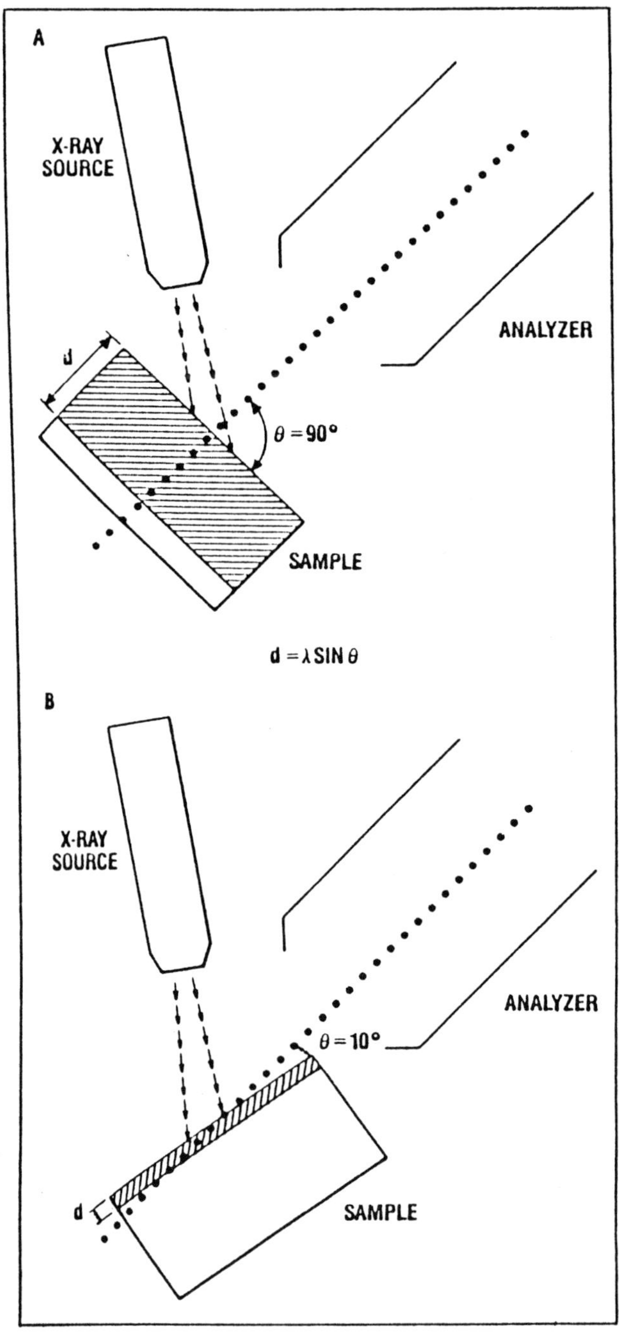

Figure 6. Schematic drawing illustrating the effect of sample tilt angle on depth analysis in angle-resolved XPS.

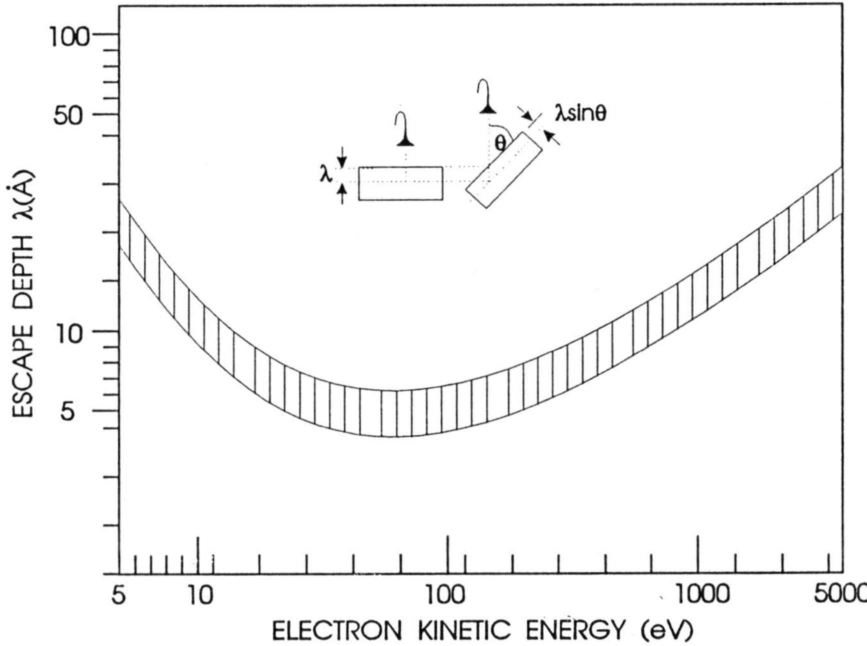

Figure 7. Escape depth plotted against electron kinetic energy.

This technique is applied frequently to polymer systems where the MFP tends to be long. By employing angle-resolved XPS one can distinguish between surface effects and bulk type effects. This technique can also be used for non-destructive depth-profiling where near-surface variations in concentration are on the order of 20 Å. It also can be applied where there are thin films such as oxides on metal surfaces.

This method can also be accomplished by the choice of different x-ray anodes. Since the KE of photoelectrons is proportional to the x-ray energy, the depth sensitivity can be increased by going to high energy x-ray sources (Ti, Zr, etc).

Typically the surface of most films are reactive with air, usually oxygen and carbon monoxide. This tends to create an oxide and carbonaceous overlayer on the surface of the film. Sputtering is often employed to remove the surface oxide and carbon prior to analysis. It is important to remove this surface layer since it attenuates the intensity of the signal from the underlying bulk of the film.

2.7 The 'Chemical Shift' in XPS

The so-called chemical shift in XPS is probably the most important feature of the method and certainly accounts for its popularity as a surface technique. It was mentioned earlier that the energies of the core levels (i.e., inner shell electron levels) are known for each element and allow an elemental identification to be made. However it has been known for some time that the core level of a particular atom does depend on the chemical environment of the atom. Thus the binding energy of the Al2s level in aluminum metal is different from the binding energy of the Al2s level in Al_2O_3 by about 3eV. The 2s level of the aluminum associated with the oxide is the more tightly bound (i.e., higher binding energy, lower kinetic energy of ejection). All elements (from and including Li upwards) in the periodic system show the chemical shift to a greater or lesser extent. It was, in fact, this chemical shift which first drew attention to XPS, or ESCA as it is sometimes known, due to the outstanding work of Siegbahn and his colleagues at the University of Uppsala (11). It was also the chemical shift which was largely responsible for the introduction of dedicated XPS instruments into the market ca. 1969 - 1970. It was hoped that it would be a sort of 'poor resolution NMR' applied to all elements above and including Li. Figure 8 shows the C1s spectrum of a fluoro-polymer and it shows that the carbon 1s level has six very distinct peaks. The peaks are shown (from left to right) in terms of increasing kinetic energy or decreasing binding energy. The peaks with the higher binding energies are due to carbon atoms in the more electronegative environments, i.e., bound to three electronegative fluorine atoms. The remaining carbons are bound to fewer fluorine atoms or less electronegative species, resulting in lower binding energy.

Despite this dramatic example, it is probably true to say XPS never quite reached its full potential in structural work. One reason is the limited sampling depth and limited resolution, but for chemical state determination of surfaces it is an extremely powerful technique—probably the most powerful of all surface techniques for chemical bonding studies.

Although it is impossible to be specific about the changes in bonding for different chemical states, on an empirical basis there are certain general rules which hold in most cases, e.g., the core levels in the metal atom of a metal oxide such as Al_2O_2, MgO, SiO_2, MoO_3 tend to be more tightly bound than in the pure metal. The same holds for metals bonded to other electronegative elements of the halogens and chalcogens. The degree of bonding for different anion/cation and covalent combinations is usually obtained from tables or the literature. Thus tables for sulphur and carbon compounds have been published by Siegbahn and his group (5)(11) while other combinations can be found in tables in Ref. 12.

Having discussed how to determine the elements present on the surface and how to determine their chemical (oxidation) state, attention is now turned on artifacts that can occur in XPS spectra which may be confusing to those uninitiated in the art. These artifacts can, under certain circumstances, yield useful information. Some of the effects have already been discussed but are discussed again in the following section.

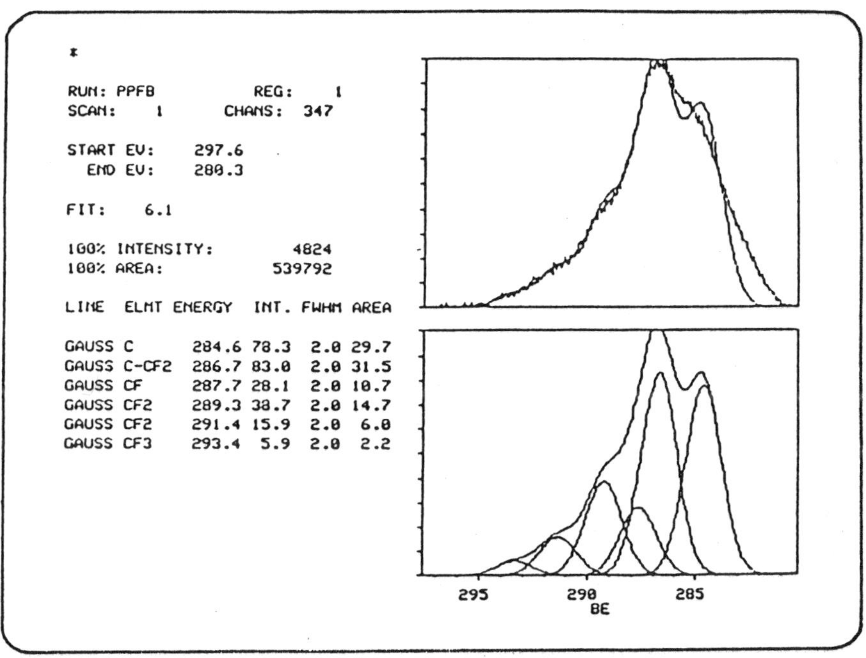

Figure 8. High resolution spectrum of the C1s peak for a fluoropolymer. Curve synthesis is employed to determine the various carbon species.

3.0 AUGER ELECTRON SPECTROSCOPY (AES)

The radiationless transition known as the Auger effect was discovered in 1925 by the French physicist Pierre Auger (2)(4). The Auger effect refers to a particular de-excitation process of an excited atom, i.e., an atom having an electron vacancy in what would otherwise constitute its ground-state electronic structure. The de-excitation proceeds by a radiationless transition in which an outer electron drops into the vacancy and the corresponding energy release causes another electron to be ejected.

The use of Auger spectroscopy did not become useful as a surface analysis technique until the 1960's. This was probably due to a general lack of adequate ultra-high vacuum technology.

Typically Auger Electron Spectroscopy (AES) is used to identify atomic species within the top 50 Å of the surface. It is sensitive to atomic concentrations on the order of 0.1%. AES is further applied in determining molecular species by observation of either small shifts in the Auger electron kinetic energies or changes in the fine structure peaks of the Auger spectrum.

3.1 General Theory

Upon exposing a sample surface with a monoenergetic beam of electrons, an energy distribution of the emitted electrons is realized as illustrated in Fig. 9. Such a spectrum can be generated with the use of electron energy analyzers as described in the previous sections. Several important features are labelled in the illustration, the elastic peak, electron loss peaks, and the Auger peaks. The elastic peak, E_o, is simply electrons of the primary beam which elastically scatter off the surface atoms with no energy losses. The electron loss peaks are associated with the absorption of energy due to molecular vibrations, much like Infrared Spectroscopy, and surface plasmons. The analysis of these peaks is known as Electron Loss Spectroscopy (ELS). ELS is outside the scope of this chapter, but is described in several journals, i.e., *Journal of Vacuum Science and Technology, Applied Surface Science,* etc.

When an atom is ionized by the incident electron beam (note: the Auger energy is independent of the ionization process), it may revert to a lower energy state by one of two processes: (1) an electron in an outer shell may drop into the lower shell vacancy with the resulting energy released in the form of a characteristic x-ray, or (2) the excess energy may result in the ejection of a electron (a radiationless transition). The latter is termed the Auger effect with reference to his identification of the process in Wilson cloud chamber studies of x-ray ionization of gases. This is a two-electron process and the ejected electron is usually referred to as an Auger electron.

The energy E of the Auger electron is determined by the ionization energy E_i of the original vacancy, the binding energy E_b of the outer shell electron that will fill the vacancy, and the ionization energy E_e of the ejected electron:

Eq. (7) $$E = E_i - E_b - E_e$$

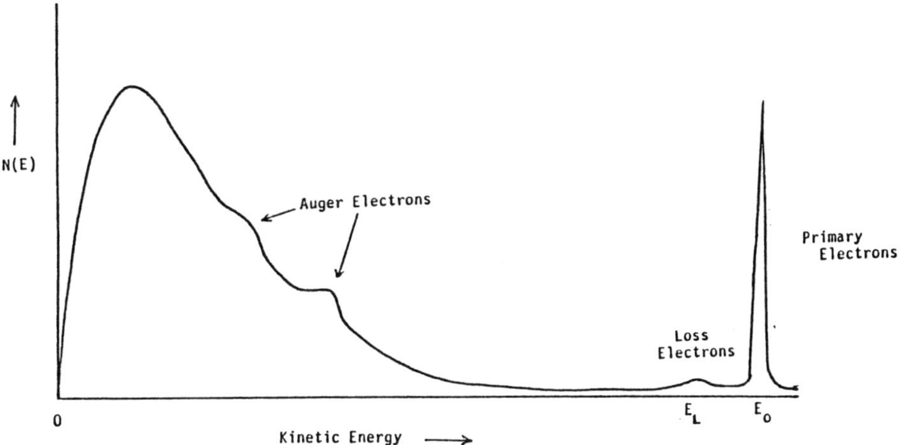

Figure 9. The backscattering spectrum for a monoenergetic beam of electrons impinging on a metallic surface.

Clearly, the Auger electron is ejected with a well defined energy that is characteristic of the original ion and may serve to identify it. An atom of large atomic number initially ionized in a low shell will have multiple transition modes and its Auger spectrum will exhibit corresponding complexity.

A schematic representation of the Auger process is shown in Fig. 10a. A particular Auger transition is characterized by three energy levels, one for the original vacancy and one each for the initial status of the two electrons. It is customary, therefore, to use three letters to identify these processes (e.g. KLL), one for each energy level.

An ion bound to the surface of a metal has additional transition for de-excitation and Auger electron ejection via the nearly inexhaustible supply of conduction electrons of the metal (Fig. 10b) (13)(14).

The electron energies ζ in the conduction band are measured with respect to the Fermi level lying at an energy ϕ below the vacuum level.

In this case, de-excitation by two conduction band electrons, the Auger electron energy is given by

Eq. (8) $$E = E_{L_{2,3}} - (\zeta_1 + \phi) - (\zeta_2 + \phi)$$

The two electrons of energy ζ_1 and ζ_2 may originate anywhere within the conduction band $0 \leq \zeta \leq \zeta_0$. Therefore, a band of energies at the Auger

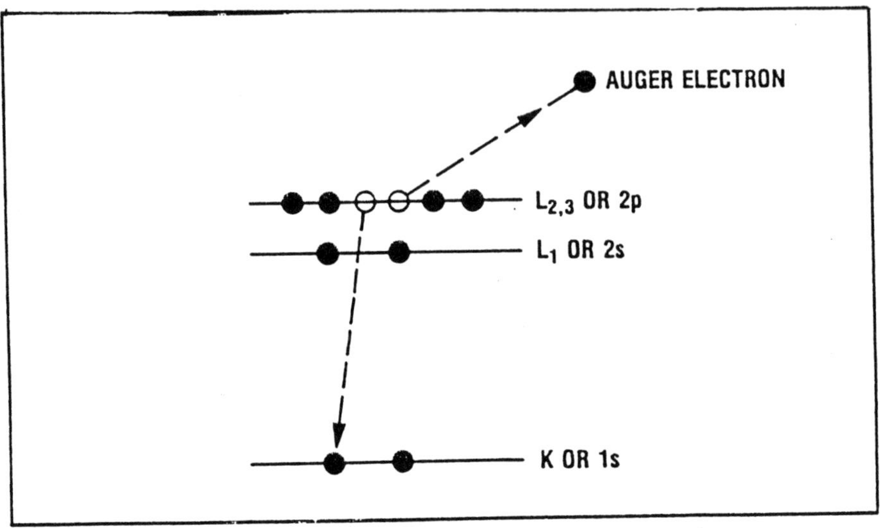

Figure 10a. Electron energy level diagram depicting the Auger process.

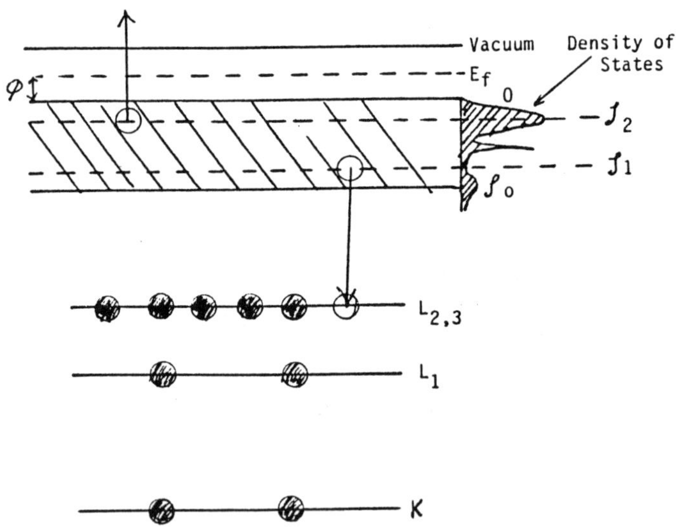

Figure 10b. Auger process involving the conduction band of a metal.

energy E is expected and is associated with the origination of both electrons from within the full width of the conduction band and in accordance with the density of states in the conduction band. As seen in the above equation, the energy band at the Auger energy is twice the width of the conduction band ($2\zeta_0$), because ζ_1 and ζ_2 both vary from 0 to ζ_0.

As already indicated, there are two parallel transitions for de-excitation of the original ion, one is the Auger transition and the other is x-ray fluorescence. They are, in effect, competitive processes since they must balance in accounting for the net rate of de-excitation. It is well known that the fluorescence yield increases with atomic number and the converse applies to the Auger yield.

A calculation of the fluorescence yield w as a function of atomic number Z gives (15).

Eq. (9) $$w = (1 + \underline{a}Z^{-4})^{-1}$$

where $\underline{a}$ is a function of the shell that is ionized, and the shells that supply electrons in the Auger transition. The value of $\underline{a}$ is on the order of 10^6.

Lastly, the primary electron beam is typically 2 to 10 keV for AES. This energy is sufficient to generate Auger electrons by ionizing various shells with a differing range of atomic numbers (16).

K shell, $3 \leq Z \leq 17$ Li-Cl

L shell, $11 \leq Z \leq 44$ Na-Ru

M shell, $19 \leq Z \leq 87$ D-Fr

The remainder of this paper concentrates on the discussion of the usefulness of AES in heterogeneous catalysis.

3.2 Sampling Depth in AES

The energy of the bombarding electron beam is of higher energy than the Auger electrons produced by it, so it is the escape depth of the Auger electrons which determine the sampling depth in AES. The energy of Auger lines is generally about the same as those in XPS, hence the sampling depth is the same as for XPS, i.e., ~50 Å (depending on energy—see Fig. 6 and 7).

3.3 The Chemical Shift in AES

There are chemical shifts associated with AES just as there are with XPS. This can be understood by considering Eqs. 7 and 8. The electron energy levels and density of states are sensitive to the chemical environment of the atom undergoing the Auger process. In some cases the shifts are smaller and in some cases larger than they would be in XPS. Both experimental and theoretical work is being done in the area of Auger chemical shifts and lineshape analysis (15)-(17). Figure 11 shows the variation in carbon Auger lineshapes of different molecular species. Basically Auger peak analyses are very powerful in determining structure, however they are very difficult to interpret. It is often useful, when possible, to compare Auger lineshapes of unknowns to those of standard materials.

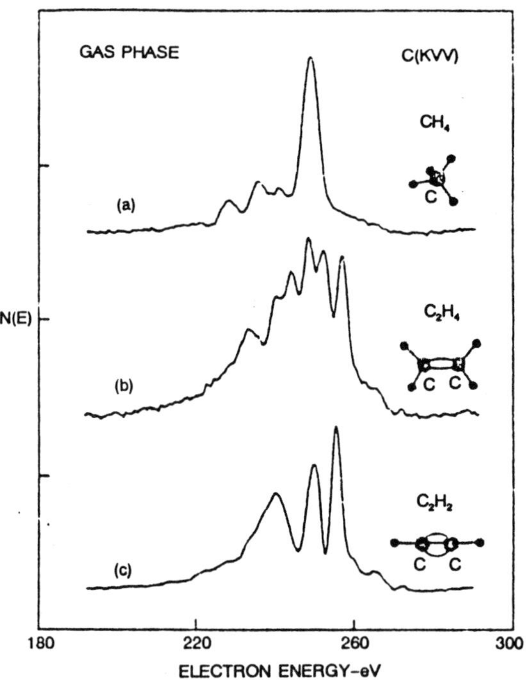

Figure 11. Gas-phase, electron excited C(KVV) spectra for (a) CH_4, (b) C_2H_4, and (c) C_2H_2.

4.0 BACKSCATTERING SPECTROMETRY (RBS and ISS)

Consider the close-impact scattering which is governed by the well-known coulomb repulsion between positively charged nuclei of the projectile and target atom. This is the underlying principle of backscattering spectrometry. Typically a monoenergetic beam of light ions (usually ^4He ions) are used as the incident probe. At high beam energies (~ 1 MeV), incident ions will penetrate several microns below the surface before elastically scattering off a target atom. This technique in this mode is typically referred to as Rutherford Backscattering Spectrometry (RBS).

At low energy (~1 keV), incident ions will predominantly scatter from surface atoms, making the technique very surface sensitive (top monolayer). In this mode, the technique is referred to as Ion Scattering Spectrometry (ISS) or Low Energy Ion Scattering (LEIS).

4.1 Kinematics of the Elastic Collision

Backscattering spectrometry involves the elastic collision between an energetic primary ion and a stationary target atom. The transfer of kinematics in elastic collisions between two isolated particles can be solved fully by applying the principles of conservation of momentum and energy. Referring to Fig. 12, the equations of motion can be written as:

Eq. (10) $$\tfrac{1}{2}M_1v^2 = \tfrac{1}{2}M_1v_1^2 + \tfrac{1}{2}M_2v_2^2$$

Eq. (11) $$M_1v = M_1v_1\cos\theta + M_2v_2\cos\phi \quad \text{(x-axis)}$$

Eq. (12) $$0 = M_1v_1\sin\theta + M_2v_1\sin\phi \quad \text{(y-axis)}$$

Eliminating v_2 and ϕ yields

Eq. (13) $$E_1/E_o = \{[(M_2^2 - M_1^2\sin^2\theta)^{0.5} + M_1\cos\theta]/(M_1 + M_2)\}^2$$

As with other spectroscopies, an accelerator for the primary ion beam and an energy analyzer are required. In LEIS, the accelerator is typically a conventional ion gun as found on most commercially available surface analysis instruments. The spectrometer is commonly an Hemispherical Energy Analyzer (HSA) used similarly for XPS and AES. In RBS, the accelerator is typically a tandetron, and the energy analyzer is a nuclear particle detector.

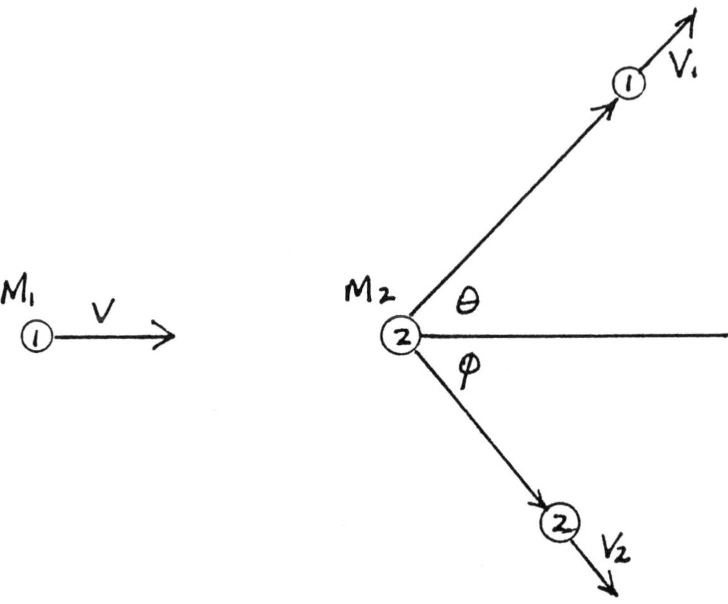

Figure 12. Schematic of an elastic collision showing incident and resulting trajectories.

In RBS, the incident ions (H^+, d^+, or He^+) are at high energies (0.5 - 5 MeV), so they can penetrate to deeper depths before elastically scattering off an atom. As the incident ions pass through the material, they lose energy through interactions with electrons which are raised to excited levels or ejected from atoms. Although these are discrete processes, macroscopically the energy loss can be considered to be continuous. A compositional depth profile can be obtained by considering the energy loss (dE/dx:eV/Å) during the passage through the solid. The energy lost in penetration is directly proportional to the thickness of material traversed so that a depth scale can be assigned directly and quantitatively to the energy spectra of detected particles.

4.2 Energy Loss

As originally done by Bohr in 1913, dE/dx can be derived by considering a heavy particle, an alpha particle or a proton, of charge Ze, mass M, and velocity v passing an atom electron of mass m at a distance b. As the heavy particle passes, the coulomb force acting on the electron changes direction continuously. The component of force in the direction perpendicular to the trajectory of the particle transfers momentum p to the electron. This problem of energy transfer can be solved by coulomb force scattering. We will not give the derivation here since it can be found in Ref. 18. However the final expression is

Eq. (14)
$$-\frac{dE}{dx} = \frac{4Z^2e^4n}{mv^2} \ln \frac{2mV^2}{I}$$

where n is the electron density (number of electrons per unit volume) and I is the excitation energy. As MeV He ions traverse the solid, they lose energy along their incident path at a rate dE/dx between 30 and 60 eV/Å. This derivation assumes that the solid material is elemental. However for compounds the energy loss is the sum of the losses of the constituent elements weighted by the abundance of elements, i.e., Bragg's Law. For a compound A_aB_b the stopping cross-section is given by

Eq. (15)
$$k^{AB} = ak^A + bk^B, \text{ and}$$

Eq. (16)
$$dE/dx = Nk^{AB}$$

where N is the number of molecules/volume.

4.3 Depth Profiles by RBS

Figure 13 shows the shape of an RBS spectrum (Y vs. E) for a single element target of finite thickness (~5000 Å). The shape of the spectrum is characterized by a leading edge at a higher energy and a trailing edge at lower energy. The leading edge reflects elastic scattering of incident ions with the target surface atoms. The energy of these ions can be calculated by Eq. 13. However the incident ions can scatter with target atoms at a depth t, resulting in a continuous energy spectrum to lower energies, until the incident ions pass through the target, hence the trailing edge. For an infinitely thick target, there would be no trailing edge and the spectrum would extend to E = 0.

The shape of the continuous portion of the spectrum is given by

Eq. (17)
$$Y(t) = \left[\frac{Z_1Z_2e^2}{4E(t)}\right]^2 Q\Omega N\Delta t$$

where

$E(t)$ = energy of detected particles that scattered at depth t
$N\Delta t$ = number of target atoms/cm^2 in the layer t
Q = measured number of incident particles
Ω = solid angle subtended by the detector

For multicomponent targets, there are leading edges for each element present, along with the associated continuous spectrum. Figures 14 and 15 are examples of RBS spectra for both homogeneous and layered thin film samples.

RBS spectra and depth profiling are readily simulated by computer programs, which is very useful in analyzing RBS data (18).

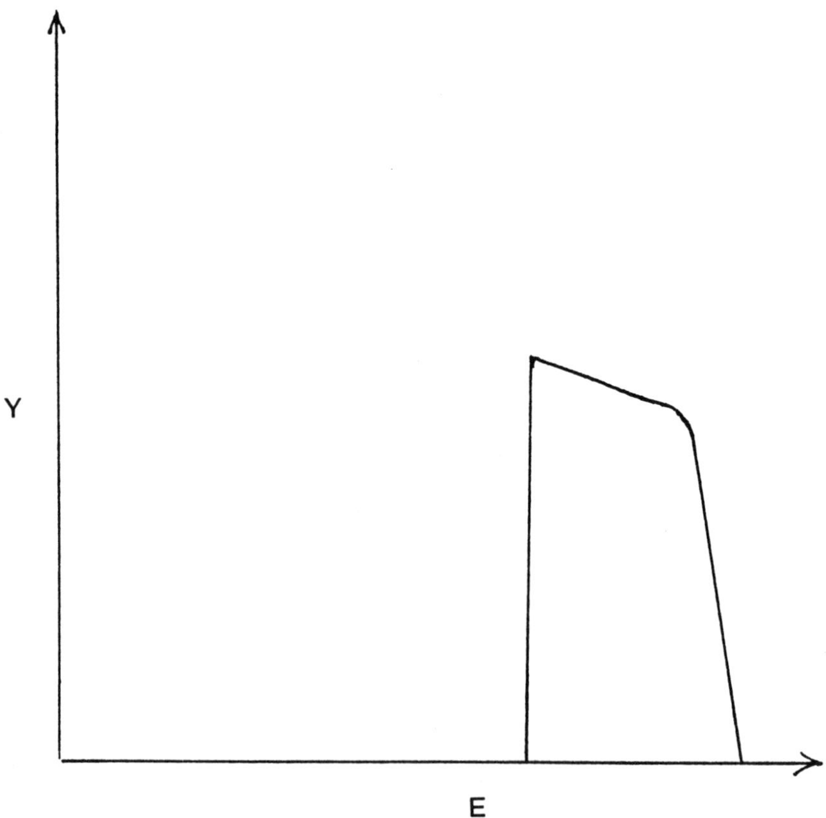

Figure 13. RBS spectrum of a single component film 5000 Å thick.

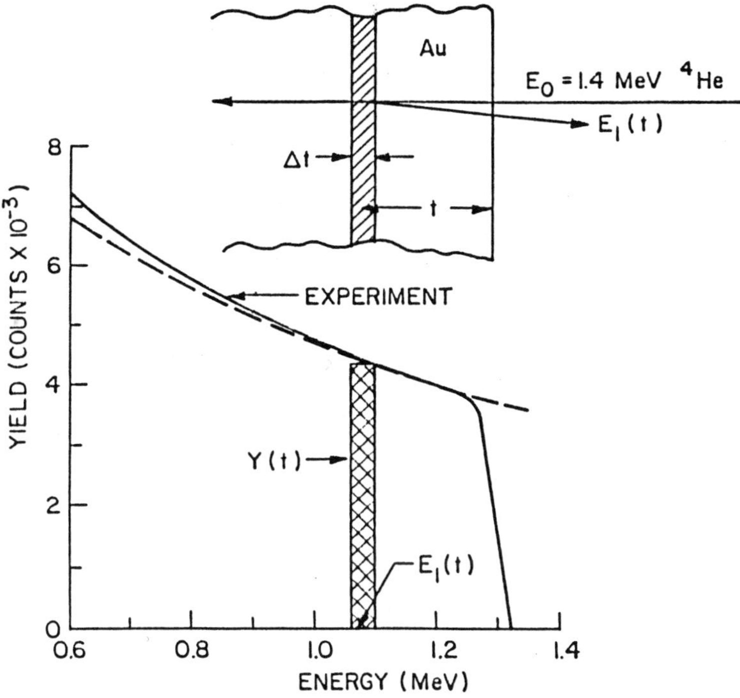

Figure 14. Backscattering spectrum for 1.4 MeV ions incident on a thick gold sample.

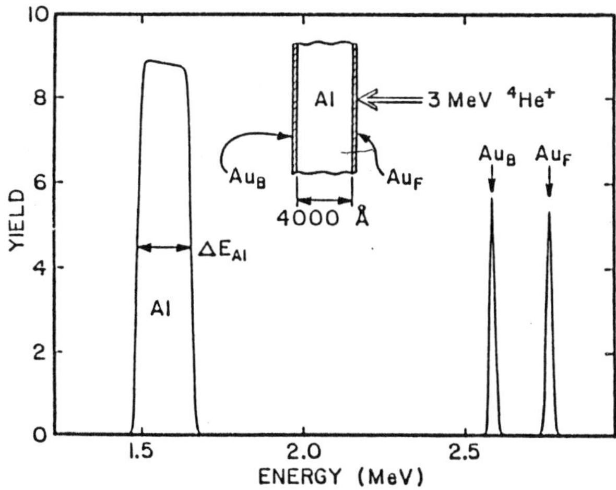

Figure 15. RBS spectrum of a layered film, 4000 Å of aluminum sandwiched by two 400 Å layers of gold.

4.4 Ion Scattering Spectroscopy

In ion scattering spectroscopy (ISS), surfaces are analyzed with low energy (200 - 2000 eV) noble gas ions, primarily He^+, Ne^+, or Ar^+. In concept, ISS is the simplest of all surface analysis techniques. The beam of positive ions is reflected with a loss of energy appropriate to the simple binary elastic collision of the beam with a particular surface atom. At any fixed scattering angle, the energy transferred is dependent only on the mass of the surface atoms causing the scattering. The equation governing the binary collision is given by

Eq. (18) $$\frac{E}{E_o} = \frac{M_p^2}{(M_p + M_+)^2} \cos \Theta + \left[\frac{M_+^2}{M_p} - \sin^2 \Theta \right]^2$$

where

E = scattered ion energy
E_o = primary ion energy
M_p = primary ion mass
M_+ = target ion mass
Θ = scattering angle

The ISS technique has been applied to a variety of problems of surface analysis of metals, semiconductors, and insulators.

A very effective application of ISS is the analysis of the surface composition of compounds. As XPS analyzes the surface and near surface composition, ISS complements the technique when there is a heterogeneity of composition in the near surface region. Frequently segregation or dissolution occurs creating surface enrichment or depletion of certain species.

5.0 SECONDARY ION MASS SPECTROSCOPY (SIMS)

The first experiments dealing with SIMS were performed in the late 1930's by Arnot and coworkers (19) and Sloane and Press (20) as part of a general study of negative ion formation resulting from ion bombardment of metal surfaces. This early beginning did not produce any immediate interest or any profound appreciation for the potential usefulness of the method. The interest in SIMS as a tool for surface and bulk solid analysis has grown steadily from the mid-1960's to the present time as is evident from the number of publications on SIMS that have appeared in the past fifteen years.

SIMS is performed by ion etching in conjunction with mass analysis of the sputtered particles. An energetic ion impinging on a solid is either back-scattered from a surface atom (a low probability event) or it enters the solid and dissipates its energy to lattice atoms through a number of collisions (Fig. 16). Sputtering takes place when the recoil atoms produced at or near the surface have enough energy to escape the solid. The sputtered atoms or fragments leave the surface in a neutral, excited, or ionic state. The escape depth for sputtered particles ranges from the surface to values greater than 20 Å and is strongly dependent on the characteristics of the collision cascade. The sputtered particles are mass analyzed usually by a quadrupole mass spectrometer (QMS). SIMS is a very useful technique for performing depth profiles.

The advantage of SIMS over the other techniques is its greater sensitivity to trace elements. Typical sensitivity values are on the order of $<10^{-7}$ of a monolayer, or 1 ppb atomic under some ideal situations. A factor that influences the sensitivity of a particular species is its ionizability or its stability as an ion. Hence, SIMS is very sensitive to alkali ions, i.e., Na, K, or Li. Furthermore, one has to perform positive and/or negative SIMS to detect either the positive or negative ions respectively. These techniques cannot be performed simultaneously, therefore depth profiles need to be carried out twice in order to get the complete picture. It is very difficult to be quantitative, since sensitivity factors depend on too many variables, i.e., differential sputtering, ionization probability, etc. A newer technique which has been recently developed and is becoming commercially available is SNMS (Sputtered Neutrals Mass Spectrometry). This technique shows a much more uniform sensitvity to a majority of species, and the hope is that SNMS can be used as a quantitative tool.

A disadvantage of SIMS is that it is a highly destructive analytical technique. To analyze an unknown sample, it is wise to begin with XPS, the least destructive technique, and use SIMS last.

SIMS can be very powerful as a complimentary technique. For example (21), a $YBa_2Cu_3O_{7-x}$ superconductor pressed from a powder in to the form of a pellet was immersed in water. XPS was used to try to identify the surface species formed. Figures 17a and 17b show XPS spectra of the Ba3d orbitals before and after exposure to water respectively. After exposure to water it is evident that a new chemical state developed at a higher binding energy. Although it is reasonable to assume that the new chemical state is due to the presence of $Ba(OH)_2$, it is inconclusive.

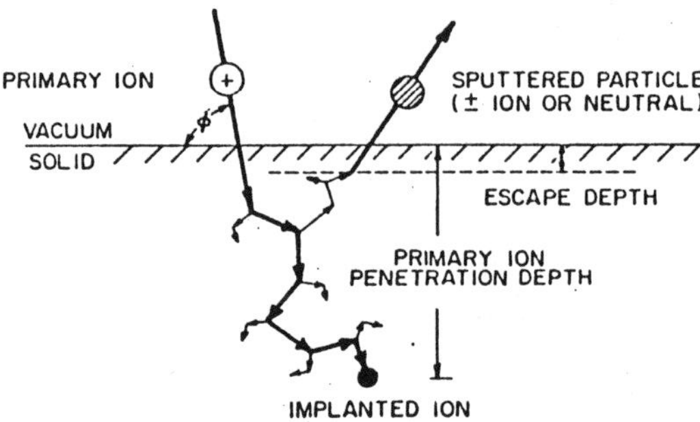

Figure 16. Schematic of ion-solid interactions and the sputtering process.

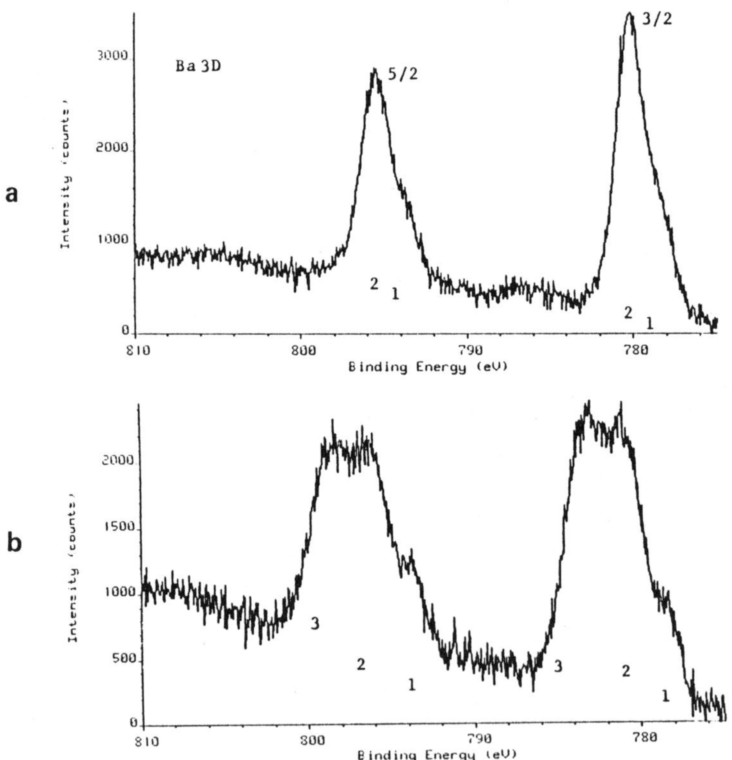

Figure 17. Ba3D spectra a) before, and b) after water exposure.

However SIMS spectra (Figs. 18a and b) of the sample before and after water exposure do indicate the formation of these species. This is evidenced by an increase in signal of the $Ba(OH)^+$ (155 amu, Fig. 18b).

When performing depth profiles with SIMS, one has to consider the crater produced in the substrate by ion etching. Figure 19 shows a profilometer scan of the crater in a layered substrate. The profile is elliptical in shape. When evaluating depth profiles the crater effects can lead to error in interpretation. Consider the example shown in Fig. 20. A layered sample of three thin films of compositions X, Y, and Z respectively, is sputtered and a crater is developed as in Fig. 19. When the ion beam is at the edge of the crater it produces ions from the topmost layer, which is commonly referred to as "crater effects". A way to get around this is to use a technique called "gating" in which the signal is electronically clipped as the ion beam approaches the edges of the crater. The signal is only acquired when the beam is in the center, the flattest portion of the crater.

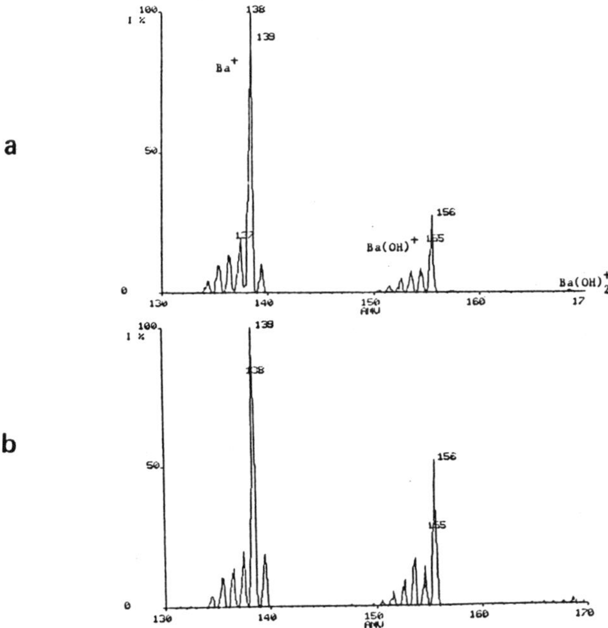

Figure 18. SIMS spectra of the barium region a) before, and b) after water exposure.

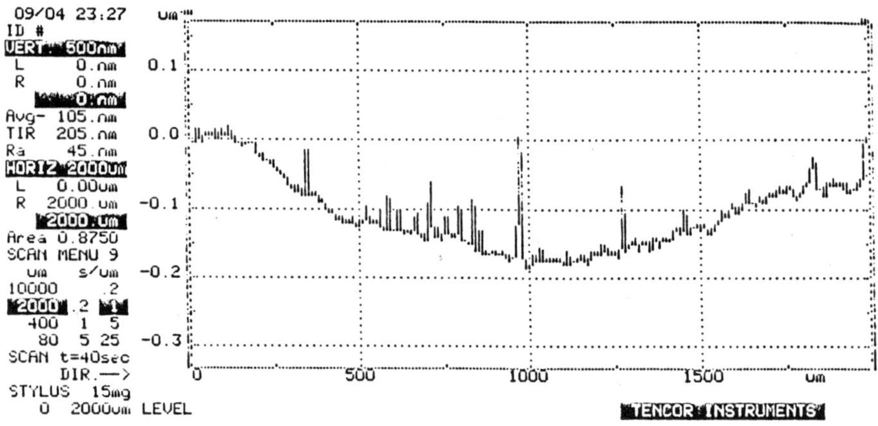

Figure 19. Profilometer scan of crater produced from ion etching.

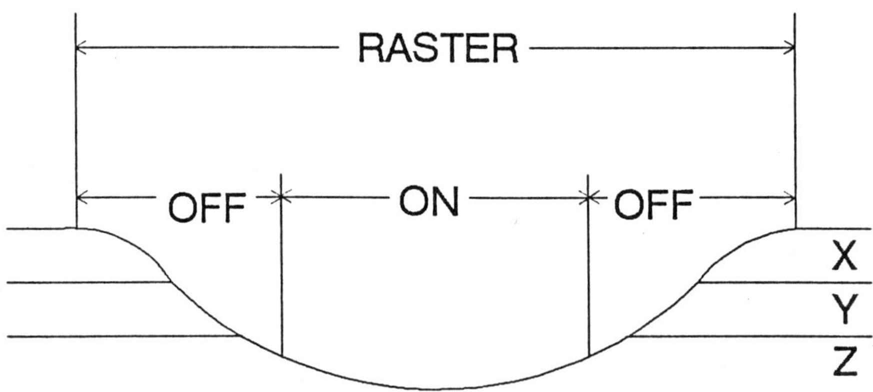

Figure 20. Schematic depicting how electronic gating minimizes crater effects during depth profiling.

6.0 SUMMARY

In this chapter we have outlined the most commonly used techniques for thin film and surface analysis. These techniques are well established and for the most part are commercially available. RBS resides mainly in large university and government laboratories.

By using these techniques, it is possible to determine the elemental composition as a function of position in the X, Y, and Z directions. This can be readily achieved with AES, RBS, ISS, and SIMS. XPS complements these techniques by providing chemical state information. This, along with SIMS, makes possible the deduction of molecular structure.

These techniques are very powerful tools and will remain so for years to come. The more recently developed techniques, such as Scanning Tunneling Microscopy, STM, and Atomic Force Microscopy, AFM, are joining the above techniques. They will provide structural information with spatial resolutions at the atomic and molecular levels.

REFERENCES

1. *Applied Surface Analysis, ASTM STP 699*, (T. L. Barr and L. E. Davis, eds.) American Society for Testing and Materials (1980)

2. *Methods of Surface Analysis*, (A. W. Czanderna, ed.) Elsevier Scientific Publishing Co. (1975)

3. Feldman, L. C. and Mayer, J. W., *Fundamentals of Surface and Thin Film Analysis,* Elsevier Scientific Publishing Co., Inc. (1986)

4. Briggs D. and Seah, M. P., *Practical Surface Analysis by Auger and X-ray Photoelectron Spectroscopy*, John Wiley and Sons (1983)

5. Siegbahn, K. et al., *ESCA - Atomic, Molecular and Solid State Structure Studied by Means of Electron Spectroscopy,* Alqvist and Wiksells, Uppsala (1967)

6. Lindberg, B. J., Hamrin, K., Johansson, G., Gelius, U., Fahlman, A., Nordling, C. and Siegbahn, K., *Molecular Spectroscopy by Means of ESCA II Sulphur Compounds. Correlation of Electron Binding Energy with Structure*. Uppsala University Institute of Physics UUIP-63B (March 1970)

7. *Handbook of X-ray Photoelectron Spectroscopy,* (C. D. Wagner, W. M. Riggs, L. E. Davis, J. F. Moulder and G. E. Mullenberg, eds.), Perkin-Elmer Corporation (1979)

8. Jen, J. S. and Kalinowski, M. R., *J. Non-Crystalline Solids*, 38 & 39, 21 (1979)

9. Caracciolo, R. and Garofalini, S. H., *J. American Ceramic Society,* C346-349 (1987)

10. Powell, C. J., *Surface Science* 44:29-46 (1974)

11. Siegbahn, K., et al., *ESCA - Atomic, Molecular and Solid State Structure Studied by Means of Electron Spectroscopy,* Alqvist and Wiksells, Uppsala (1967)

12. *Handbook of X-ray and Ultraviolet Photoelectron Spectroscopy* (D. Briggs, ed.), Heyden, London (1977)

13. Chang, C. C., *Surface Science,* 25:53-79 (1971)

14. Bishop, E. H. S., Asaad, W. N., *Advances in Atomic and Molecular Physics,* 8:163-284 (1972)

15. Caracciolo, R. and Schmidt, L. D., *Applied Surface Science,* 25:95-106 (1986)

16. Sickafus, E. N., Bonzel, H. P., *Progress in Surface and Membrane Science,* 4:115-230 (1971)

17. Rye, R. R., Madey, T. E., Houston, J. E. and Holloway, P. H., *J. Chem. Phys.,* 69(4) pp. 1504-1512 (1978)

18. Evans, R. D., *The Atomic Nucleus,* McGraw Hill Book Co., NY (1955)

19. Arnot, F. L. and Milligan, J. C., *Proc. Roy. Soc. Ser. A,* 156:538 (1936)

20. Sloane, R. H. and Beckett, C., *Proc. Roy. Soc. Ser. A,* 168:284 (1938)

21. Caracciolo, R., Parkhe, V., Safari, A. and Wachtman, J. B., presented at *Second Annual Symposium of Laboratory for Surface Modification,* Rutgers University, (December 10, 1987)

11

High T_c Superconducting Thin Films

X. D. Wu, A. Inam, T. Venkatesan

1.0 INTRODUCTION

High temperature superconductors (HTSC) are metal oxide based materials. The metal oxide system has been very important for micro- and opto-electronics for properties other than superconductivity (1). Properties such as ferroelectricity, optical nonlinearities, high optical transparency, relatively large controllable refractive indices, etc., have made metal oxides very useful for various technological applications (Table 1). The metal oxides can be doped with transition metal ions significantly affecting their optical properties; e.g., Ti doping of $LiNbO_3$ to form waveguides and Cr doping of Al_2O_3 to form light emitters. Technologies such as ion implantation can be effectively utilized to modify the surface properties of metal oxides (2). With the discovery of HTSC in metal oxide systems, the importance of these materials significantly escalated. This is probably the only system where, by modifying the oxygen composition, the materials can be tailored from a perfect dielectric to a superconductor. As a result, films of these materials have potential novel applications in advanced technologies such as micro- and opto-electronics.

Since one of the potentially important applications is in microelectronics it may be a worthwhile digression to speculate on how electronics, photonics and HTSC devices could coexist in a single device or system. The electron-electron interaction is strong, and this qualifies electronics for efficient switching devices, whereas the photon-photon interaction is weak, which qualifies photonics for information transmission with minimum cross-talk

Table 1. A brief list of the properties and examples of metal oxides that have applications in micro- and opto-electronics.

Property	Example of metal oxide system	Application
High optical transparency	MgO, ZrO	Mirror Coatings
Low loss diectric with electro-optic effect	$LiNbO_3$, $LiTaO_3$	Optical wave guides and integrated optics
Piezoelectricity	$BaTiO_3$, $PbTi_{.46}Zr_{.54}O_3$	Transducers
Ferromagnetism	γ-Fe_2O_3	Magnetic tape memories
Optical norlinearity	Nb_2O_5-SiO_2-Na_2O-Ba_2O_3-TiO_2	All optical switching devices
High optical gain	Nd^{3+} doped $Y_3Al_5O_{12}$	Lasers
Tansparent conductors	$InSnO_x$	Novel device coatings
Superconductivity	Y-Ba-Cu-O Bi-Sr-Ca-Cu-O, Tl-Ba-Ca-Cu-O	SQUID Superconducting electronics

(further low absorption and dispersion media for photons exist as well). However, superconductors exhibit properties of both electrons and photons; there is a strong interaction between the basic quanta, and superconducting materials exhibit low absorption losses and dispersion in propagating signals. Hence, while superconductors will not replace photonics in terms of their attractiveness for data transmission with low cross talk, a hybrid evolution of superconductors coexisting with electronics and photonics seems highly likely for eventual high bandwidth applications. An example of a futuristic large bandwidth hybrid system (3) is shown in Fig. 1, where high bit rate information arrives at a compound semiconductor interface where photons get converted to electrons; the electronic signal is processed by VLSI Si-based chips as well as specialized HTSC chips before being retransmitted as photons via a compound semiconductor interface. Hybrid superconducting/semiconductor VLSI based device technology has tremendous future potential.

Table 2 illustrates the requirements for the thin film HTSC materials in order to integrate the superconductors with micro- and opto-electronics.

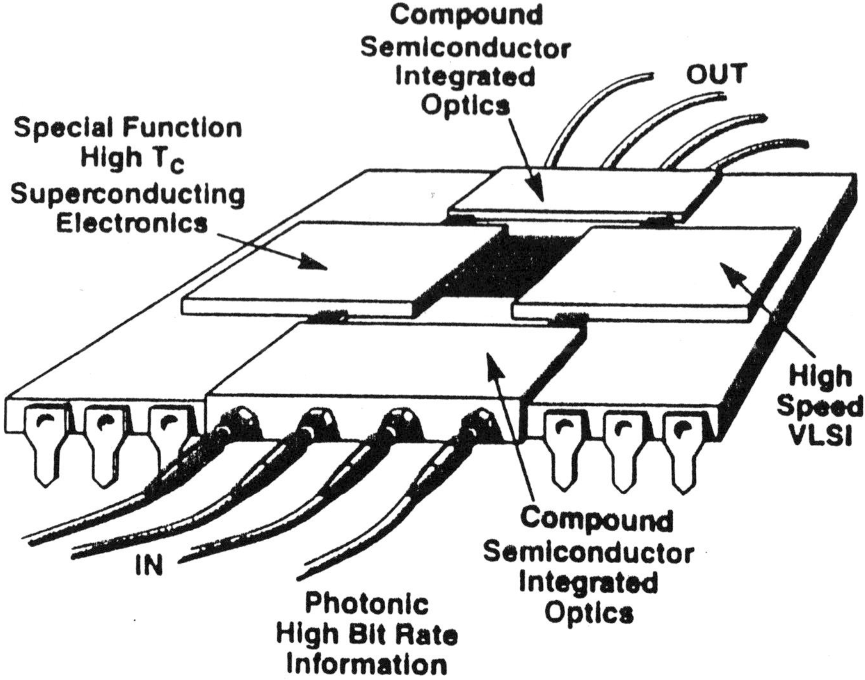

Figure 1. A futuristic chip combining opto-, micro- and superconductor electronic technologies (3).

Figure 2 shows the molecular and complex crystal structure for the various HTSC materials (4), and the increasing complexity of the materials with increasing T_c illustrates the difficulties inherent in the process for synthesizing these films in order to accomplish the requirements listed in Table 2. A viable film fabrication technique must be able to:

(a) produce smooth films with the appropriate composition,
(b) get the right crystal phase, and
(c) produce the right oxygen stoichiometry in the film.

The number of thin film deposition techniques demonstrated to date vary in terms of their ease in meeting the above criteria in producing good quality films. The various deposition techniques could be generically divided into two classes, multiple sources and a single source for depositing the different elements. Figure 3a shows a typical geometry for deposition from

Table 2. Needs for Device Fabrication.

1. High T_c (R=0)
2. Small ΔT (transtion width)
3. High J_c (critical current density)
4. Surface smoothness
5. Stable film on substrates such as Si
6. Sharp interfaces

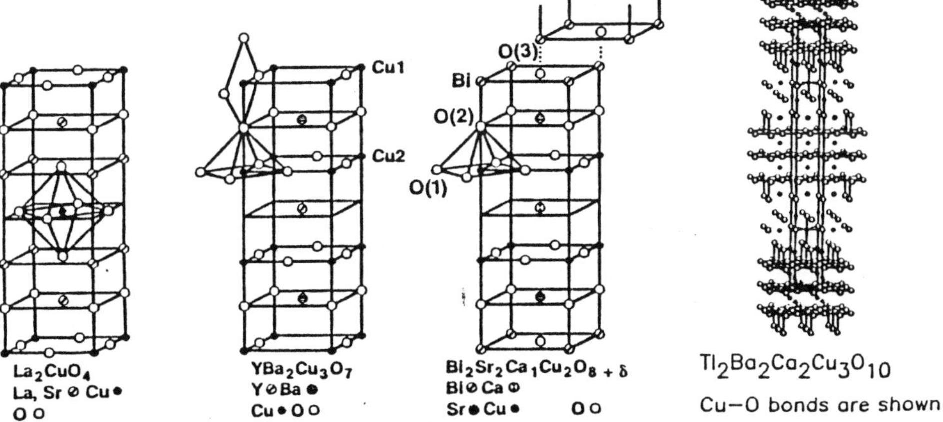

Figure 2. Approximate crystal structure of the various high T_c superconductors (courtesy of J. M. Tarascon and M. A. Subramanian).

different sources. The elements are ejected from these sources using heat, electrons, photons or ions. The generic problems with these systems are:

– the relative ejection rate of the elements must be monitored (in mTorr oxygen ambient) and kept constant,
– the non-overlap of the atomic trajectories from the three or more sources must be overcome either with planetary manipulation of the substrate holder, or by mounting the sources in a radial configuration.

On the other hand, the use of a single target (Fig. 3b) has its associated problems:

– the ejection rate of the three elements is not the same (e.g., ion sputtering yields are different for Cu, Y and Ba),
– the sticking coefficients of these elements on the substrate are also different.

As a result, the target will not have the same composition as that of the deposited films, and the optimum target composition will also depend on the deposition parameters.

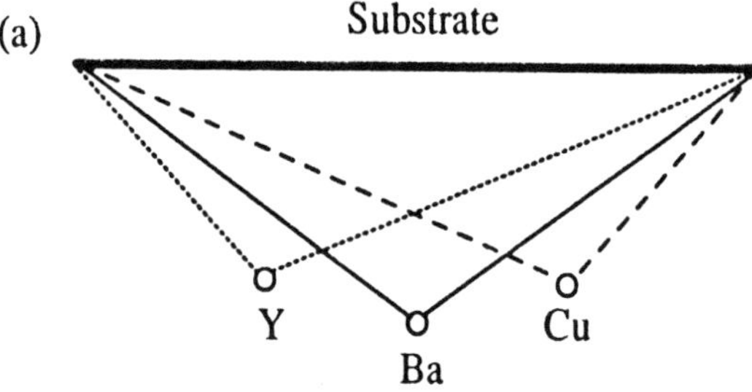

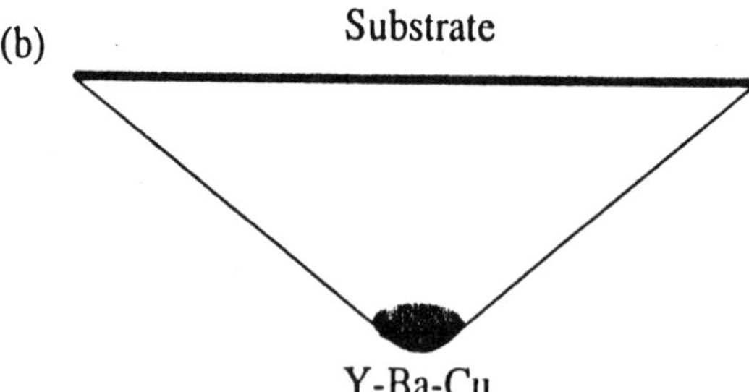

Figure 3. Generic deposition systems: (a) multiple sources and (b) single source.

It should be realized that no new thin film fabrication techniques have been invented after the discovery of the high T_c superconductors. But remarkable advances have been made in the last few years in the technology of thin film processes for deposition of the high T_c superconducting thin films. The technology falls into two categories: vacuum deposition and non-vacuum deposition. The discussion in this chapter shows that it is easy to make thin films by using non-vacuum methods, but the film quality is not always as good as those made by vacuum deposition techniques.

There are at least two approaches to making good superconducting thin films. One is to deposit the film in some random phases at a relatively low temperature and follow this up with a high temperature (800 - 900°C) anneal in oxygen. However, a more elegant way to accomplish this is to deposit the films directly in the right superconducting phase at a temperature of 650 - 750°C in an oxygen ambient. From a technological point of view, in order to produce sharp interfaces and minimize film-substrate interaction and stresses in the film, a low temperature process is absolutely essential. Moreover, if the as-deposited film is a random or disordered phase, after a high temperature anneal, the final film will be probably be polycrystalline in nature. On the other hand, with the second approach, epitaxial film with superior superconducting properties can be prepared since the films are grown layer by layer.

2.0 VACUUM DEPOSITION

2.1 Thermal and Electron Beam Evaporation

The thermal evaporation technique is widely used for the preparation of thin films of many materials. In order to evaporate materials, a vapor source is required that will support the evaporant and supply heat for vaporization while allowing the charge of evaporant to reach a temperature sufficiently high to produce the desired vapor pressure, hence, rate of evaporation.

In order to prevent the contamination of the evaporant from the support materials, two types of materials can be used: either refractory metals or certain non-metallic materials such as oxides or fluorides. Wire and metal foil structures are widely used for a variety of evaporants; the material from these sources can be evaporated by either resistive or induction heating.

It is, however, possible to cause vaporization of materials by using electron bombardment. An electron beam is accelerated up to few keV and focused on to the evaporant surface. Temperatures exceeding a few thousand degrees can be obtained, enabling a variety of otherwise non-evaporable materials, such as refractory metals, to be used. Since only a very limited portion of the evaporants are heated, any interaction between the evaporant and the support materials is reduced considerably, thus electron beam evaporation is a technique for preparation of very pure films, and is particularly attractive for materials which evaporate only at very high temperatures.

Since the vapor pressure for barium, yttrium and copper, in the case of $YBa_2Cu_3O_{7-\delta}$ (YBCO) superconductor, are different, the change of composition in the films with time during the evaporation process is a major disadvantage. The problem can be overcome by using multiple sources with the rates of evaporation separately controlled in order to obtain the desirable composition in the deposited films, or by a technique known as flash thermal evaporation (5)(6) where a small amount of evaporant is evaporated at once or the evaporant is fed to the source slowly. Although flash thermal evaporation has been used to prepare superconducting YBCO thin films, the composition of the thin film is still not the same as the evaporant. At present, for 1-2-3 (the composition ratio for the cations such as Y, Ba and Cu) superconductors, the configuration used in the evaporation system is primarily three separated sources.

The first high T_c superconducting thin film with T_c over 77 K and critical current density (J_c) over 1 x 10^5 A/cm^2 at 77 K was made on $SrTiO_3$ substrates by electron beam evaporation with a subsequent annealing at high temperatures of 900°C (7). The reported J_c was a great breakthrough for the application of the new oxide superconductor since only very low critical current density was obtained in ceramic superconductor samples. Subsequently, the Stanford group (8) reported similar results using the same technique. Those early films were prepared by evaporating yttrium, barium and copper metals in the presence of a partial oxygen pressure of 10^{-4} Torr, and the as-deposited films were not stable in air. The use of BaF_2 rather than the reactive Ba metal, an idea pioneered by a group from AT&T Bell Labs (9), resulted in stable as-deposited and partially annealed films. Films prepared by this technique could be patterned, exposed to room air indefinitely, or even immersed in water before annealing, without affecting the superconducting properties. However, to form the superconducting phase, the fluorine had to be removed using a wet oxidation method in which water is added to the oxygen stream by bubbling oxygen through a water bath during a high temperature thermal anneal. The annealed films on $SrTiO_3$ made using the BaF_2 process have critical temperatures as high as 92 K, and critical current densities over 10^6 A/cm^2 at 77 K at zero field (9). It should be noted that all the films with high critical current densities have to be properly aligned on the substrates with the c-axis of the films perpendicular to the substrate surfaces. The preferential orientation strongly depends on the substrate as well as composition of the film (8). While the earlier films prepared with the BaF_2 source material have excellent transport properties, the surface of these films are far from being smooth. By fine adjusting the annealing process and utilizing excellent lattice matches with the substrates, quality epitaxial YBCO films can be obtained on substrates such as $LaAlO_3$ and $SrTiO_3$ (10). Though these films have excellent crystallinity in terms of ion beam channeling minimum yields (as low as 2 - 3%, which is comparable to the value for single crystals), the critical current densities of the films at 77 K and zero field are still 2 - 3 times less than the values from the best films deposited in situ. The BaF_2 method is a very simple process for preparation of single layer large area or double-sided superconducting films. It cannot be used for depositing epitaxial multi-layer structures.

As-deposited YBCO superconducting thin films on ZrO_2 (stabilized with 9.5% Y) and $SrTiO_3$ were first prepared by a combination of electron beam evaporation and thermal evaporation in 0.65 mTorr oxygen pressure at 600 - 750°C (11). Apparently these films are oxygen deficient during deposition, resulting in a superconducting transition temperature of only about 82 K. To overcome the disadvantage of the low oxygen pressure in the vacuum system (since the electron guns will not work under high oxygen pressures), a group from Kyoto University (12) used an oxygen plasma, which produced more reactive atomic oxygen between the source and substrate. The as-deposited films made by this group showed T_c and J_c as high as 90 K and 3 x 10^6 A/cm^2 at 77 K, respectively. A film of only 100 Å thickness had a zero resistance temperature of 82 K, reported first by the same group (13). In order to oxide the cations at a low oxygen pressure, pure ozone was used to prepare as-deposited superconducting Dy-Ba-Cu oxide films (14). The advantage is that films can be made under higher vacuums. The disadvantage is the explosive nature of ozone.

High T_c and J_c YBCO films have been successfully prepared by electron beam and thermal evaporation techniques. Both techniques are relatively simple, but in order to obtain as-deposited superconducting thin films, RF plasma (12)(15), or differential pumping (16), or ozone (14) has to be used, because the biggest problem in both techniques is that high oxygen pressure cannot be introduced into the vacuum system as the films are grown. This is in addition to the generic problems of multi-source systems discussed earlier.

2.2 Sputtering

If a surface is bombarded with energetic particles, it is possible to cause ejection of the surface atoms, a process known as sputtering (17). These ejected atoms can be condensed on a substrate to form a thin film. In most cases, positive ions of heavy neutral gases such as argon are used to bombard the surface of target materials by making the surface the cathode in an electrical circuit. High melting point materials can be used as easily as low melting point ones, and, using RF sputtering technique, both metals and insulators can be deposited.

There are a number of types of DC and RF sputtering systems. The most popular sputtering system used for high T_c superconducting thin films is magnetron sputtering, while other sputtering systems such as triode sputtering are also used.

Sputtering from three separated sources gives good superconducting thin films (18). Due to generic problems of multi-source systems, more and more groups have started using single target sputtering systems, either DC or RF. If the target is conductive, DC can be used while RF is preferred when using a nonconducting target. Normally, in order to obtain 1-2-3 composition, the target used is often off-stoichiometric with a typical composition of $Y_1Ba_2Cu_{4.5}O_x$ (19), compensating for the deficiency of a particular element such as Cu. Problems are caused by preferential sputtering effects, different sticking coefficients, and negative ion resputtering effects (20). The negative ion resputtering effect is due to the negative oxygen ion bombardment of the deposited films in the conventional target/substrate face-to-face geometry. The effect is minimized by depositing the films at a fairly high pressure or placing the substrate outside the glow discharge region (21)(23). The most popular solution is *off-axis sputtering* (24), where the substrate holder is rotated about 90 degrees with respect to the target as compared to the face-to-face setup.

Using sputter deposition, Enomoto et al. (25) made superconducting YBCO thin films on (110) $SrTiO_3$ after a high temperature post annealing. With a-axis normal to the substrate surface, the anisotropy of critical current density in the films was observed for the first time by a transport measurement. A high critical current density of 2 - 3 x 10^6 A/cm^2 in a superconducting YBCO thin film made by sputtering was obtained after a high temperature anneal

(26). As-deposited superconducting thin film with T_c of 85 K by sputtering was first reported by Adachi et al. (19). Results by in situ sputter deposition using a small, high pressure RF magnetron sputtering system called *cylindrical magnetron sputtering* are quite impressive (T_c over 90 K and J_c ~ 5 - 6 x 10^6 A/cm^2 at 77 K) (27). In cylindrical magnetron sputtering, the target is in the form of a cylinder, and the substrate is held perpendicular to the cylinder major axis. The sputter gas and oxygen are allowed to enter the cylinder from the end of the cylinder behind the anode, and the jet of gases pushes the plasma etched materials from the surface of the cylindrical target to the substrate holder. Due to the geometry, deposition rates in excess of a few Å/s have been demonstrated. However, the anode is coated during the deposition, and the anode surface becomes non-conducting. Frequent removal of non-conducting layer on the anode limits the capability of continuous deposition of the films. Using off-axis geometry for the substrate and a conventional magnetron sputtering gun, in situ high quality superconducting YBCO films are obtained using stoichiometric targets. Films up to 5 cm in diameter have been deposited using this method (28)(29). Due to the simple setup, this method is the most popular sputtering method. The best films obtained by this method are typically deposited under a high total argon and oxygen pressure, however the deposition rate decreases as the pressure increases. A typical deposition rate is 500 - 1000 Å per hour. At low pressures (<100 mTorr), in situ superconducting YBCO films have T_c of typically 88 K and less probably due to oxygen deficiency during deposition. Off-axis sputtering is simple but has a limited deposition rate.

2.3 Ion Beam Deposition

Two basic configurations can be used in ion beam deposition of thin films. Primary ion beam deposition utilizes low energy (-100 eV) ion beams consisting of the desired film material which is deposited directly on a substrate. In secondary ion beam deposition, much higher energy (hundreds to thousands of electron volts) ion beams of an inert or reactive gas are directed at a target of the desired materials. The target is sputtered and collected on to substrates. The latter method is also called ion beam sputter (IBS) deposition.

Due to the complicated composition of oxide superconductors such as YBCO, no encouraging results have been reported using primary ion beam deposition. Using IBS deposition and high temperature annealing, superconducting YBCO films with T_c ~ 90 K were obtained (30)-(32). Good as-grown epitaxial YBCO films were prepared using ion beam sputtering (33).

In ion beam deposition, greater isolation of the substrate from the ion generation process is obtained compared to conventional plasma sputtering. This makes it possible to exercise independent control over the substrate temperature, ion beam current and energy, and angle of deposition as well as the ambient gas pressure. At present, for high T_c oxide superconducting films, more time is needed to assess the potential of these techniques.

2.4 Pulsed Laser Deposition

In the pulsed laser deposition (PLD) technique, a high intensity pulsed laser beam is focused onto a target. The optical energy from the laser is turned instantaneously into heat at the target surface where the light is absorbed. Once the surface temperature exceeds the vaporization temperature, the surface materials are vaporized making it possible to deposit thin films. Although this technique is not well understood, it has been used for over twenty-five years in the preparation of thin films of semiconductors, dielectrics and metals. More recently, a pulsed ruby laser was used to deposit superconducting thin films of Ba-Pb-Bi oxide (34). Earlier results of laser-solid interaction were summarized by J. F. Ready in 1971 (35), and Cheung and Sankur (36) gave an excellent review of laser deposition of thin films just before the discovery of high temperature superconductors.

A good review of pulsed laser deposition (PLD) of HTSC thin films can be found in "Laser Ablation for Material Synthesis," in the *Proceedings of the Materials Research Society* (37). Typically, a PLD system consists of a vacuum system with a face-to-face target/substrate setup. A laser beam is focused on the target surface at an incident angle of typically 45 degrees. To remove the material uniformly from the target, a rotating target or a rastering laser beam is used. Though there are various pulsed lasers for thin film deposition, excimer lasers are the choice for multi-elemental materials such as HTSC due to their short wavelengths (less than 300 nm), and the great progress in excimer laser technology in the last ten years.

The report of the first high T_c superconducting thin films using the PLD technique came from a Bellcore/Rutgers team (38). At about the same time other groups used the same technique to make Y-Ba-Cu oxide and La-Sr-Ca oxide films (39). However, these films were not electrically superconducting. The Bellcore/Rutgers team was also the first to report a low temperature preparation of YBCO thin films by laser deposition using high oxygen pressure (40). Since then many groups have made as-deposited YBCO superconducting thin films (41)-(46). The highest T_c and J_c reported for YBCO thin films made by PLD were about 90 K and 4 - 5 x 10^6 A/cm^2 at 77

K in a zero field, respectively, and even YBCO films with thickness less than 100 Å were prepared by PLD (47).

The advantage of the PLD technique is that deposition is from a single target. More importantly films can be made with a composition very close to that of the target; in other words, the stoichiometry of the target can be replicated in the films, which significantly simplifies the deposition process. The composition of the film is not effected by the change of oxygen pressure in the deposition system or the substrate temperature. So far, no other technique competes with PLD in the area. It has been shown that parameters such as deposition angle, laser energy density, and oxygen pressure are critical parameters for obtaining as-deposited superconducting thin films (48)(49).

For PLD, a typical deposition rate of 0.1 - 1 Å/pulse can be obtained depending on the substrate and target distance. It has been shown that a deposition of 600 $Å \cdot cm^2/sec$ is possible with a 30 watt excimer laser (50), which is significantly higher than that of sputtering. It should be noted that the YBCO films made at the high deposition rate (up to 150 Å/s) by PLD were still high quality films (50). Using PLD, uniform, large area (up to 7.5 cm in diameter) and double-sided YBCO thin films were prepared (51)(52). PLD also has a natural advantage for multilayer fabrications since different targets can be easily placed in the laser beam for deposition.

However, the process suffers from the problems of particulate incorporation on the film surface and target modification after long laser exposure (53). In almost every YBCO film made by PLD, spherical particulates are observed. By using high density (over 95% of the theoretical density) targets, the particle density is greatly reduced but not totally eliminated. After cumulative laser exposure, column formation on the target surface is observed (53). These columns are responsible for the observed decrease in film deposition rate.

So much progress in practical applications of in situ PLD has been made in a relatively short time, but fundamentally, the laser-target interaction, resulting plasma processes, and subsequent thin film growth are still poorly understood owing to the highly nonlinear, nonequilibrium nature of the PLD process. More work is indeed needed in this area, and one may anticipate that, as our understanding of the PLD process improves, further development of this emerging technology will be forthcoming.

2.5 Molecular Beam Epitaxy

Molecular beam epitaxy (MBE) is a thin film deposition process in which thermal beams or atoms or molecules are deposited on crystalline substrates to form epitaxial films. The molecular beams are commonly created through the use of Knudsen effusion cells for MBE growth. In order to grow films with high purity and excellent crystal quality, a vacuum better than 10^{-10} torr is needed. Although the original focus of the MBE process was to grow group III-V semiconductors, the process is now widely used in the growth of other semiconductors such as the group IV and II-VI families, metals such as Co and Al, and insulators such as fluorides (CaF_2, SrF_2 and BaF_2).

At first, the MBE technique was used as molecular beam deposition for the preparation of high T_c 123 oxide thin films because all the as-deposited films were not epitaxially grown, and in fact were insulating (54)(55). Superconducting thin films could only be obtained after a high temperature anneal. It was realized that copper is not oxidized by molecular oxygen at pressure compatible with MBE (56). Using an oxygen plasma source, epitaxial growth of YBCO films was demonstrated (56)-(58). In fact, the growth conditions for the YBCO films by MBE are similar to those of in situ electron beam evaporation of the films (59)(12).

The advantage of the MBE process is that it is possible to layer the new oxide superconductors (which have a layered structure) with precise thickness control. Getting enough oxygen into the YBCO films seems to be the biggest problem. Since these films seem to favor growth under a high oxygen partial pressure, most of the advantages of the MBE system are muted, and there does not seem to be any overwhelming reason to pursue this process.

Recent work on growth of Bi-based oxide superconducting thin films by MBE and utilizing sequence shuttering of various elements indicates that MBE may be useful for layer-by-layer growth of the materials (60)(61), although layer-by-layer growth can be obtained using other techniques such as PLD (62), sputtering and e-beam evaporation.

2.6 Chemical Vapor Deposition (CVD) and Organometallic CVD

In the chemical vapor deposition (CVD) process, the deposited phase is produced via chemical reactions. Thin films are deposited from gaseous precursors (typically metal halides), which are flowed with a nonmetal source gas such as H_2 or an oxidant to the CVD reactor on a heated substrate. The deposition reaction can be pyrolysis or oxidation or hydrolysis or combinations of these, and may even be catalyzed by the substrate. There are a variety of CVD processes such as low pressure CVD (LPCVD), plasma-assisted CVD (PACVD), organometallic CVD (OMCVD or MOCVD) and laser CVD (LCVD). CVD has been used for deposition of metals, carbides, nitride silicides, sulfides and semiconductors. In principle, OMCVD requires at least one of the sources to be transported by an organometallic compound. The chemical reaction in OMCVD is irreversible in contrast to CVD, though the basics are the same for both processes.

Although superconducting YBCO thin films have been made by CVD methods (63)(64), the transport properties of the films are still not as good as the best superconducting thin films made by e-beam evaporation, sputtering and laser deposition. The in situ CVD films made by a group from Tohoku University showed T_c of 84 K and J_c of 2×10^4 A/cm^2 at 77 K (63). Later, the T_c and J_c had been improved by the same group to 89 K and over 1×10^6 A/cm^2 at 77 K (65). Other groups have used oxygen plasma to enhance the oxidation process or different geometries, and obtained YBCO films with even better quality (66)-(68).

For most of groups using MOCVD, three individual sources, typically b-diketone of $Y(thd)_3$, $Ba(thd)_3$ and $Cu(thd)_3$ for Ba, Y and Cu in the case of YBCO are adjusted to desired temperatures (between 100 - 300°C for YBCO) to achieve appreciable vaporization. The recent work by Hiskes et al. (68) showed that the demand on precise control of the individual sources can be eliminated by using a single source MOCVD technique, which simplifies the whole MOCVD process. The in situ YBCO films made by this technique have excellent superconducting properties (68).

In principle, CVD offers the advantages of excellent film uniformity, high deposition rates, and large scale processing. The results of the best films made by MOCVD process are very encouraging, though more work has to be done to optimize the CVD process for depositing large area films. Efforts are needed in search of volatile and thermally stable precursors.

3.0 NON-VACUUM DEPOSITION

3.1 Liquid Phase Epitaxy

The liquid phase epitaxy (LPE) process for 123 superconductors is based on the pseudo-binary Y_2BaCuO_5-$YBa_2Cu_3O_y$ phase diagram. The idea is to melt the starting composite materials and then allow the precipitation of $YBa_2Cu_3O_7$ crystals upon cooling. These crystals will grow on the substrate immersed into the melt solution. Once the liquid solution is separated, thin epitaxial films are obtained on the substrates.

So far, these are few reports on the growth of the 123 superconducting films by LPE (69)(70). The problems are that the detailed phase diagram is not known, and the oxide superconductors melt incongruently. In LPE, high temperatures (close to the melting point of the superconductors) can not be avoided.

3.2 Plasma Spray

In plasma spray deposition (71)(72), YBCO powders are injected into a plasma, which provides both the energy to heat the powders and the gas velocity for accelerating the powder particles. The energetic powder particles are deposited on the substrates. All deposition can be done at atmospheric pressures.

This method has advantages of high deposition rate (few mm/s), composition control, and the ability to coat many shapes and large areas. Apparently, however it is difficult to make quality in situ or post annealed thin films by the technique.

3.3 Solution Method

Solution precursors of Y, Ba and Cu such as fluoroacetates, nitrates and carboxylates (all these could be called metalorganic compounds) can be used for preparation of high T_c thin films. The solutions of Y, Ba and Cu are atomically mixed, and sprayed or spun onto substrates. Then, the mixed solution is thermally decomposed to remove the residual solvent at temperatures of 100 - 500°C. Finally, superconducting thin films are obtained by subsequent annealing at high temperatures (900°C) in oxygen.

There are a number of reports on the preparation of high T_c YBCO thin films using the solution method (73)-(75). So far, transition temperatures as high as 94 K for YBCO films have been obtained (75), but the J_c is low.

The greatest advantage of the technique is its simplicity. Any shape can conceivably be coated. But the trade-off is the difficulty of making in situ superconducting thin films.

4.0 COMPARISON OF THE THIN FILM TECHNIQUES

For a given vacuum deposition technique, good superconducting thin films can be obtained if detailed research is carried out. So far, the YBCO thin films made by most of the techniques have comparable structural and superconductive properties. As examples, Fig. 4 and Fig. 5 show typical

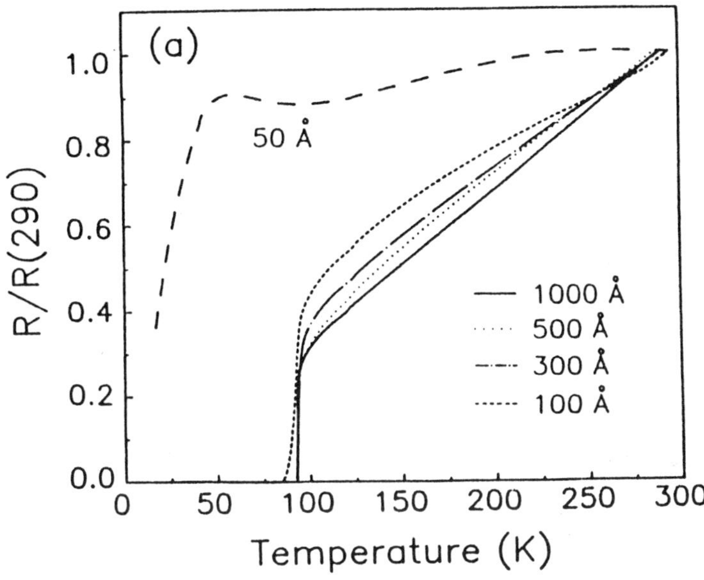

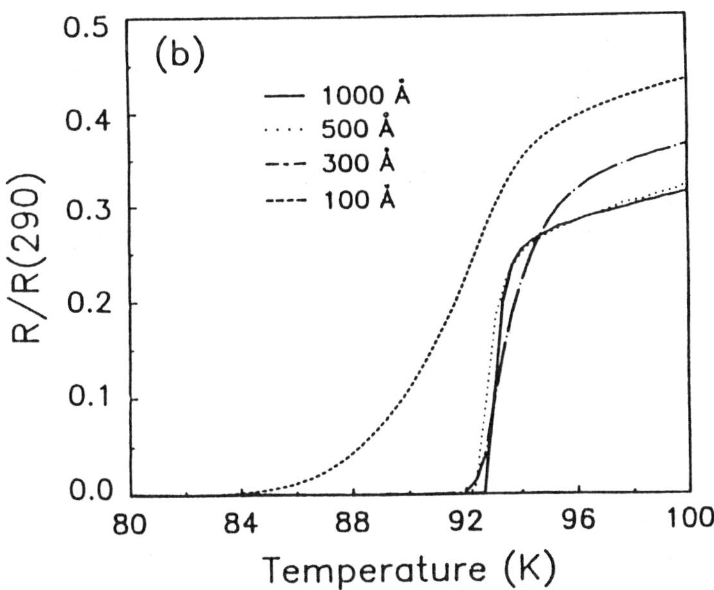

Figure 4. Resistance vs. temperature for YBCO films on (100) $SrTiO_3$ with various thicknesses.

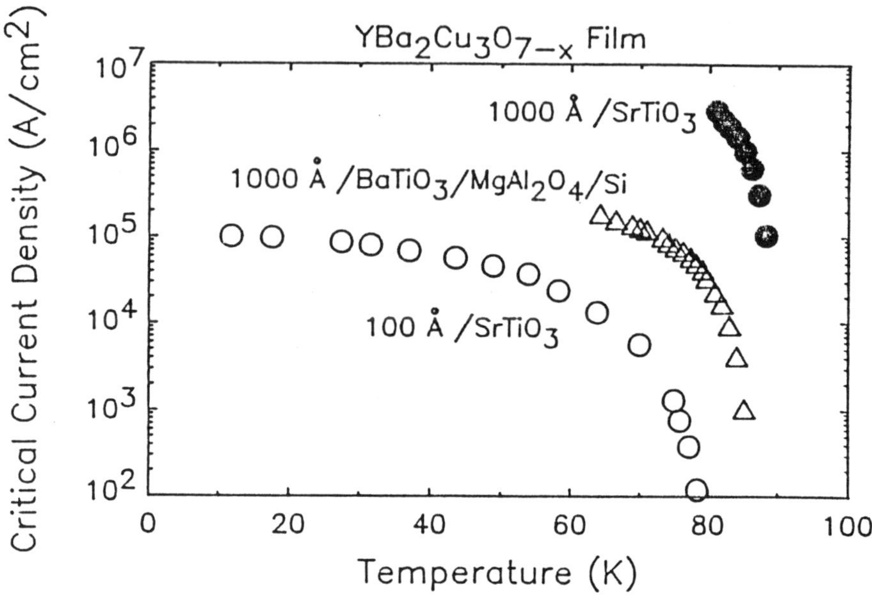

Figure 5. J_c vs. temperature for 1000 Å YBCO/$SrTiO_3$, 1000 Å YBCO on Si with buffer layers, and 100 Å YBCO/$SrTiO_3$.

resistance transitions for YBCO thin films with various thicknesses, and J_c as a function of temperature, respectively (47).

Based on the results published in the last few years, a few conclusions can be drawn for the comparison. For vacuum deposition, sputtering, PLD and MOCVD are better than the other methods. Off-axis sputtering is a simple but slow process. Cylindrical magnetron sputtering needs to be further developed in order to solve the problem of anode coating. PLD is an expensive (the cost of the laser system), simple and versatile process. It is best for research development, and for multi-layer deposition. Single source MOCVD technique shows tremendous potential, and further research is needed to explore the technique.

The non-vacuum thin film deposition techniques will not be useful for depositing films for electronic applications. These techniques may be needed for coating large or odd shaped objects for applications where a high J_c is not required.

5.0 OTHER HIGH T_c SUPERCONDUCTORS

We have limited the discussions on the high temperature superconductors to the 123 compounds only. As seen from Fig. 3, there are other superconductors with transition temperatures over 100 K such as the Bi-based or Tl-based oxides. So far, much of the research done on growth of Bi-based superconducting thin films was carried out in Japan. There are few Bi-based or Tl-based compounds which show superconductivity at a temperature ranging from 10 K to 110 K for the Bi oxides or 10 K to 125 K for the Tl oxides, which create difficulties in terms of stabilization of the phases with the highest superconducting transition temperatures.

For Bi-based superconductors, in order to obtain transition temperatures over 100 K in the films, a post-annealing process has to be used. The highest transition temperature of in situ grown Bi-based superconducting thin films is over 95 K but less than 100 K (76).

Due to its toxicity, only few groups are working on Tl-based oxide superconducting thin films. Since there is substantial loss of Tl at elevated temperatures, in situ growth of the films is not yet realized.

6.0 SUBSTRATE

Obviously, the choice of the substrates is a very important step for making a quality thin film. Table 3 contains a short list of substrates which are used YBCO thin films or other HTSC films. The following key points should be kept in mind when selecting a substrate for high T_c superconducting thin films.

6.1 Lattice Constant

It is well known in the field of semiconductors that a substrate enables epitaxial growth of the top layer if the substrate has a good lattice match with the over-layer. In another words, the lattice mismatch should be as small as possible to promote easier nucleation of the film. Due to a poor lattice match, multiple in-plane orientations were observed in such films as YBCO on (100) MgO substrates (77). As the result, the films had low critical current densities. The distribution of the multiple orientations largely depends on the growth temperature.

Because the basal plane of YBCO unit cell is nearly cubic (a = 3.82 Å and b = 3.89 Å), it is also important to find an orientation arrangement for the

Table 3. Substrates for high T_c superconductors; α and ε are thermal expansion coefficient and dielectric constant at room temperature respectively.

	Crystal system	Lattice constant	α	ε
$YBa_2Cu_3O_{7-x}$	Othorhomic	a=3.82 Å b=3.89 Å c=11.68 Å	$14x10^{-6}/K$ $12x10^{-6}/K$ $25x10^{-6}/K$	
$SrTiO_3$	cubic	a=3.905 Å	$10.8x10^{-6}/K$	~ 300
YSZ	cubic	a=5.16 Å	$10x10^{-6}/K$	27
MgO	cubic	a=4.203 Å	$13.8x10^{-6}/K$	10
$LaGaO_3$	orthorhombic	a=5.519 Å b=5.494 Å c=7.77 Å	$10.55x10^{-6}/K$	25
$LaAlO_3$	cubic	a=3.792 Å		15
$BaTiO_3$	tetragonal	a=3.99 Å c=4.03 Å	$13x10^{-6}/K$	high
$MgAl_2O_4$	cubic	a=8.086 Å	$7.6x10^{-6}/K$	-
Si	cubic	a=5.43 Å	$2.6x10^{-6}/K$	12
GaAs	cubic	a=5.65 Å	$6.86x10^{-6}/K$	13

superconductor growth. For example, $LaGaO_3$ (78) has an orthorhombic crystal structure with a = 5.519 Å, b = 5.494 Å and c = 7.77 Å. One of the possible orientation arrangements will be that the a and b axes of YBCO lie above the Ga-O planes in $LaGaO_3$ where the Ga-O-Ga distance is 3.894 Å, though the lattice constants of $LaGaO_3$ are much larger than the those of the 123 superconductors.

6.2 Thermal Expansion Coefficient

At present, the best HTSC superconducting thin films are made in situ at temperatures of 600 - 800°C. It seems very difficult to reduce the temperature to less than 500°C and maintain the quality of the films. Moreover, the high T_c thin films will be tested or used at low temperatures such as 77 K. Because of the difference in the thermal expansion coefficients of the superconductors and the substrates, the films are always strained upon cooling or heating. As a result, cracks will be formed in the films, therefore, a good match in the thermal expansion coefficients is necessary. Certainly, any structural phase changes in the substrates within the range of the process and measurement temperatures are not desired.

6.3 Reactivity

In order to reduce reaction between the film and substrate, the substrate should be chemically compatible with the superconductors. In the case of Si and sapphire, buffer layers of YSZ or CeO_2 were successfully used to prevent chemical reactions between YBCO films and the substrates (79)-(81).

6.4 Dielectric Constant

One of the applications of the superconducting thin films will be in high frequency devices. The dielectric constant should be as small as possible to avoid the high frequency loss in the substrate. It should be noted that the dielectric constant is a function of temperature. For $SrTiO_3$, the constant is -300 at room temperature, over 1000 at 77 K and over 18000 at 4 K. Unlike $SrTiO_3$, $LaAlO_3$ has a reasonable dielectric constant (~15) and low loss tangent (1 x 10^4 at 300 K and 10 GHz), making it the most popular substrate for microwave devices at present (82). However, $LaAlO_3$ is always twined; as a result, its physical properties are varied not only from substrate to substrate but also in the same substrate. Therefore, it will be difficult to control the device properties such as frequency in a microwave resonator.

So far, the most popular substrates are $SrTiO_3$, MgO, ZrO_2 (Y-stabilized) (YSZ), $A1_2O_3$ (sapphire), $LaGaO_3$ and $LaAlO_3$. All the substrates used for the films have one or another limitation depending on applications. Many oxide compounds with cubic or pseudo-cubic structures have lattice constants between 3.8 - 4 Å. The possibility of using these compounds as substrates needs to be explored.

7.0 APPLICATIONS

Wires and Tapes. Since the fabrication of long superconducting wires and tapes with high J_c is still a problem (particularly in the presence of a magnetic field) though very encouraging critical current densities were reported for Bi-based superconducting tapes (83). Few groups have been working on depositing HTSC films on fibers or tapes of appropriate materials. For example, if a 1 mm thick film with a J_c of 5×10^6 A/cm^2 at 77 K could be fabricated on a 100 mm fiber, the resulting wire would have a critical current density of 10^5 A/cm^2 at 77 K, an acceptable number for these configurations. To achieve a J_c of 5×10^6 A/cm^2 at 77 K seems very difficult since epitaxial growth of the films is almost impossible for the configuration.

Electromagnetic Shielding. Since high quality thin films could be prepared over large areas by scaling up many of the deposition processes, fabrication of shielding panels for sensitive electronic circuits is a possibility.

Passive Microwave Components. High frequency components such as inters, resonators and strip lines could be made from superconductors with properties better than those prepared with the best of conventional conductors, say Cu. Devices made from YBCO films have shown performance better than those made from Cu at least by 10 times at 10 GHz and 77 K (see papers published in *IEEE Trans. Mag.*, Vol 27, #2). This is one of areas where HTSC has offered a practical application. So far, the expected surface resistances of the superconducting thin films at low temperatures (say 4 K) are much higher than expected values. This is due to the so called residual surface resistance (84), which is not fully understood in terms of the nature of the origin. Currently, researchers are working on improvement of the films and microwave systems based on HTSC.

Discrete Devices. Due to their short coherence lengths, no superconductor-insulator-superconductor (SIS) junctions have been realized based on HTSC. As a consequence, no superconducting quantum interference devices (SQUIDs) based on SIS have been made from the HTSC materials. However, in the last few years, significant progress has been made in the area of bi-crystal junctions (85)(86), multi-layer junctions (87)-(90), natural weak link junctions (91)(92), step junctions (93)-(95), and bi-epitaxial junctions (96). All the junctions show a resistively shunted Josephson (RSJ) junction behavior. SQUIDs fabricated on bi-crystals have demonstrated good device performance in terms of sensitivity (97). These devices can only be made along the boundary line of two crystals, which limits circuit possibilities. It should be pointed that these devices are certainly applicable where only a single SQUID is needed. So far, the performance of SQUIDs based on multi-layer junctions is still relatively poor. Weak links are difficult to control and not reproducible. Currently devices based on the step and bi-epitaxial junctions are the two most promising approaches for superconducting circuits. Recently, a monolithic DC SQUIDs magnetometer (SQUIDs and input coil together) operating above 77 K have been realized using multi-layer deposition and bi-epitaxial junctions (98). These results indicate a very promising future for HTSC SQUIDs.

Results of bolometers (99) and microbolometers (100) make the application of these for near and far IR very attractive indeed.

Interconnection. The results of high J_c films on silicon using a suitable buffer layer (79) looks quite promising for interconnections based on this technology. Multilayer structures based on HTSC have been realized (98). Devices such as MOSFETs utilizing a superconducting gate have been demonstrated (101).

8.0 CONCLUSION

The status of high T_c thin films after the few years since the original discovery of the high T_c superconductors is very impressive. The superconducting properties of films on a number of different substrates are currently adequate for a number of applications. However, further work is needed to fabricate smooth, reproducible and quality films, and multilayer structures.

REFERENCES

1. *Handbook of Materials Science*, Vol III, (C. T. Lynch, ed.) CRC Press, OH

2. White, C. W., Boatner, L. A., Sklad, P. S., McHargue, C. J., Rankin, J., Farlow, G. C. and Aziz, M. J., *Proceedings of the Radiation Effects in Insulators*, Lyon, France, (July 6 - 10, 1987)

3. Venkatesan, T., Wu, X. D., Inam, A., Hegde, M. S., Chase, E. W., Chang, C. C., England, P., Hwang, D. M., Krchnavek, R., Wachtman, J. B., McLean, W. L., Levi-Setti, R., Chabala, J. and Wang, Y. L., *Chemistry of High Temperature Superconductors II,* Ch. 19, (D. L. Nelson and T. F. George, eds.) American Chemical Society, Washington, DC (1988)

4. Courtesy of J. M. Tarascon and M. A. Subramanian

5. Hatou, T., Takai, Y. and Hayakawa, H., *Jpn. J. Appl. Phys.* 27:L617 (1988)

6. Terasaki, I., Nakayama, Y., Uchinokura, K., Maeda, A., Hasegawa, T. and Tanaka, S., *Jpn. J. Appl. Phys.* 27:L1480 (1988)

7. Chaudhari, P., Koch, R. H., Laibowitz, R. B., McGuire, T. R. and Gambino, R. J., *Phys. Rev. Lett.* 58:2684 (1987)

8. Naito, M., Hammond, R. H., Oh, B., Hahn, M. R., Hsu, J. W. P., Rosenthal, P., Marshall, A. F., Beasley, M. R., Geballe, T. H. and Kapitulnik, A., *J. Mater. Res.* 2:713 (1988)

9. Mankiewich, P. M., Scofield, J. G., Skocpol, W. J., Howard, R. E., Dayem, A. H. and Good, E., *Appl. Phys. Lett.* 51:1753 (1987)

10. Siegal, M. P., Phillips, J. M., Hebard, A. F., van Dover, R. B., Farrow, R. C., Tiefel, T. H. and Marshall, J. H., *J. Appl. Phys.* 70:4982 (1991)

11. Lathrop, D. K., Russek, S. E. and Buhrman, R. A., *Appl. Phys. Lett.* 51:1554 (1987)

12. Terashima, T., Iijima, K., Yamamoto, K., Bando, Y. and Mazaki, H., *Jpn. J. Appl. Phys.* 27:L91 (1988)

13. Bando, Y., Terashirna, T., Enma, K., Yamamoto, K., Hirata, K. and Mazaki, H., in *Proceedings of the Symposium of the Materials Research Society,* Tokyo (1988)

14. Berkley, D. D., Johnson, B. R., Anand, N., Beauchamp, K. M., Conroy, L. E., Goldman, A. M., Maps, J., Mauersberger, K., Mecartney, M. L., Morton, J., Tuominen, M. and Zhang, Y. J., *Appl. Phys. Lett.* 53:1973 (1988)

15. Missert, N., Hammond, R., Mooij, J. E., Matijasevic, V., Rosenthal, P., Geballe, T. H., Kapitulnik, A., Beasley, M. R., Laderman, S. S., Lu, C., Garwin, E. and Barton, R., *IEEE Trans. Mag.* 25:2418 (1989)

16. Silver, R. M., Berean, A. B., Wendman, M. and de Lozanne, A. L., *Appl. Phys. lett.* 52:2174 (1988)

17. *Thin Film Processes,* (J. L. Vossen and W. Kern eds.) Academic Press, New York (1982)

18. Char, K., Kent, A. D., Kapitulnik, A., Beasley, M. R. and Geballe, T. H., *Appl. Phys. Lett.* 51:130 (1987)

19. Adachi, H., Hirochi, K., Setsune, K., Kitabatake, M. and Wasa, K., *App. Phys. Lett.* 51:2263 (1988)

20. Rossnagel, S. M. and Cuomo, J. J., in *Thin Film Processing and Characterization of High Temperature Superconductors,* (J. M. E. Harper, R. J. Colton and L. C. Feldman, eds.) p.106, AIP, New York (1988)

21. Lee, W. Y., Salem, J., Lee, V., Retiner, C. T., Lim, G., Savoy, R. and Deline, V., in *Thin Film Processing and Characterization of High-Temperature Superconductors,* (J. M. E. Harper, R. J. Colton and L. C. Feldman, eds.) p.94, AIP, New York (1988)

22. de Vries, J. W. C., Dam, B., Heijman, M. G. J., Stollman, G. M., Gijs, M. A. M., Hagen, C. W. and Griessen, R. P., *Appl. Phys. Lett.* 52:1904 (1988)

23. Sandstrom, R. L., Gallagher, W. J., Dinger, T. R., Koch, R. H., Laibowitz, R. B., Kleinsasser, A. W., Gambino, R. J., Bumble, B. and Chishohn, M. F., *Appl. Phys. Lett.* 53:444 (1988)

24. Eom, C. B., Sun, J. Z., Yamamoto, K., Marshall, A. F., Luther, K. E., Geballe, T. H. and Laderman, S. S., *Appl. Phys. Lett.* 55:595 (1989)

25. Enomoto, Y., Murakami, T., Suzuki, M. and Moriwaki, K., *Jpn. J. Appl. Phys.* 26:L1248 (1987)

26. Tanaka, S. and Itozaki, H., *Jpn. J. Appl Phys.* 27:L622 (1988)

27. Xi, X. X., Link, G., Meyer, O., Nold, E., Obst, B., Ratzel, F., Smithey, R., Strhlau, B., Weschenfelder, F. and Geerk, J., *Z. Phys.* B74:13 (1989)

28. Newman, N., Cole, B. F., Garrison, S. M., Char, K. and Taber, R. C., *IEEE Trans. Mag.* 27:1276 (1991)

29. Talvacchio, J., Forrester, M. G., Gavala, J. R. and Braggins, T. T., *IEEE Trans. Mag.* 27:978 (1991)

30. Hebard, A. F., Eick, R. H., Fiory, A. T., White, A. E. and Short, X. T., in *Processing and Application of High T_c Superconductors: Status and Prospects,* TMS Northeast Regional Meeting, Rutgers University, Piscataway, NJ (May 9 - 11, 1988)

31. Yotsuya, T., Suzuki, Y., Ogawo, S., Kuwahara, H., Otani, K., Emoto, T. and Yamamoto, J., Presented in *5th International Workshop on Future Electron Devices,* Miyagi-Zoa, Japan (June 2 - 4, 1988)

32. Gao, J., Zhang, Y. Z., Zhao, B. R., Out, P., Yuan, C. W. and Li, L., *Appl. Phys. Lett.* 53:2675 (1988)

33. Fujita, J., Yoshitake, T., Kaniijo, A., Satoh, T. and Igarashi, H., in *Extended Abstracts of High Temperature Superconductors II,* (D. W. Capone II, W. H. Butler, B. Batlogg and C. W. Chu, eds.) MRS, Pittsburgh, p.109 (1988)

34. Zaitsev-Zotov, S. V., Martynyuk, A. N. and Protasov, E. A., *Sov. Phys. Solid State* 25:100 (1983)

35. Ready, J. F., *Effects of High-Power Laser Radiation*, Academic Press, New York (1971)

36. Cheung, J. T. and Sankur, H., CRC *Critical Reviews: Solid State Materials Sciences* 15:63 (1988)

37. Wu, X. D., Venkatesan, T., Inam, A., Xi, X. X., Li, Q., Chang, C. C., Ramesh, R., Hwang, D. M., Nazar, L., Wilkens, B., Schwarz, S. A., Ravi, R. T., Martinez, J. A., Barner, J. B., England, P., Rogers, C. T., Tarascon, J. M., Muenchausen, R. E., Foltyn, S., Dye, R. C., Garcia, A. R. and Nogar, N. S., *Proc.of MRS* 191:129 (1990)

38. Dijkkamp, D., Venkatesan, T., Wu, X. D., Shaheen, S. A., Jisrawi, N., Min-Lee, Y. H., McLean, W. L. and Croft, M., *Appl. Phys. Lett.* 51:619 (1987)

39. Moorjani, K., Bohandy, J., Adrian, F. J., Kim, B. F., Shull, R. D., Chiang, C. K., Swartzendruber, L. J. and Bennet, H., *Phys. Rev.* B36:4036 (1987)

40. Wu, X. D., Inam, A., Venkatesan, T., Chase, E. W., Barboux, P., Tarascon, J. M. and Wilkeses, B., *Appl. Phys. Lett.* 52:754 (1988)

41. Inam, A., Hegde, M. S., Wu, X. D., Venkatesan, T., England, P., Miceli, P. F., Chase, E. W., Chang, C. C., Tarascon, J. M. and Wachtman, J. B., *Appl. Phys. Lett.* 53:908 (1988)

42. Roas, B., Schultz, L. and Endres, G., *Appl. Phys. Lett.* 53:1557 (1988)

43. Witanachchi, S., Kwok, H. S., Wang, X. W. and Shaw, D. T., *Appl. Phys. Lett.* 53:234 (1988)

44. Koren, G., Polturak, E., Fisher, B., Cohen, D. and Kimel, G., *Appl. Phys. Lett.* 53:2330 (1988)

45. Golovashikin, A. I., Eklinov, E. V., Krasnosvobodtsev, S. I. and Pechen, E. V., *Physica* C153-155:1455 (1988)

46. Frohlingsdrof, J., Zander, W. and Stritzker, B., *Solid State Comm.* 67:965 (1988)

47. Venkatesan, T., Wu, X. D., Dutta, B., Inam, A., Hegde, M. S., Hwang, D. M., Chang, C. C., Nazar, L. and Wilkeses, B., *Appl. Phys. Lett.* 59:581 (1989)

48. Venkatesan, T., Wu, X. D., Inam, A., and Wachtman, J. B., *Appl. Phys. Lett.* 62:1193 (1988)

49. Wu, X. D., Dutta, B., Hegde, M. S., Inam, A., Venkatesan, T., Chase, E. W., Chang, C. C. and Howard, R., *Appl. Phys. Lett.* 54:179 (1989)

50. Wu, X. D., Muenchausen, R. E., Foltyn, S., Estler, R. C., Dye, R. C., Garcia, A. R., Nogar, N. S., England, P., Ramesh, R., Hwang, D. M., Ravi, T. S., Chang, C. C., Venkatesan, T., Xi, X. X., Li, Q. and Inam, A., *Appl. Phys. Lett.* 57:523 (1990)

51. Greer, J. A. and Hook, J. V., *SPIE Proc.* 79:1377 (1990); J. A. Greer, presented at *38th National American Vacuum Society Symposium,* Seattle, WA (Nov 11 - 15, 1991)

52. Foltyn, S. R., Muenchausen, R. E., Dye, R. C., Wu, X. D., Luo, L., Cooke, D. W. and Taber, R. C., *Appl. Phys. Lett.* 59:1374 (1991)

53. Foltyn, S. R., Dye, R. C., Ott, K. C., Peterson, E., Hubbard, K. M., Hutchinson, W., Muenchausen, R. E., Estler, R. C. and Wu, X. D., *Appl. Phys. Lett.* 59:594 (1991)

54. Webb, C., Weng, S. L., Eckstein, J. N., Missert, N., Char, K., Schlom, D. G., Hellman, E. S., Beasley, M. R., Kapitulnik A. and Harris, J. S. Jr., *Appl. Phys. Lett.* 51:1191 (1987)

55. Kwo, J., Hsieh, T. C., Fleming, R. M., Hong, M., Liou, S. H., Davidson, B. A. and Feldman, L. C., *Phys. Rev.* B36:4039 (1987)

56. Schlom, D. G., Eckstein, J. N., Hellman, E. S., Webb, C., Turner, F., Harris, J. S. Jr., Beasley, M. R. and Geballe, T. H., in *Extended Abstracts of High-Temperature Superconductors II,* (D. W. Capone II, W. H. Butler, B. Batlogg and C. W. Chu, eds.) MRS, Pittsburgh, p.197 (1988)

57. Spah, R. J., Hess, H. F., Stormer, H. L., White, A. E. and Short, K. T., *Appl. Phys. Lett.* 53:441 (1988)

58. Kwo, J., Hong, M., Trevor, D. J., Fleming, R. M., White, A. E., Farrow, R. C., Kortan, A. R. and Short, K. T., *Appl. Phys. Lett.* 53:2683 (1988)

59. Kwo, J., Fulton, T. A., Hong, M., and Gammel, P. L., *Appl. Phys. Lett.* 56:788 (1990)

60. Eckstein, J. N., Bozovic, I., von Dessonneck, K. E., Schlom, D. G., Harris, J. S. Jr. and Baumann, S. M., *Appl. Phys. Lett.,* 57:931 (1990)

61. Eckstein, J. N., Bozovic, I., Schlom, D. G. and Harris, J. S. Jr., *Appl. Phys. Lett.* 57:1049 (1990)

62. Kanai, M., Kawai, T. and Kawai, S., *Appl. Phys. Lett.* 58:771 (1991)

63. Yamane, H., Masumoto, H., Hirai, T., Iwasaki, H., Watanabe, K., Kobayashi, N., Muto, Y. and Kurosawa, H., *Appl. Phys. Lett.* 53:1548 (1988)

64. Panson, A. J., Charles, R. G., Schmidt, D. N., Szedon, J. R., Machiko, G. J. and Braginski, A. I., *Appl. Phys. Lett.* 53:1756 (1988)

65. Watanabe, K., Yamane, H., Kurosawa, H., Hirai, T., Kobayashi, N., Iwasaki, H., Noto, K. and Muto, Y., *Appl. Phys. Lett.* 54:575 (1989)

66. Zhao, J., Chern, C. S., Li, Y. Q., Norris, P., Gallois, B., Kear, B., Wu, X. D. and Muenchausen, R. E., *Appl. Phys. Lett,* 58:2839 (1991)

67. Schulte, B., Maul, M., Becker, W., Schlosser, E. G., Elschner, S., Haussler, P. and Adrian, H., *Appl. Phys. Lett.* 59:869 (1991)

68. Hiskes, R., DiCarolis, S. A., Young, J. L., Laderman, S. S., Jacowitz, R. D. and Taber, R. C., *Appl. Phys. Lett.* 59:606 (1991)

69. Yue, A. S. and Yang, C. S., in *Extended Abstracts of High-Temperature Superconductors II,* (D. W. Capone II, W. H. Butler, B. Batlogg and C. W. Chu, eds.) p. 85, MRS, Pittsburgh (1988)

70. Chen, H. S., Liou, S. H., Kortan, A. R. and Kimerling, L. C., *Appl. Phys. Lett.* 53:705 (1988)

71. Cuomo, J. J., Guamieri, C. R., Shivashankar, S. A., Roy, R. A., Yee, D. S. and Rosenberg, R., *Adv. Ceram. Mater.* 2:422 (1987)

72. Tachikawa, K., Watanabe, I., Kosuge, S., Kabasawa, M., Suzuki, T., Matsuda, Y. and Shinbo, Y., *Appl. Phys. Lett.* 52:1011 (1988)

73. Rice, C. E., van Dover, R. B. and Fisanick, G. J., *Appl. Phys. Lett.* 51:1842 (1987)

74. Gross, M. E., Hong, M., Liou, S. H., Gallagher, P. K. and Kwo, J., *Appl. Phys. Lett.* 52:106 (1987)

75. Gupta, A., Jagannathn, R., Cooper, E. I., Giess, E. A., Landman, J. I. and Hussey, B. W., *Appl. Phys. Lett.* 52:2077 (1988)

76. Ohbayashi, K., Anma, M., Takai, Y. and Hayakawa, H., *Jpn. J. Appl. Phys.* 29:L2049 (1990)

77. Hwang, D. M., Ravi, T. S., Ramesh, R., Chan, S. W., Chen, C. Y., Nazar, L., Wu, X. D., Inam, A. and Venkatesan, T., *Appl. Phys. Lett.* 57:1690 (1990)

78. Sandstrom, R. L., Giess, E. A., Gallagher, W. J., Segmuller, A., Copper, E. I., Chishohn, M. F., Gupta, A., Shinde, S. and Laibowitz, R. B., *Appl. Phys. Lett.* 53:1874 (1988)

79. Fork, D. F., Fenner, D. B., Connell, G. A. N., Phillips, J. M. and Geballe, T. H., *Appl. Phys. Lett.* 57:1137 (1990)

80. Wu, X. D., Muenchausen, R. E., Nogar, N. S., Pique, A., Edwards, R., Wilkens, B., Ravi, T. S., Hwang, D. M. and Chen, C. Y., *Appl. Phys. Lett.* 58:304 (1991)

81. Wu, X. D., Dye, R. C., Muenchausen, R. E., Foltyn, S. R., Maley, M., Rollett, A. D., Garcia, A. R. and Nogar, N. S., *Appl. Phys. Lett.* 58:2165 (1991)

82. Simon, R. W., Platt, C. E., Lee, A. E., Lee, G. S., Galy, K. P., Wire, M. S., Luine, J. A. and Urbanik, M., *Appl. Phys. Lett.* 53:2677 (1988)

83. Enomot, N., Kikuchi, H., Uni, N., Kumakura, N., Togkno, K. and Watanabe, N., *Jpn. J. Appl. Phys.* 29:L447 (1990)

84. Miller, D., Richards, P. L., Etemad, S., Venkatesan, T., Nazar, L., Inam, A., Dutta, B., Wu, X. D., Eom, C. B., Spielman, S. R., Geballe, T. H., Newman, N. and Cole, B. F., *Appl. Phys. Lett.* 59:2326 (1990)

85. Dimos, D., Chaudhari, P., Mannhart, J. and LeGoues, F. K., *Phys. Rev. Lett.* 60:1653 (1988)

86. Ivanov, Z. G., Nilsson, P. A., Winkler, D., Alarco, J. A., Claeson, T., Stepantsov, E. A. and Tzalenchuk, A. Y., *Appl. Phys. Lett.* 59:3030 (1991)

87. Roger, C. T., Inam, A., Hegde, M. S., Dutta, B., Wu, X. D. and Venkatesan, T., *Appl. Phys.Lett.* 55:2032 (1989)

88. Gao, J., Aarnink, W. A. M., Gerritsma, G. J., Veldhuis, G. and Rogalla, H., *IEEE Tran. Mag.* 27:3062 (1991)

89. Laibowitz, R. B., Koch, R. H., Gupta, A., Koren, G., Gallagher, W. J., Foglietti, V., Oh, B. and Viggiano, J. M., *Appl. Phys. Lett.* 56:1156 (1990)

90. Chin, D. K. and Van Duzer, T., *Appl. Phys. Lett.* 58:753 (1991)

91. Koch, R. H., Gallagher, W. J., Bumble, B. and Lee, W. Y., *Appl. Phys. Lett.* 54:951 (1989)

92. Russek, S. E., Lathrop, D. K., Moeckly, B. H., Buhrman, R. A., Shin, D. H. and Silcox, J., *Appl. Phys. Lett.* 57:1155 (1990)

93. Daly, K. P., Dozier, W. D., Burch, J. F., Coons, S. B., Hu, R., Platt, C. E. and Simon, R. W., *Appl. Phys. Lett.* 58:543 (1991)

94. Dilorio, M. S., Yoshizami, S., Yang, K. Y., Zhang, J. and Maung, M., *Appl. Phys. Lett.* 58:2552 (1991)

95. Ono, R. H., Beall, J. A., Cromar, M. W., Harvey, T. F., Johansson, M. E., Reintsema, C. D., Rudman, R. A., *Appl. Phys. Lett.* 59:1126 (1991)

96. Char, K., Colclough, M. S., Lee, L. P. and Zaharchuk, G., *Appl. Phys. Lett.* 59:2177 (1991)

97. Gross, R., Chaudhari, P., Kawasaki, M., Ketchen, M. B. and Gupta, A., *Appl. Phys. Lett.* 57:727 (1990)

98. Lee, L. P., Char, K., Colclough, M. S. and Zaharchuk, G., *Appl. Phys. Lett.* 59:3051 (1991)

99. Verghese, S., Richards, P. L., Char, K. and Sachijen, S. A., *IEEE Trans. Mag.* 27:3077 (1991)

100. Nahum, M., Hu, Q., Richards, P. L., Sachijen, S. A., Newman, N. and Cole, B. F., *IEEE Trans. Mag.* 27:3081 (1991)

101. Xi, X. X., Li, Q., Doughty, C., Kwon, C., Bhattacharya, S., Findikoglu, A. T. and Venkatesan, T., *Appl. Phys. Lett.* 39:3470 (1991)

12

Chemical Vapor Deposited Diamond*

Albert Feldman, Edward N. Farabaugh, and Lawrence H. Robins

1.0 INTRODUCTION

The development of chemical vapor deposition (CVD) techniques that can deposit diamond over large areas has generated considerable interest in using diamond for numerous new applications. These new applications make use of combinations of superior properties that diamond possesses. These extreme properties, which are based on measurements in bulk single-crystal diamond, include greatest hardness, highest elastic moduli, and highest thermal conductivity at room temperature of any material. Other important properties include optical transparency over an extensive wavelength range from the ultraviolet through the far infrared, high electrical resistivity, dopability to form a semiconductor, low permeability to diffusion, chemical inertness, and low coefficient of friction.

The principal uses of bulk diamond, either natural or man made by the high-pressure/high-temperature process, have been abrasives from diamond powders and cutting tools from polycrystalline diamond compacts, machining with diamond crystal points, heat dissipating substrates for small electronic components and laser diodes, windows for specialized applications, and diamond scalpel blades. CVD diamond is already being used in cutting tools for nonferrous materials, x-ray windows, and loud speakers. Other uses expected soon are large area, heat dissipating substrates for electronics and mask supports for x-ray lithography.

2.0 HISTORICAL BACKGROUND

Researchers have attempted to employ synthetic means to produce diamond for many years. Graphite is the stable form of carbon at room temperature and at one atmosphere (10^5 Pa) of pressure and, under these conditions, diamond is a metastable phase of carbon. Diamond exists as a stable phase only at extremely high pressures and temperatures. Graphite can be converted directly to diamond but this process requires pressures that are extremely high (>1.2×10^{10} Pa). An explosive shock method was developed to produce diamond directly from graphite; however, the diamond is produced as small particles ranging in size from submicrometer to several micrometers. The first commercially successful method to synthesize diamond was a high-pressure/high-temperature process developed at the General Electric Company (1). In this process diamond is produced by precipitation from a solution of graphite and a metal catalyst such as iron, nickel, or cobalt. The process is conducted at pressures above 4.5 GPa and temperatures greater than 1100°C. This process is generally used to produce polycrystalline diamond from graphite. By employing temperature gradients and charges of polycrystalline diamond in the high-pressure/high-temperature apparatus, it is possible to produce gem quality diamonds. However, the maximum dimension of these diamonds has been limited to less than 2 cm. Most recently, a diamond crystal of isotopically pure ^{12}C has been produced which shows excellent optical quality and which has a thermal conductivity that exceeds the thermal conductivity of type II diamond by 50% (2). Prior to this work, natural type II diamond was considered the diamond of highest purity with a room temperature thermal conductivity of about 21 W/cm/K, the highest of any material.

Methods for depositing diamond from the gas phase have been under investigation for many years. At about the time that General Electric developed its high-pressure/high-temperature process, Eversole at Union Carbide was able to produce diamond particles from carbon monoxide gas (3). At about the same time, work on depositing diamond from the gas phase began in the Soviet Union under Deryagin (4). These early gas phase processes had two principal problems, graphitic material was deposited simultaneously with the diamond, and the growth rate of diamond was very low. A key discovery made at Case Western Reserve (5) eventually led to high rate growth methods. A study of diamond growth from a feed gas containing methane found that atomic hydrogen etched the graphitic impurities while leaving behind the diamond. The Deryagin group made use of this information to grow diamond at practical growth rates by using large quantities of hydrogen in its gas mixtures (6). Diamond deposition depended on employing an energetic process for converting molecular hydrogen to atomic hydrogen, which is now recognized as essential for diamond growth from the gas phase in most CVD processes that employ hydrocarbon gases. Recently, a group at Rice University found that halogenated gases can also be used to produce diamond by a CVD method that does not require an energetic activation process (7).

The first widely used methods for depositing diamond were developed at the National Institute for Research in Inorganic Materials (NIRIM) in Japan. These included the hot filament CVD method (8) and the microwave plasma CVD method (9). Both of these methods employed feed gas mixtures of methane in hydrogen. The deposition rates are low by today's standards, although many researchers are using these methods in order to understand the deposition process. More recently, hot plasma (10) and oxy-acetylene torch (11) methods have been developed that exhibit very high growth rates and produce material of high quality. All of the above methods produce polycrystalline diamond.

Recent publications report that single-crystal diamond can also be grown by ion implantation of carbon into a copper surface followed by an annealing process (12).

3.0 METHODS OF DEPOSITION

The earliest method of producing diamond at a reasonable deposition rate was developed in the Soviet Union (13). Figure 1 is a schematic diagram of the experimental apparatus. The method is based on transport of carbon from a graphite susceptor to the substrate by means of the hydrogen catalysis. The graphite was heated by optical means to a temperature of about 2000°C. A fraction of the hydrogen gas in contact with the graphite was converted to atomic hydrogen, which etched the graphite, resulting in a number of hydrocarbon gas species such as methane and acetylene. The hydrocarbons diffused to the cooler substrate, held at about 1000°C, where it reacted to deposit diamond.

The hot filament method was the first practical method to produce diamond in a systematic way because of a greater degree of process control (8). Figure 2 shows a schematic diagram of a hot filament reactor at NIST. A hydrogen and methane feed gas mixture is allowed to pass over a hot filament. Typical deposition conditions are: substrate temperature, 600 to 950°C, filament temperature, 1800 to 2100°C; gas pressure, (2.5 to 13) x 10^3

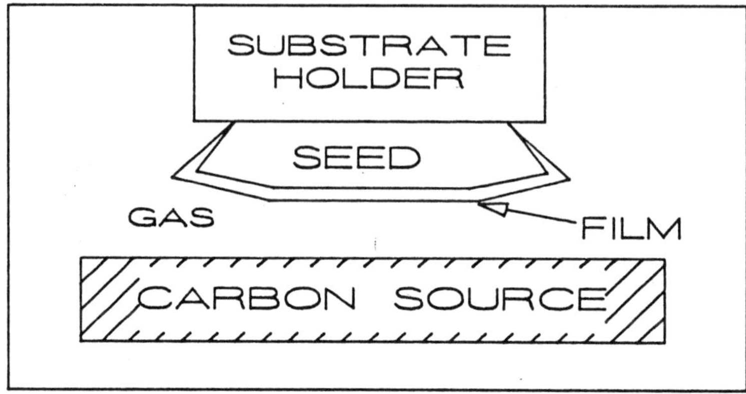

Figure 1. Schematic diagram of diamond deposition as described by Spitsyn (13). The gas consists of hydrogen, atomic hydrogen, methane, acetylene, and other hydrocarbons.

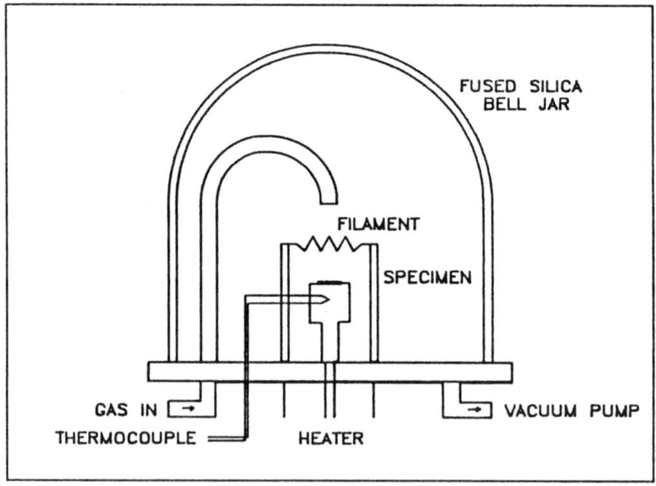

Figure 2. Schematic diagram of hot filament CVD reactor for depositing diamond.

Pa (20 to 100 torr); flow rate, 40 to 100 cm^3/min; and, methane fraction in the feed gas, 0.1 to 5%. The quality of the diamond produced, as revealed by Raman spectroscopy, improves with decreasing methane fraction in the feed gas.

The microwave plasma CVD method was the next practical method to be developed (9). Figure 3 shows a schematic diagram of a microwave system. A heated substrate is placed below a plasma ball sustained by a microwave discharge. The substrate can be heated by the plasma alone or with a separate heating source. The deposition parameters are similar to those in the hot filament reactor. Adding oxygen to the feed gas mixture improves the quality of the diamond (14). Most commercial microwave systems operate at a frequency of 2.45 GHz.

Modulation of the gas composition with time in the microwave reactor acts to increase the growth rate and to improve the diamond quality (15). For example, a gas mixture of 5% methane and 95% hydrogen flowed in the chamber for four minutes, followed by a gas mixture of 1% oxygen and 99% hydrogen in the chamber for two minutes would represent one period of a cyclical process. The process periodically acts to deposit diamond and then etch away any non-diamond impurities that may deposit with the diamond.

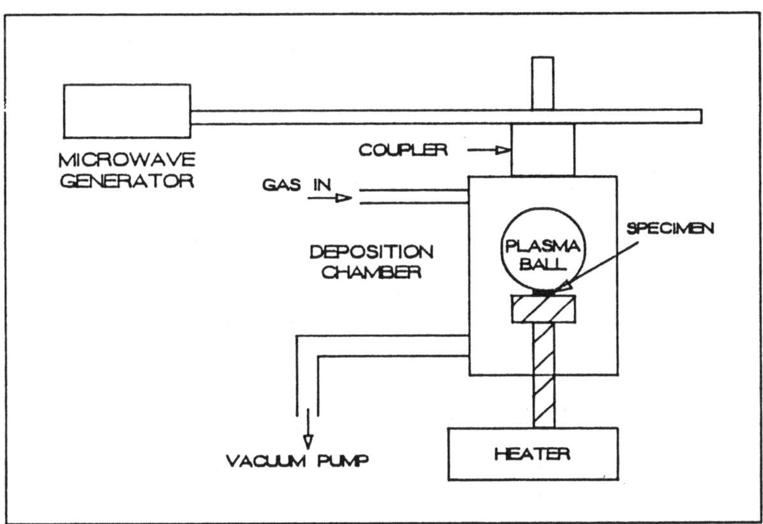

Figure 3. Schematic diagram of microwave plasma CVD reactor for depositing diamond.

The hot filament and microwave plasma methods are low deposition rate methods with deposition rates ranging from 0.1 to 1 µm/hr for reasonably high quality diamond. Other deposition methods have been developed that result in much higher deposition rates. All of these methods rely on the use of thermal plasmas at high temperatures or combustion processes. These include DC plasma deposition (16), radio frequency (RF) plasma deposition (17), and the oxy-acetylene torch (11). The high deposition rates in these systems are attributed to the large amount of atomic hydrogen generated at the high plasma temperatures. However, because of the large amounts of heat generated by these techniques, extensive use of water cooling is required.

Growth rates as high as 930 µm/hr have been reported for the DC plasma torch method (18). These systems contain a nozzle equipped with several gas inlets that allow for various mixtures of Ar, H_2, hydrocarbons, and oxygen-containing organic compounds. Deposition is usually conducted in a chamber below atmospheric pressure. DC plasma deposition is used for some of the commercially produced CVD diamond. One problem with this type of system is electrode erosion which can lead to contamination of the deposited diamond.

Deposition rates of about 200 µm/hr have been achieved with an RF plasma torch. The RF generator typically operates at a frequency of 4 MHz with a power output up to 50 kW. Gas mixtures of Ar, H_2 and methane are utilized at flow rates of tens of liters per minute. Problems with this technique include plasma instabilities, power transfer inefficiencies, and nonuniform depositions.

The oxy-acetylene torch method is receiving considerable attention as a means for depositing high quality diamond at high deposition rates. Diamond has been found to deposit in the reducing region of an oxygen-poor flame. The greater the amount of oxygen in the flame, the higher the quality of the diamond; however, the growth rate decreases with increasing oxygen content in the flame. An advantage of this method is that diamond can be grown in the open atmosphere. Recent reports indicate this method can be used to significantly increase the size of diamond seed crystals at rapid growth rates. This is accomplished at substrate temperatures as high as 1600°C, which is considerably higher than the highest temperature normally expected for diamond growth (1000°C). It is believed that growth occurs at these high temperatures because the abundance of atomic hydrogen in the flame prevents the graphitization of the material.

Halogenated compounds have been used to grow diamond by a direct CVD process without the need for an activating process or for atomic hydrogen. Furthermore, deposition of diamond has been observed at temperatures as low as 300°C (7). Mixtures of hydrogen and fluorine containing gases such as CF_4 flow through a monel tube containing a thermal gradient. The highest temperature at the center of the tube is ~950°C, decreasing to 250°C near the tube ends. One problem that must be addressed with this method is the removal of gases, such as HF, that are toxic and corrosive. The quality of the diamond produced by this method is yet to be evaluated.

Electronic applications will require large area, single crystal diamond coatings; this is one of the major goals of CVD diamond research. Recently, thin single-crystal diamond has been deposited onto a copper substrate (12). Carbon ions were initially implanted into the surface of a single-crystal copper substrate. Carbon has negligible solubility in copper. Upon heating with a pulsed high-power laser beam, the copper at the surface melts. Due to the high thermal conductivity of copper, the heat rapidly dissipates into the bulk of the copper crystal causing rapid solidification. The solid-liquid interface moves rapidly toward the crystal surface expelling the implanted carbon. Due to the rapid cooling, the carbon has insufficient time to crystallize into the stable phase, which is graphite, but does crystallize into the diamond phase. This latter process is not a CVD process; however, the CVD process can be used to greatly increase the thickness of the single-crystal diamond film, once a large area single-crystal surface has been created.

4.0 GROWTH AND QUALITY OF CVD DIAMOND

Angus and Hayman (19) have discussed the fundamental processes leading to nucleation and growth of diamond. In practice, the substrate upon which the diamond is grown is usually rubbed, scratched or polished with diamond powder. It is not yet clear how this process promotes nucleation. It may be due to exposure of chemically active nucleation sites on the surface of the substrate, or it may be due to residual diamond remaining on the substrate as seeds for diamond growth. Diamond appears to grow best on carbide forming substrates such as silicon or molybdenum. Diamond can also be grown on other substrates such as silicon carbide, silicon nitride, mullite, fused silica or sapphire. Most researchers use silicon as the substrate material. Silicon has the advantages of being readily available and of having a linear thermal expansion that is close to that of diamond from room temperature to 950°C. Large differences in the thermal expansion over this temperature range usually lead to fracture of the diamond or the substrate or to delamination of the diamond from the substrate. Figure 4 compares the linear thermal expansion of diamond with several substrate materials. Silicon nitride has a good thermal expansion match and for this reason is being used as a substrate material for diamond coated cutting tools.

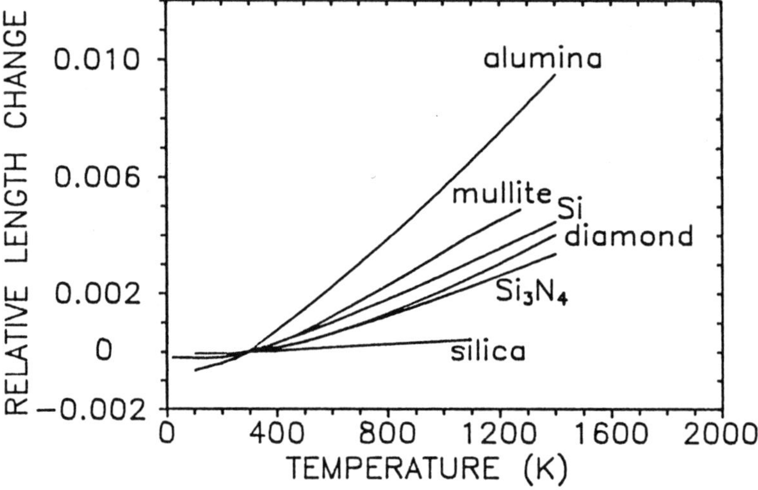

Figure 4. Linear thermal expansion of diamond and several other materials.

Diamond nucleates as discrete particles on the surface of the substrate. When the diamond particle size approaches the interparticle separation distance, the particles merge to form a continuous layer. Thus, the surface of a diamond film is rough. The morphology of the film depends upon growth conditions. The film also shows a preferential orientation. Let us consider diamond produced in a hot filament or a microwave reactor using methane and hydrogen as the feed gas. At the lowest methane concentrations, the films produced show a triangular morphology where the triangular faces are {111} planes. The films show preferential orientation of the <111> direction normal to the film surface. At higher methane concentrations, a pyramidal morphology is observed which is indicative of preferential orientation of <110> directions normal to the film surface. At even higher methane concentrations, the films show predominantly square faceting, corresponding to {100} planes, and the <100> directions are preferentially oriented normal to the film surface. At even higher methane concentrations, the surfaces have a cauliflower appearance and faceting is not observed. Figure 5 shows examples of the different morphologies observed in a scanning electron microscope.

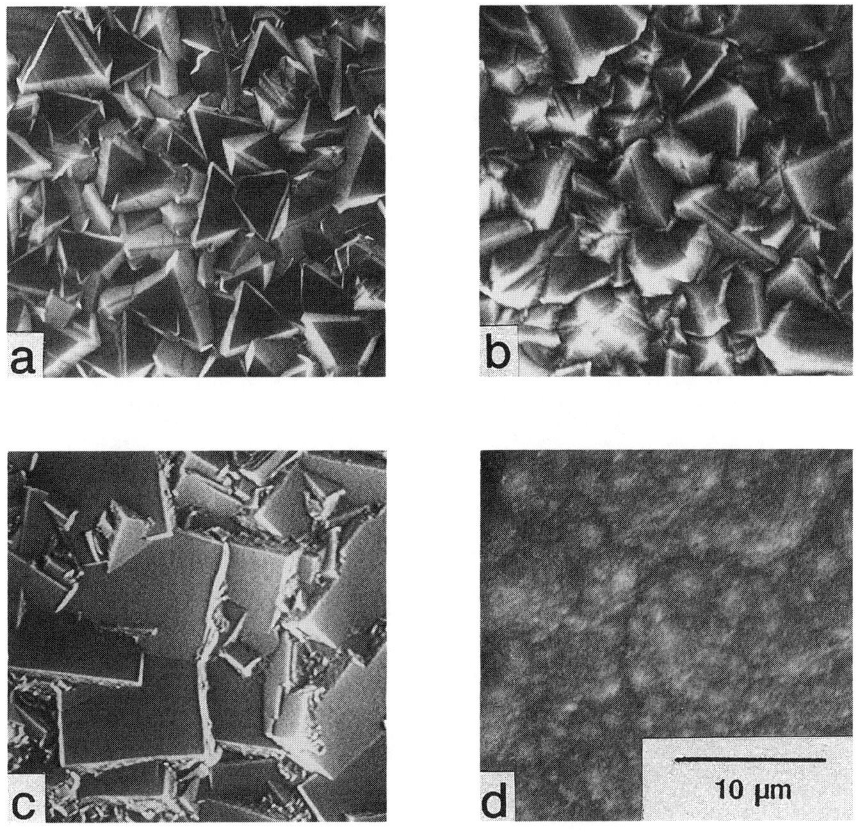

Figure 5. Morphologies of films grown in a hot filament CVD reactor: a) triangular {111} morphology; b) {110} morphology; c) {100} morphology; d) "cauliflower" morphology.

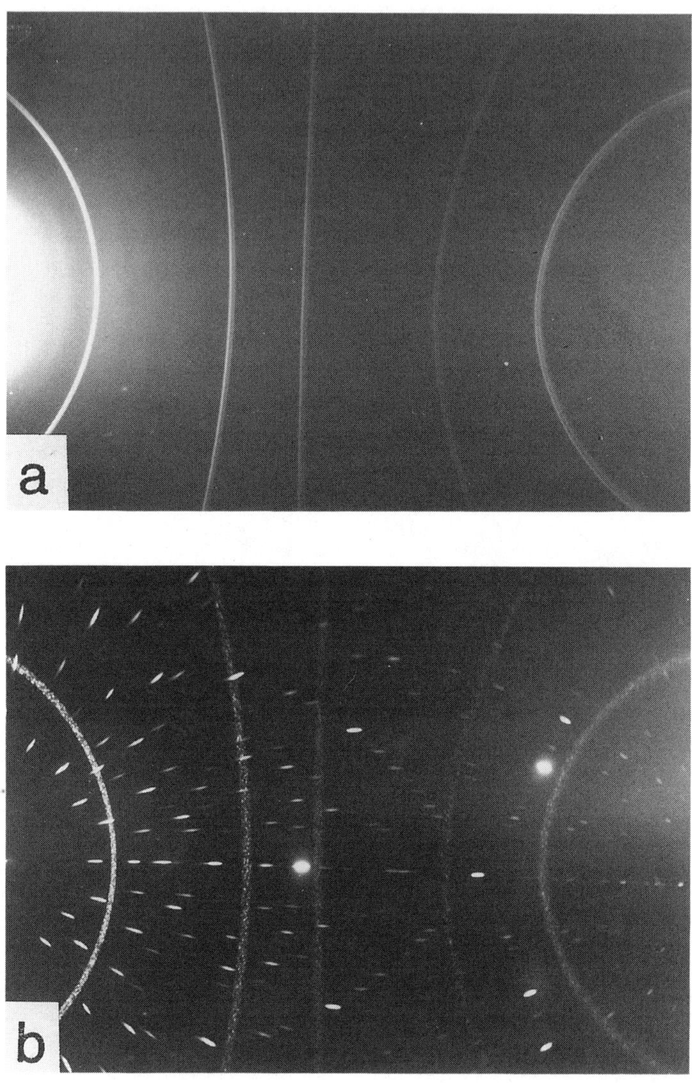

Figure 6. X-ray diffraction patterns of: a) natural diamond powder and b) of CVD diamond produced at NIST.

Several techniques are used to evaluate the diamond produced. The simplest is x-ray diffraction which is used to verify the presence of the crystalline diamond phase. Figure 6 compares an x-ray pattern of natural diamond powder with an x-ray pattern of a diamond film produced at NIST. A modified Debye-Scherrer wide film technique was used to obtain these patterns. The discrete spots seen in the diamond film pattern are due to diffraction from the single-crystal silicon substrate. A disadvantage of this type of x-ray diffraction is its insensitivity to the presence of amorphous carbon phases that are usually present as impurities in the diamond.

Auger spectroscopy is another technique that has been used to examine diamond. Distinct Auger spectra are observed from graphite, diamond, and amorphous carbon (20)(21). However, since this technique is sensitive only to the first few atomic layers on the surface of the specimen, care must be taken to avoid erroneous identifications when confirming the presence of diamond.

Diamond films may also be characterized by electron energy loss spectroscopy (EELS) (21). As with Auger spectroscopy, diamond and graphite possess distinct EELS spectra. Recently, EELS has been useful in determining the relative amounts of diamond chemical bonding (sp^3) and graphitic bonding (sp^2) in diamond-like hydrogenated carbon films (22).

Raman spectroscopy has been accepted as the method of choice for evaluating the quality of the diamond produced. The Raman spectrum of diamond consists of a single sharp peak located at 1332 cm^{-1} wavenumber shift relative to the exciting laser source (23). This line comes from scattering of a photon from the transverse optical phonon of diamond. However, when one examines the Raman spectrum of CVD diamond, one usually observes a more complicated spectrum composed of several spectral features. Figure 7 shows a typical Raman spectrum of CVD diamond excited by the 514.5 nm line of an argon-ion laser. In addition to the sharp line at 1334 cm^{-1}, (small deviations in the diamond line position are believed to be due to internal stress) there is a broad peak centered near 1500 cm^{-1} that is attributed to graphitic or sp^2 bonding, a broad luminescence background of undetermined origin, and a luminescence band at 5890 cm^{-1} associated with point defects in the material. The intensity of the graphitic band correlates positively with the intensity of the background luminescence, and to a lesser extent, with the width of the diamond Raman peak (24). Figure 7 also shows the Raman spectrum of a CVD diamond particle. The spectrum is dominated by the diamond Raman peak. Raman spectra from diamond particles usually show much less fluorescence background and smaller graphitic carbon peaks than diamond films suggesting that these

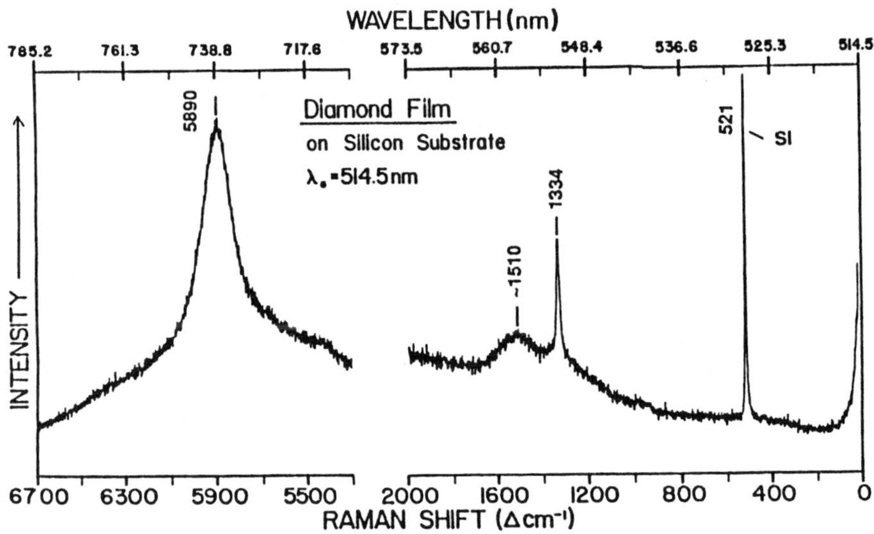

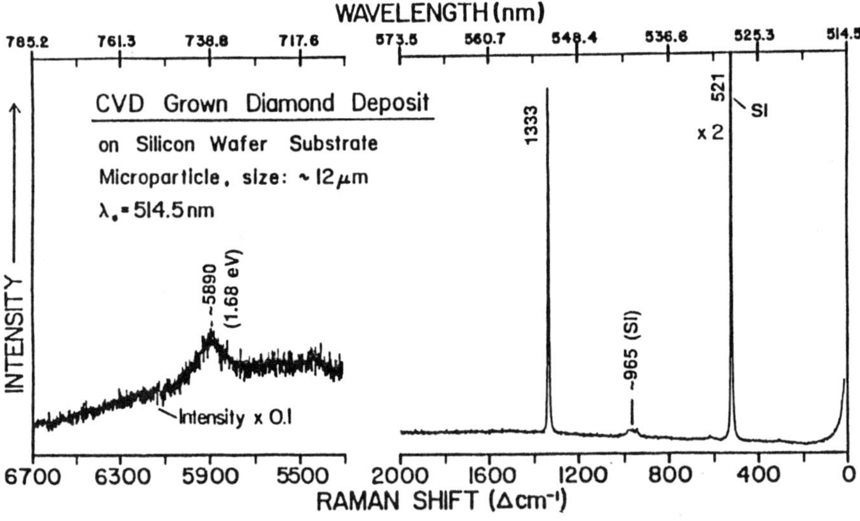

Figure 7. Raman spectra of a CVD diamond film and a CVD diamond particle.

spectral features are associated with the grain boundaries in the polycrystalline films. The luminescence feature seen at 5890 cm^{-1} is believed to be due to silicon incorporated in the diamond lattice (25).

An interesting means of examining diamond on a microscopic scale is cathodoluminescence imaging and spectroscopy. In this method, the specimen is placed in a scanning electron microscope and the optical radiation emitted by the specimen is collected by a photodetector. The optical signal arises when valence band electrons are excited above the fundamental energy gap of the diamond into the conduction band by the energetic electron beam (>10 kV) in the SEM. The electrons in the conduction band can decay to defect states within the band gap. The electrons can lose energy from these defect states by emitting optical radiation whose spectral features are characteristic of the defect center.

The cathodoluminescence provides an optical image of the specimen as the electron beam scans over the specimen. This image can be compared to the secondary electron image customarily observed with the electron microscope. Figure 8 shows a cathodoluminescence image and an SEM image of diamond particles deposited by CVD. The image provides information regarding the distribution of luminescent defects in the diamond. However, the interpretation of the image in terms of defect densities is not necessarily straightforward because the intensity of the luminescence from a particular defect species depends not only on the number density of that species but also on the densities of other species that give rise to competing decay processes.

The nature of the defect centers can be deduced by analyzing the cathodoluminescence spectrum. This identification is based on a large body of work in which the luminescence spectra of many defects have been identified. Figure 9 is the cathodoluminescence spectrum of a diamond film that shows several spectral features often observed in CVD diamond. These features have been associated with particular defects: a sharp line at 1.68 eV believed to be due to a silicon impurity introduced during deposition (this feature is identical to the luminescence line observed at 5890 cm^{-1} in the Raman experiment); a line at 2.156 eV, with an associated vibronic band centered near 2 eV, due to a nitrogen-vacancy (N-V) complex; a line at 2.326 eV due to a different N-V complex; and, a broad violet band centered at 2.85 eV, due to a dislocation related defect. A line at 3.188 eV due to a nitrogen interstitial-carbon complex has also been observed in some CVD diamond films.

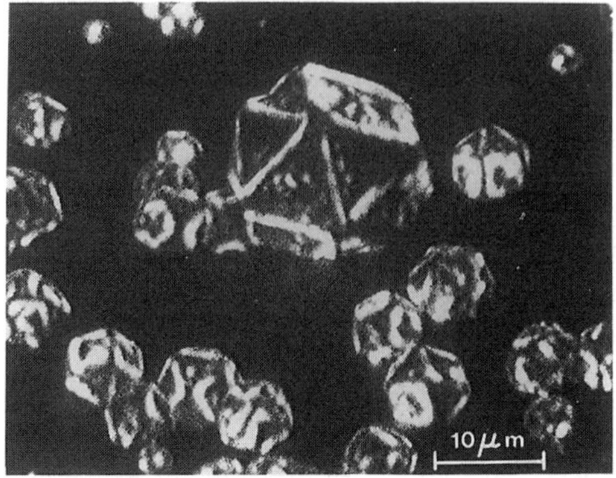

CL IMAGE

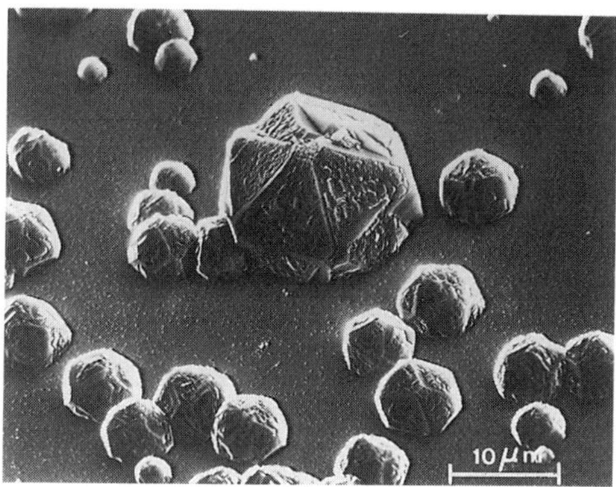

SECONDARY-ELECTRON IMAGE

Figure 8. Cathodoluminescence image of diamond particles and the corresponding SEM image.

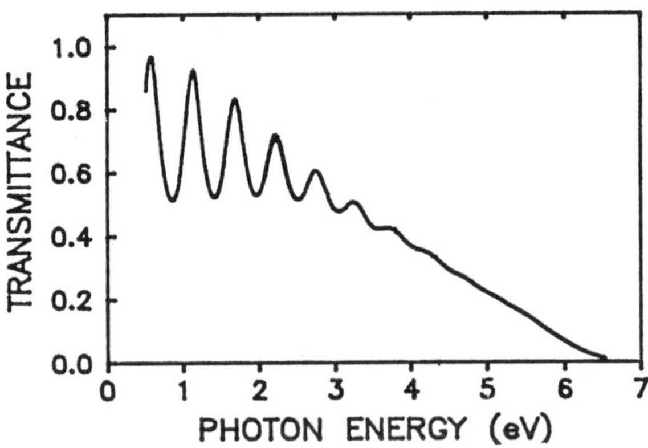

Figure 9. Cathodoluminescence spectrum of CVD diamond film showing principal spectral features.

5.0 THERMAL PROPERTIES OF CVD DIAMOND

High thermal conductivity makes diamond very desirable for heat dissipation applications. However, even in bulk diamond, the thermal conductivity can vary considerably, depending on the presence of defects and impurities. The question arises whether the thermal conductivity of CVD diamond can be as high as the that of the best bulk diamond. Furthermore, the measurement of thermal conductivity in specimens with high values is difficult, especially when the specimens are in thin film form.

Ono et al. (26) have systematically measured the thermal conductivities of a series of diamond film specimens prepared in a microwave reactor. They measured thermal conductivity as a function of the methane/hydrogen ratio used in the specimen preparation which varied from 0.1% to 3%. The specimens, which consisted of bare diamond strips coated with black paint, were supported in vacuum between two heated posts. Radiative cooling resulted in a temperature profile that varied approximately parabolically with distance from the posts. The temperature profile was measured with a thermograph. By considering the radiative cooling and the heat flow in the strips, the authors calculated the thermal conductivities of the films. The specimen prepared with 0.1% methane/hydrogen ratio showed the highest thermal conductivity, 10 W/cm/K. This is about one half that value of type IIa diamond but is still respectably high. The thermal conductivity decreased rapidly with increasing methane/hydrogen ratios and this was found to correlate well with an increasing sp^2 component in the Raman spectrum.

Morelli et al. (27) used a steady state four probe technique to measure the thermal conductivity of two freestanding films of CVD diamond as a function of temperature between 10 K and 300 K. The low temperature values were as much as two orders of magnitude lower than the values for type IIa diamond; however, the values significantly increased with increasing temperature so that at 300 K the values were comparable to the highest values obtained by Ono et al. The small values of thermal conductivity at low temperatures were attributed to phonon scattering from grain boundaries.

Albin et al. (28) measured the thermal diffusivity of two CVD diamond films. Thermal diffusivity, α, is related to thermal conductivity, κ, by $\alpha = \kappa/(\rho C)$, where ρ is the density and C is the specific heat. The authors focused a repetitively pulsed Nd:YAG laser onto a specimen consisting of CVD diamond on a silicon substrate. The time dependence of the temperature distribution across the specimen along a line that passed through the heated spot was measured with an infrared camera as a set of successive images. The effective thermal diffusivity of the specimen was calculated from the computed phase and amplitude of the temperature profile away from the heated spot. If one assumes that the substrate thickness, d_s, and film thickness, d_f, were much less than the thermal diffusion lengths in both materials at the modulation frequency used, the effective thermal diffusivity, α_e, is given by

$$\text{Eq. (1)} \qquad \alpha_e = \frac{\alpha_f \alpha_s (\kappa_f d_f + \kappa_s d_s)}{\alpha_s \kappa_f d_f + \alpha_f \kappa_s d_s}$$

Using this equation, the authors computed α_f to be about 8 cm^2/s for two CVD diamond films 16 μm and 32 μm thick. The corresponding thermal conductivity is 14 W/cm/K if the bulk values $\rho = 3.5$ g/cm^3 and $C = 0.51$ J/g/K are assumed to hold for CVD diamond. This is the largest value of thermal conductivity reported for CVD diamond. This method measures the component of the thermal diffusivity parallel to the surface.

Feldman et al. (29) have used photothermal radiometry to measure the thermal diffusivity of a CVD diamond plate 0.24 mm thick. The method is similar to that of Albin et al. A modulated laser beam from an argon-ion laser is focused onto the surface of the diamond plate which has a black carbon coating for increasing the surface absorptance at the laser wavelength and the surface emittance in the infrared. The infrared radiation emitted by the heated surface is detected with an indium-antimonide detector. The phase of the thermal signal as a function of modulation frequency will depend on the thermal diffusivity of the specimen. The thermal conductivity obtained assuming bulk values for ρ and $C = 5.5$ W/cm/K.

Most recently Lu and Swann (30) have used the method of Cielo (31) to measure the thermal diffusivity of CVD diamond. The method is similar to that of Albin et al. except that the beam of a ruby laser is focused to a ring of light on the diamond plate with an axicon lens. The radius of the ring was 11 mm, thus the method is applicable to large area specimens. A mercury-cadmium-telluride detector placed behind the specimen was used to detect the thermal radiation emitted by the back surface of the specimen at the center of the ring. For a thin specimen, the temperature at the center of the heated ring is given by

$$\text{Eq. (2)} \qquad T = E/(4\pi\rho\alpha t)\exp[-r^2/(4\alpha t)]$$

where E is the absorbed energy per unit thickness, r is the focused ring radius, and t is the time after the laser pulse. The maximum temperature at the center of the ring occurs at a time, $t = r^2/(4\alpha)$, after the laser pulse. For a CVD diamond plate several hundred micrometers thick, the authors obtained a transverse thermal conductivity of 12 W/cm/K. The authors checked the accuracy of their measurements by performing the measurements on specimens of copper, silver, aluminum, and aluminum nitride. Agreement with published values was better than 8%.

The recent discovery of high thermal conductivity in isotopically pure ^{12}C diamond was based on photothermal deflection measurements (2). During the measurement the surface of the specimen is heated with a modulated laser beam in a manner similar to that in the photothermal radiometry method mentioned above. A second, low power probe laser beam is made to skim the surface of the specimen in the vicinity of the heated spot. Due to transfer of heat from the specimen surface to the air above the specimen, a modulated thermal gradient will be present in the air that acts to periodically deflect the probe beam. The deflection is measured as a function of the distance of the probe beam from the heating beam. An analysis of the phase and amplitude of the probe beam deflection allows for calculating the thermal diffusivity of the specimen. Because the diamond specimen was very transparent, it was necessary to coat the surface of the specimen with an absorbing layer in order to perform the measurements. A value of 34 W/cm/K was obtained for the thermal conductivity of the isotopically pure diamond.

6.0 OPTICAL PROPERTIES

Pure diamond has the widest transmission range of any solid material. It is transparent from the electronic absorption edge at 225 nm through the far infrared except for a region of absorption between 3 and 6 µm. Most crystalline materials absorb infrared radiation in particular wavelength regions due to the excitation of lattice vibrations (or phonons) by the infrared radiation. Because of the symmetry of the diamond lattice, no absorption should occur due to excitation of single phonons. Absorption does occur due to excitation of two or more phonons, but this absorption is relatively weak. This is the process responsible for the region of absorption between 3 and 6 µm observed in all diamond crystals. Lattice defects can disrupt the perfection of the diamond lattice, leading to infrared absorption due to single phonon excitations. This absorption process is very weak and great care must be exercised for its observation in thin specimens. In applications requiring thick optical components, these weak absorption processes can produce large absorptances that can be deleterious. Absorption can also occur due to the presence of impurities in the diamond. Both nitrogen and boron, impurities occurring naturally in diamond, increase the wavelength range of absorption in natural diamond. The classification of diamond into types I and II is based on absorption due to nitrogen impurities. Type II diamonds do not display the characteristic ultraviolet and infrared absorption bands associated with nitrogen impurities.

Optical transmission measurements have been made on CVD diamond films principally in the infrared part of the spectrum (32)-(39). The infrared region was examined because the growth surface of CVD diamond is usually too rough to transmit light at visible wavelengths without excessive optical scatter. Typical average roughnesses, R_a, are 0.1 to 0.5 μm. Several authors have published transmission data showing typical spectral features: a decrease in the transmittance with increasing wavenumber above 1000 cm^{-1}, an absorption band due to C-H stretching near 2800 cm^{-1}, and an oscillatory transmittance due to beam interference in the specimen. The two photon absorption region usually coincided with the region of optical scatter and, due to the small thicknesses of the specimens, was not strongly evident. However, transmittance measurements in thicker specimens have shown significant absorption effects. Gatesman et al. (39) were able to measure absorption that was attributed to free carriers in a film 76.2 μm thick. Figure 10 shows their fit to optical transmission data obtained in a Fourier transform infrared spectrometer. The fit includes models for free carrier absorption and surface roughness.

An important region of the spectrum for optical applications of CVD diamond is between 8 and 12 μm. As mentioned above, diamond should be transparent over this wavelength range. Until recently, CVD diamond had not been available with sufficient quality and thickness to observe the limits of optical absorption between 1250 and 833 cm^{-1} due to weak single phonon processes. In a recent study, Klein et al. (40) have reported on infrared transmission measurements performed on good quality diamond films 0.2 to 0.4 mm thick grown by microwave plasma and hot filament CVD methods. Spectra taken in a Fourier transform spectrometer over the wavenumber range 500 to 4000 cm^{-1} show three distinct absorption regions associated with one phonon, two phonon, and three phonon absorption. A small absorption is found in the region where one phonon absorption should occur suggesting that symmetry-breaking defects are indeed present. By comparing the peaks in the absorbance spectra with expected critical points in the phonon spectrum determined by neutron diffraction (41), Klein et al. were able to attribute spectral features in the absorbance spectrum to particular phonons.

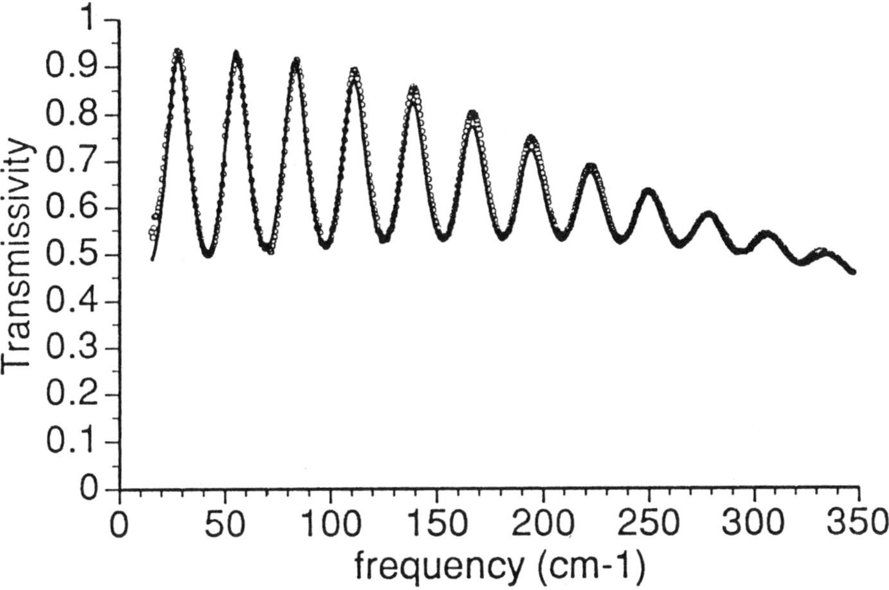

Figure 10. Transmission spectrum of CVD diamond measured by Gatesman et al. (39). (Reprinted with permission of the author.)

By careful preparation of the substrate prior to diamond deposition, it is possible to produce reasonably smooth diamond films that are transparent in the visible and near ultraviolet region of the spectrum. The substrate, typically silicon, is rubbed for approximately one minute against a one micrometer diamond powder placed on a glass plate. This process causes a high density of diamond nucleation sites on the substrate; thus, the diamond particles that nucleate on the substrate merge to form a continuous film at thicknesses significantly less than one micrometer. The resultant films show root-mean- squared surface roughnesses ~0.02 µm as long as the film thickness is not much greater than 1 µm. Transmittance measurements are conducted on unsupported diamond film specimens in which the substrate material is etched away; reflectance measurements are also made. Figure 11 shows the transmittance spectrum of one such diamond film 0.8 µm thick (42). The film is transparent to photons with energies less than the absorption edge at 5.45 eV (225 nm). The oscillations in the spectrum are due to interference effects. By fitting transmittance and reflectance data to appropriate models, one can obtain surface roughness, refractive index, thickness, and absorption coefficients of the films.

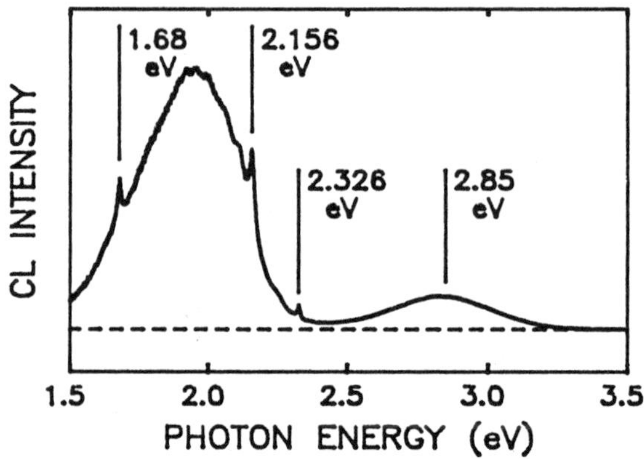

Figure 11. Transmission spectrum of an optically transparent diamond film.

Refractive index is an important optical property of a material. Diamond has a refractive index that is high for an ultraviolet transmitting material. The refractive index of bulk diamond has been determined in the ultraviolet, the visible and in the infrared. Table 1 lists some values at selected wavelengths (43)-(45). These values have been partially verified for CVD diamond.

Table 1. Refractive Index of Diamond, n, vs. Wavelength (43)-(45)

Wavelength µm	Index n	Wavelength µm	Index n	Wavelength µm	Index n
0.250	2.6333	0.700	2.4062	7.00	2.3761
0.300	2.5407	2.50	2.3786	10.00	2.3756
0.400	2.464	3.00	2.3782	15.00	2.3752
0.500	2.4324	4.00	2.3773	20.00	2.3750
0.600	2.4159	5.00	2.3767	25.00	2.3749

7.0 MECHANICAL PROPERTIES

Large area bearing-surfaces of diamond are now possible because of the CVD process. Hardness and low friction coefficient (<0.1) make diamond very desirable for this application. The last statement must be qualified because the coefficient of friction of crystalline diamond is large

(0.9) in a vacuum environment when an adsorbed layer of hydrogen is absent (46). In addition, the wear and coefficient of friction depend on crystallographic orientations of the diamond surface and the sliding direction (47). Isotropic wear would be one advantage of a randomly oriented CVD diamond surface.

Jahanmir et al. (48) have compared the friction and wear characteristics of a silicon carbide ball rubbed against a bare SiC plate and rubbed against a SiC plate coated with CVD diamond. The diamond films were deposited in a hot filament reactor in thicknesses of 2.6 and 4.3 μm. Diamond grain sizes varied from 0.5 to 2.5 μm. The experiments were conducted in a ball-on-three flat arrangement with a ball rotation rate of 100 rev/min (sliding speed of 0.038 m/s) at room temperature in air without lubrication. During a test, the normal force exerted by the ball against the flats was increased in 4 N increments; the force at each increment was kept constant for 10 min. Figure 12 shows the friction coefficient of SiC against SiC and of SiC against CVD diamond as a function of contact load. After the initial load, the coefficient of friction of the diamond coated specimen against SiC was 0.08±0.02 which is one order of magnitude less than the coefficient of friction of SiC on SiC. The wear rate on the diamond coated specimen was four orders of magnitude lower than the wear rate on the uncoated specimen.

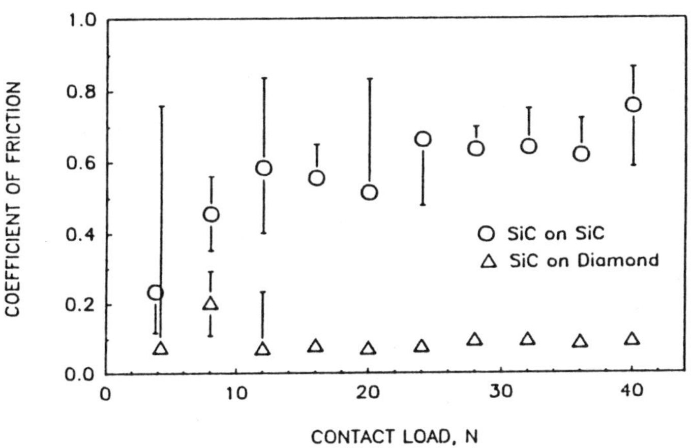

Figure 12. Friction coefficient as a function of contact load for SiC against SiC and SiC against diamond ball-on-flat tests.

Gardos and Ravi (49) have conducted controlled-environment tests of friction and wear on CVD diamond films. Pin-on-flat tests were conducted at low pressure (1.33×10^{-3} Pa or 10^{-5} torr) and at high pressure (13.3 Pa or 0.1 torr) over a range of temperatures from room temperature to 800°C. Pins of SiC and SiC coated with diamond were rubbed against flat silicon substrates coated with CVD diamond by the DC plasma method. The authors found that oxygen or water adsorbates on the films lead to high friction coefficients (0.5 to 0.8) while hydrogen adsorbates lead to low friction coefficients (0.1). Experiments conducted at high temperatures at low pressure lead to increased friction coefficients due to loss of hydrogen from the film surface.

Bulge tests to determine the biaxial modulus and the residual stresses have been made on CVD diamond films (50). The biaxial modulus is related to Young's modulus, E, and Poisson's ratio, ν, by the relationship, $E/(1-\nu)$; this coefficient is the ratio of planar stress to planar strain in an isotropic medium. The test specimen was a diamond film prepared by microwave plasma CVD at a gas pressure 4000 Pa (30 torr). The feed gas ratios H_2:O_2:CH_4 by flow rate were 0.965:0.03:0.005 with a total flow rate of 500.5 cm^3/min. The film thickness was 9.61 μm. The film, which was polycrystalline, showed some (220) texture. The amount of hydrogen in the film was estimated to be 0.95%. The biaxial modulus was 960 GPa with a standard deviation of 4.3%. Depending on the value chosen for the Poisson ratio of polycrystalline CVD diamond, the authors found Young's modulus to be 864 GPa ($\nu = 0.1$) or 893 GPa ($\nu = 0.07$). These are reasonably close to the values of Young's modulus for crystalline diamond which are 1,053 GPa for <100> uniaxial stress and 1207 GPa for <111> uniaxial stress, calculated from the elastic constant values of McSkimin et al. (51).

8.0 POLISHING CVD DIAMOND

CVD diamond films usually grow with rough surfaces that would be undesirable for many applications. Smooth diamond films can be made if the nucleation density during growth is high; however, the thickness of such films is limited to one micrometer or less, as the roughness increases with increasing film thickness.

Methods of polishing diamond films are being developed to produce smooth films. Because CVD diamond is polycrystalline and hard, it is very difficult to polish; polishing by means conventionally used to polish diamond is a very slow process. Wang et al. (37) have polished CVD diamond films on a cast iron scaife heated to 350°C. Six weeks of polishing were required to obtain a mirror-like surface. To increase the polishing rate a sample was annealed in an atmosphere of 0.01% oxygen in argon at 1000°C for 4 hours; the film surface turned black. In this case, the time for polishing was reduced to one week. Polishing with potassium nitrate also increased the polishing rate; however, the specimen had to be carefully monitored to avoid destruction. Polishing decreased the peak-to-valley surface roughness from 1.2 μm to less than 0.1 μm.

Yoshikawa (52) has pioneered a thermochemical method for polishing diamond at high rates. In his method, a rotating polishing plate of iron or nickel is held at an elevated temperature inside an environmental chamber capable of supporting a vacuum. The CVD diamond surface is polished by holding it in contact with the rotating plate. In an atmosphere of hydrogen, iron produced the highest polishing rate and nickel produced nearly as high a polishing rate. No polishing action was observed with molybdenum or with cast iron plates and no polishing was observed at 700°C or lower. At 750°C and above, the polishing rate increased with increasing temperature. At 950°C, the entire surface was polished after 20 min. The polishing rate also increased with applied pressure; however, excessively high pressures made the polishing process unstable. Increasing the lapping speed also increased the polishing rate. The average roughness, R_a, obtained on a 7 mm square specimen was 2.7 nm.

Frequently, the diamond surface is too rough for polishing directly. Yoshikawa has planed the surface of the specimen prior to polishing by irradiating the specimen with a Q-switched Nd-doped yttrium aluminum garnet (Nd:YAG) laser in one atmosphere of oxygen. The laser operated in the TEM_{00} mode with a pulse repetition rate of 1 kHz and a peak power of 23 kW. The laser beam skimmed the diamond surface at an inclination angle of 7° and was focused at different depths relative to the surface. A peak-to-valley roughness of 3 μm could be obtained by this process. Several authors have used variations of Yoshikawa's method to polish CVD diamond (53)(54)

Protrusions that sometimes grow on the diamond surface must be removed prior to polishing. Harker et al. (53) have used reactive plasma etching to remove such protrusions. In order to etch only the protrusions and not the surrounding material, a nonreactive gold coating was applied to the entire surface. The protrusions were then exposed for reactive plasma etching with oxygen.

9.0 CONCLUSION

The combination of superior properties that diamond possesses make this material desirable for many applications. Synthesis of diamond from the gas phase makes it possible to take advantage of these properties because diamond can be deposited over large areas and with thicknesses previously not available. At present, this material is polycrystalline in nature; however, recent results suggest that large area single-crystal diamond may soon be produced. The implications are important for many diverse technical areas: mechanical, electrical, optical, electronic, thermal, etc. The economic impact is expected to grow as new advances are made. The rapid advances being made in diamond processing technology are expected to soon bring many commercial products to the market.

ACKNOWLEDGEMENT

The work was supported in part by the United States Office of Naval Research.

REFERENCES

1. Bundy, F. P., Hall, H. T., Strong, H. M. and Wentorf, R. H., *Nature* 176:51- 55 (1955)

2. Anthony, T. R., Banholzer, W. F., Fleischer, J. F., Wei, L., Kuo, P. K., Thomas, R. L. and Pryor, R. W., *Phys. Rev. B* 42:1104-1111 (1990)

3. Eversole, W. G., U.S. Patents 3030187 and 3030188 (1962); Kiffer, A. D., *Synthesis of Diamond from Carbon Monoxide,* Tonowanda Laboratories, Linde Air Products Co., Tonowanda NY, 1956; see footnote 16 in Angus J. C. and Hayman, C., *Science* 241:913-921 (1988)

4. Deryagin, B. V. and Fedoseev, D. V., *Russian Chemical Reviews* 39:783-788 (1970)

5. Poferi, D. J., Gardner, N. C. and Angus, J. C., *J. Appl. Phys.* 44:1428-1434 (1973)

6. Varnin, V. P., Deryagin, B. V., Fedoseev, D. V., Teremetskaya, I. G. and Khodan, A. N., *Sov. Phys. Crystallogr.* 22:513-515 (1977)

7. Patterson, D. E., Bai, B. J., Chu, C. J., Hauge, R. H. and Margrave, J. L., in: *New Diamond Science and Technology* (R. Messier, J. T. Glass, J. E. Butler, and R. Roy, eds.), pp. 433-438, Materials Research Society, Pittsburgh, PA (1991)

8. Matsumoto, S., Sato, Y., Kamo, M. and Setaka, N., *Jpn. J. Appl. Phys.* 21:L183-L185 (1982)

9. Kamo, M., Sato, Y., Matsumoto, S. and Setaka, N., *J. Cryst. Growth* 62:642-644 (1983)

10. Matsumoto, S., in: *Symp. Proc. International Symposium Plasma Chem., 7th,* pp. 79-84 (1985)

11. Hirose, Y. and Kondo, N., *Program and Book of Abstracts,* p.434, Japan Applied Physics 1988 Spring Meeting (March 29, 1988); Hanssen, L. M., Carrington, W. A., Butler, J. E. and Snail, K. A., *Materials Letters* 7:289-292 (1991)

12. Narayan, J., Godbole, V. P. and White, C. W., *Science* 252:416-418 (1991)

13. Spitsyn, B. V., Bouilov, L. L. and Derjaguin, B. V., *J. Cryst. Growth* 52:219-226 (1981); Spitsyn, B. V., in *Applications of Diamond Films and Related Materials,* (Y. Tzeng, M. Yoshikawa, M. Murakawa and A. Feldman, eds.), pp. 475-482, Elsevier, Amsterdam, (1991)

14. Chiang, C. P., Flamm, D. L., Ibbotson, D. E. and Mucha, J. A., *J. Appl. Phys.* 63:1744-1748 (1988)

15. Ravi, K. V., in: *Proceedings of the Second International Symposium on Diamond Materials,* The Electrochemical Society, Pennington, NJ, 91-98:31-38 (1991)

16. Suzuki, K., Yasuda, J. and Inuzuka, T., *Appl. Phys. Lett.* 50:728-729 (1987)

17. Matsumoto, S., Hino, M. and Kobayashi, T., *Appl. Phys. Lett.* 51:737-739 (1987)

18. Ohtake, N., Tokura, H., Kuriyama, Y., Mashimo, Y. and Yoshikawa, Y., in: *Diamond and Diamond-Like Films*, Proceedings Volume 89-12 (J. P. Dismukes, A. J. Purdes, K. E. Spear, B. S. Meyerson, K. V. Ravi, T. D. Moustakas, and M. Yoder, eds.), pp. 93-105, The Electrochem. Society, Pennington, NJ (1989)

19. Angus J. C. and Hayman, C., *Science* 241:913-921 (1988)

20. Moravec, T. J. and Orent, T. W., *J. Vac. Sci. Technol.* 18:226-228 (1981)

21. Lurie, P. G. and Wilson, J. M., *Surface Science* 65:476-498 (1977)

22. Wang, Y., Chen, H., Hoffman, R. W. and Angus, J. C., *J. Mater. Res.* 5:2378-2386 (1990)

23. Solin S. A. and Ramdas, A. K., *Phys. Rev. B* 1:1687-1689 (1970)

24. Robins, L. H., Farabaugh, E. N. and Feldman, A., *J. Mater. Res.* 5:2456-2468 (1990)

25. Badzian, A. R., Badzian, T., Roy, R., Messier, R. and Spear, K. E., *Mat. Res. Bull.* 23:531-548 (1988)

26. Ono, A., Baya, T., Funamoto, H. and Nishikawa, A., *Jap. J. Appl. Phys* 25:L808-L810 (1986)

27. Morelli, D. T., Beetz, C. P. and Perry, T. A., *J. Appl. Phys.* 64:3063-3066 (1988)

28. Albin, S., Winfree, W. and Scott, B. S., in: *Diamond Optics II* (A. Feldman and S. Holly, eds), 1146:85-94, SPIE, Bellingham, WA (1990)

29. Feldman, A., Frederikse, H. P. R. and Norton, S. J., in: *Diamond Optics III* (A. Feldman and S. Holly, eds.), 1325:304-314, SPIE, Bellingham, WA (1990)

30. Lu, G. and Swann, W. T., *Appl. Phys. Lett.* 59:1556-1558

31. Cielo, P., Utracki, L. A. and Lamontagne, M., *Can. J. Phys.* 64:1172-1177 (1986)

32. Johnson C. E. and Weimer, W. A., in: *Diamond Optics II* (A. Feldman and S. Holly, eds.), 1146:188-191, SPIE, Bellingham, WA (1990)

33. Feng, T., in: *Diamond Optics II* (A. Feldman and S. Holly, eds.), 1146:159-165, SPIE, Bellingham, WA (1990)

34. Snail, K. A., Hanssen, L. M., Morrish, A. A. and Carrington, W. A., in: *Diamond Optics II* (A. Feldman and S. Holly, eds.), 1146:144-151, SPIE, Bellingham, WA (1990)

35. Cong, Y., Collins, R. W., Epps, G. F. and Windischmann, H., *Appl. Phys. Lett.* 58:819-821 (1991)

36. Akerman, M. A., McNeely, J. R. and Clausing, R. E., in: *Diamond Optics III* (A. Feldman and S. Holly, eds), 1325:178-186, SPIE, Bellingham, WA (1990)

37. Wang, X. H., Pilione, L., Zhu, W., Yarbrough, W., Drawl, W. and Messier, R., in: *Diamond Optics III* (A. Feldman and S. Holly, eds), 1325:160-167, SPIE, Bellingham, WA (1990)

38. Bi, X. X., Eklund, P. C., Zhang, J. G., Rao, A. M., Perry, T. A. and Beetz, C. P. Jr., *J. Mater. Res.* 5:811-817 (1990); Bi, X. X., Eklund, P. C., Zhang, J. G., Rao, A. M., Perry, T. A. and Beetz, C. P. Jr., in: *Diamond Optics II* (A. Feldman and S. Holly, eds), 1146:192-200, SPIE, Bellingham, WA (1990)

39. Gatesman, A. J., Giles, R. H., Waldman, J., Bourget, L. P. and Post, R., in: *Diamond Optics III* (A. Feldman and S. Holly, eds.), 1325:170-177, SPIE, Bellingham, WA (1990)

40. Klein, C., Hartnet, T., Miller, R. and Robinson, C., in: *Proceedings of the Second International Symposium on Diamond Materials,* The Electrochemical Society, Pennington, NJ, 91-98:435-442 (1991)

41. Warren, J. L., Yarnell, J. L., Dolling, G. and Cowley, R. A., *Phys. Rev.* 158:805-808 (1967)

42. Robins, L. R., Farabaugh, E. N. and Feldman, A., in: *Diamond Optics IV* (A. Feldman and S. Holly, eds), Vol. 1534:105-116, SPIE, Bellingham, WA (1991)

43. Rösch, von S., *Opt. Acta* 12:253-260 (1965)

44. Peter, F., *Z. Physik* 15:358-368 (1923)

45. Edwards, D. F. and Ochoa, E., *J. Opt. Soc. Amer.* 71:607-608 (1981); D. F. Edwards and H. R. Philipp, in: *Handbook of Optical Constants of Solids* (E. D. Palik, ed.) pp. 665-673, Academic Press, Orlando, FL (1985)

46. Bowden, F. P. and Hanwell, A. E., *Proc. R. Soc. Lond.* A295:233-243 (1966)

47. Tabor, D., *J. Lubr. Technol.* 103:169-179 (1981)

48. Jahanmir, S., Deckman, D. E., Ives, L. K., Feldman, A. and Farabaugh, E., *Wear* 133:73-81 (1989)

49. Gardos, M. N. and Ravi, K. V., in: *Diamond and Diamond-Like Films*, Proceedings Volume 89-12 (J. P. Dismukes, A. J. Purdes, K. E. Spear, B. S. Meyerson, K. V. Ravi, T. D. Moustakas, and M. Yoder, eds.) pp. 475-493, The Electrochem. Society, Pennington, NJ (1989)

50. Cardinale, G. F. and Tustison, R. W., in: *Diamond Optics III* (A. Feldman and S. Holly, eds), 1325:90-98, SPIE, Bellingham, WA (1990)

51. McSkimin, H. J., Andreatch, P. Jr. and Glynn, P., *J. Appl. Phys.* 43:985-987 (1972)

52. Yoshikawa, M., in: *Diamond Optics III* (A. Feldman and S. Holly, eds), 1325:210-221, SPIE, Bellingham, WA (1990)

53. Harker, A. B., Flintoff, J. and DeNatale, J. F., in: *Diamond Optics III* (A. Feldman and S. Holly, eds), 1325:222-229, SPIE, Bellingham, WA (1990)

54. Thorpe, T. P., Morrish, A. A., Hanssen, L. M., Butler, J. E. and Snail, K. A., in: *Diamond Optics III* (A. Feldman and S. Holly, eds), 1325:230-237, SPIE, Bellingham, WA (1990)

Index